Nuclear Waste Management

ACS SYMPOSIUM SERIES **943**

Nuclear Waste Management

Accomplishments of the Environmental Management Science Program

Paul W. Wang, Editor
Concurrent Technologies Corporation

Tiffany Zachry, Editor
Concurrent Technologies Corporation

Sponsored by the
ACS Division of Nuclear Chemistry and Technology

American Chemical Society, Washington, DC

Library of Congress Cataloging-in-Publication Data

Nuclear waste management : accomplishments of the Environmental Management Science Program / Paul W. Wang, editor, Tiffany Zachry, editor ; sponsored by the ACS Division of Nuclear Chemistry and Technology.

    p. cm.—(ACS symposium series ; 943)

Developed from a symposium sponsored by the Division of Nuclear Chemistry and Technology at the 226th National Meeting of the American Chemical Society in New York, Sept. 7–11, 2003.

Includes bibliographical references and index.

ISBN 13: 978–0–8412–3947–0 (alk. paper)

ISBN 10: 0–8412–3947–9

    1. Radioactive waste disposal—Congresses. 2. Radioactive substances—Congresses. 3. Radiochemistry—Congresses.

    I. Wang, Paul W. II. Zachry, Tiffany. III. American Chemical Society. Division of Nuclear Chemistry and Technology. IV. American Chemical Society. Meeting (226th : 2003 : New York, N.Y.) V. Series.

TD897.85.N835   2006
621.48′38—dc22
                                                            2006042791

The paper used in this publication meets the minimum requirements of American National Standard for Information Sciences—Permanence of Paper for Printed Library Materials, ANSI Z39.48–1984.

Copyright © 2006 American Chemical Society

Distributed by Oxford University Press

All Rights Reserved. Reprographic copying beyond that permitted by Sections 107 or 108 of the U.S. Copyright Act is allowed for internal use only, provided that a per-chapter fee of $33.00 plus $0.75 per page is paid to the Copyright Clearance Center, Inc., 222 Rosewood Drive, Danvers, MA 01923, USA. Republication or reproduction for sale of pages in this book is permitted only under license from ACS. Direct these and other permission requests to ACS Copyright Office, Publications Division, 1155 16th Street, N.W., Washington, DC 20036.

The citation of trade names and/or names of manufacturers in this publication is not to be construed as an endorsement or as approval by ACS of the commercial products or services referenced herein; nor should the mere reference herein to any drawing, specification, chemical process, or other data be regarded as a license or as a conveyance of any right or permission to the holder, reader, or any other person or corporation, to manufacture, reproduce, use, or sell any patented invention or copyrighted work that may in any way be related thereto. Registered names, trademarks, etc., used in this publication, even without specific indication thereof, are not to be considered unprotected by law.

PRINTED IN THE UNITED STATES OF AMERICA

Foreword

The ACS Symposium Series was first published in 1974 to provide a mechanism for publishing symposia quickly in book form. The purpose of the series is to publish timely, comprehensive books developed from ACS sponsored symposia based on current scientific research. Occasionally, books are developed from symposia sponsored by other organizations when the topic is of keen interest to the chemistry audience.

Before agreeing to publish a book, the proposed table of contents is reviewed for appropriate and comprehensive coverage and for interest to the audience. Some papers may be excluded to better focus the book; others may be added to provide comprehensiveness. When appropriate, overview or introductory chapters are added. Drafts of chapters are peer-reviewed prior to final acceptance or rejection, and manuscripts are prepared in camera-ready format.

As a rule, only original research papers and original review papers are included in the volumes. Verbatim reproductions of previously published papers are not accepted.

ACS Books Department

Contents

Preface xi

Overview

1. Nuclear Waste Management: Accomplishments
   of the Environmental Management Science Program 2
   Paul W. Wang and Roland F. Hirsch

Characterization, Monitoring, and Analysis Techniques

2. Fluorophores as Chemosensors Based on Calix[4]arenes
   and Three Different Fluorescence Reporters 12
   Gudrun Goretzki, Peter V. Bonnesen, Reza Dabestani,
   and Gilbert M. Brown

3. Spectroscopic Properties and Redox Chemistry
   of Uranium in Borosilicate Glass 34
   Zuojiang Li, Shannon M. Mahurin, and Sheng Dai

4. Surface-Enhanced Raman Scattering of Uranyl-Humic
   Complexes Using a Silver-Doped Sol-Gel Substrate 53
   Lili Bao, Hui Yan, Shannon M. Mahurin, Baohua Gu,
   and Sheng Dai

5. Analysis of Electrochemical Impedance Data for Iron
   in Borate Buffer Solutions 64
   Jun Liu, Brian M. Marx, and Digby D. Macdonald

6. Investigating Ultrasonic Diffraction Grating Spectroscopy
   and Reflection Techniques for Characterizing Slurry Properties 100
   M. S. Greenwood, A. Brodsky, L. Burgess, and L. J. Bond

7. Novel Chemical Detection Strategies for Trichloroethylene
   and Perchloroethylene 133
   Andrew C. R. Pipino, Johan P. M. Hoefnagels,
   John T. Woodward, Curtis W. Meuse, and Vitalii Silin

Separations Chemistry and Technology

8. Separation of Fission-Products Based on Room-Temperature
   Ionic Liquids 146
   Huimin Luo, Sheng Dai, Peter V. Bonnesen,
   and A. C. Buchanan, III

9. Supercritical Fluid Extraction of Radionuclides:
   A Green Technology for Nuclear Waste Management 161
   Chien M. Wai

10. Fundamental Chemistry of the Universal Extraction Process
    for the Simultaneous Separation of Major Radionuclides
    (Cesium, Strontium, Actinides, and Lanthanides)
    from Radioactive Wastes 171
    R. Scott Herbst, Thomas A. Luther, Dean R. Peterman,
    Vasily A. Babain, Igor V. Smirnov, and Evgenii S. Stoyanov

11. Dynamics of Switch-Binding by a Linear Ligand That
    Transforms to a Macrocycle upon Chelation to a Metal Ion:
    Synthesis, Kinetics, and Equilibria 186
    Mansour M. Hassan, Chi Zhang, Jong-ill Lee, K. Mani Bushan,
    Anne McCasland, Richard S. Givens, and Daryle H. Busch

12. Organofunctional Sol–Gel Materials for Toxic Metal Separation 223
    Hee-Jung Im, Terry L. Yost, Yihui Yang, J. Morris Bramlett,
    Xianghua Yu, Bryan C. Fagan, Leonardo R. Allain, Tianniu Chen,
    Craig E. Barnes, Sheng Dai, Lee E. Roecker, Michael J. Sepaniak,
    and Zi-Ling Xue

Facility Inspection, Decontamination, and Decommissioning: Materials Science

13. Investigation of Nanoparticle Formation During Surface
    Decontamination and Characterization by Pulsed Laser 240
    Meng-Dawn Cheng and Doh-Won Lee

14. Recent Progress in the Development of Supercritical Carbon Dioxide-Soluble Metal Ion Extractants: Solubility Enhancement through Silicon Functionalization 250
    Mark L. Dietz, Daniel R. McAlister, Dominique Stepinski, Peter R. Zalupski, Julie A. Dzielawa, Richard E. Barrans, Jr., J. N. Hess, Audris V. Rubas, Renato Chiarizia, Christopher Lubbers, Aaron M. Scurto, Joan F. Brennecke, and Albert W. Herlinger

15. Investigation of SOMS and Their Related Perovskites 268
    Yali Su, Liyu Li, Tina M. Nenoff, May D. Nyman, Alexandra Navrotsky, and Hongwu Xu

Modeling and Waste Treatment Chemistries

16. Solubility of $TcO_2 \cdot xH_2O$(am) in the Presence of Gluconate in Aqueous Solution 286
    Nancy J. Hess, Yuanxian Xia, and Andrew R. Felmy

17. Behavior of Technetium in Alkaline Solution: Identification of Non-Pertechnetate Species in High-Level Nuclear Waste Tanks at the Hanford Reservation 302
    Wayne W. Lukens, David K. Shuh, Norman C. Schroeder, and Kenneth R. Ashley

18. Effects of Sodium Hydroxide and Sodium Aluminate on the Precipitation of Aluminum Containing Species in Tank Wastes 319
    Shas V. Mattigod, David T. Hobbs, Kent E. Parker, David E. McCready, and Li-Qiong Wang

Indexes

Author Index 339

Subject Index 341

Preface

Nuclear waste, as referred to in this volume, includes all waste streams from our past nuclear weapons production and development in the United States. These waste streams consist of radioactive and hazardous waste, waste-contaminated soil and groundwater, and surplus facilities used in nuclear weapons activities. Fundamental research for enabling solutions is needed to significantly reduce the costs, risks, and schedules associated with every stage of management of these waste streams—including safe storage, characterization, treatment, disposal, environmental cleanup, and long-term monitoring. A significant portion of this research has been sponsored by the U.S. Department of Energy (DOE) Environmental Management Science Program (EMSP). Since 1996, the EMSP has funded hundreds of scientific research projects aiming to achieve long-term impact on nuclear waste management approaches and applications.

To highlight EMSP achievements in the area of nuclear waste management, an American Chemical Society (ACS) symposium was organized through sponsorship by the Division of Nuclear Chemistry and Technology. Four technical sessions under this symposium were held in September 2003, at the 226th ACS National Meeting in New York City. Based on the leading-edge research presented and the peer-reviewed manuscripts of the selected presentations, this proceedings volume was proposed and approved by the ACS.

This volume provides a cross-section of scientific and technical advancements from the EMSP-supported projects in the high-level waste and the deactivation and decommissioning areas. The chapters in this volume are organized into four sections, in accordance with the symposium's four technical sessions, namely: (1) characterization, monitoring, and analysis techniques; (2) separations chemistry and technology; (3) facility inspection, decontamination, and decommissioning, as well as materials science; and (4) modeling and waste treatment chemistries. This volume also serves as a companion to ACS Symposium Series Volume 904 (May 2005), which highlights EMSP achievements relating to subsurface contamination remediation.

We acknowledge support from the DOE Idaho Operations Office and the DOE Office of Environmental Remediation Sciences Division. We greatly appreciate this support, which extended from the organization of the symposium sessions through this volume publication. This volume would not be possible without the contributions of the EMSP principal investigators and their collaborators, who provided us with clear, concise, and high-quality manuscripts for this publication. Lastly, we thank the many technical peer reviewers, who provided insightful review and feedback for each manuscript. The combined effort of these authors and reviewers has made this volume a sound, successful representation of research accomplishments, and thus, we hope, a useful resource for the scientific community.

Paul Wang
Principal Technical Advisor
Concurrent Technologies Corporation
425 Sixth Avenue, 28th Floor
Pittsburgh, PA 15219

Tiffany Zachry
Senior Technical Editor
Concurrent Technologies Corporation
P.O. 350
Millville, UT 84326

Nuclear Waste Management

Overview

Chapter 1

Nuclear Waste Management: Accomplishments of the Environmental Management Science Program

Paul W. Wang[1] and Roland F. Hirsch[2]

[1]Concurrent Technologies Corporation, 425 Sixth Avenue, 28th Floor, Pittsburgh, PA 15219
[2]Office of Biological and Environmental Research, SC–23.4 Germantown Building, Office of Science, U.S. Department of Energy, 1000 Independence Avenue, SW, Washington, DC 20585–1290

   The legacy of nuclear weapons development and production in the United States has resulted in considerable environmental contamination at many of the research, manufacturing, and testing facilities collectively called the nuclear weapons complex. The contamination—radioactive, hazardous, or both—poses potentially significant exposure risks to human health and the environment. In 1989, the U.S. Department of Energy (DOE) established the Office of Environmental Management (EM) with the mission of safe, accelerated risk reduction and cleanup of the legacy waste and environmental contamination.
   The scope of the EM cleanup is broad and complex, encompassing vast amounts of contained wastes, contaminated subsurface water and soils, contaminated facilities and equipment, as well as materials in inventory. Many cleanup problems are deemed to be intractable without long-term, science-based research solutions. In 1996, the Environmental Management Science Program (EMSP) was established to support scientific research focusing on the EM cleanup needs. Among the EM need areas, subsurface contamination (i.e., groundwater and soils) and high-level waste are the two areas where scientific research was determined to have the greatest potential impact. Thus, a significant majority of the EMSP portfolio of projects falls in these two areas, while a much smaller portion falls into the area of deactivation and decommissioning of facilities.

This volume presents a cross section of scientific and technical progress in the EMSP-supported projects in the high-level waste and the deactivation and decommissioning areas. The project reports were initially presented at the symposium "Environmental Management Science Program Symposium on Nuclear Waste Management," at the 226th National Meeting of the American Chemical Society (ACS) in September 2003. This volume is a companion to ACS Symposium Series 904 (March 2005), titled *Subsurface Contamination Remediation: Accomplishments of the Environmental Management Science Program (1)*, which was developed from the namesake symposium at the 225th ACS National Meeting. Publication of these two volumes is intended to provide scientific and technical communities with information about the problems of and research solutions for subsurface contamination, high-level waste, and deactivation and decommissioning of facilities, where scientific breakthroughs are needed most to accomplish the EM cleanup mission.

The budget for the EMSP, originally in the DOE Office of Environmental Management, has been in the DOE Office of Science since 2003. This has provided an opportunity to align the EMSP with the Natural and Attenuated Bioremediation Research Program (NABIR), a major basic research program also addressing EM needs. At the start of fiscal year 2006, both EMSP and NABIR have been integrated into the Environmental Remediation Sciences Program of the Office of Biological and Environmental Research in the Office of Science.

The High-Level Waste Problem and Science Research Agenda

High-level waste is the highly radioactive material resulting from the production of nuclear materials and from the reprocessing of spent nuclear fuel and irradiated target assemblies. The inventory of DOE high-level waste consists of approximately 91 million gallons of liquid waste stored in underground tanks and approximately 4,000 m^3 of calcined waste solids stored in bins. Most of this waste is stored at three locations: the Savannah River Site in South Carolina, the Hanford Site in Washington State, and the Idaho National Laboratory. The overall radioactivity is estimated to be 784 million curies, over 99% of which results from relatively short-lived fission products—cesium-137, strontium-90, and their daughter products, yttrium-90 and barium-137. The remaining 1% of radioactivity results from long-lived isotopes of plutonium, americium, uranium, and their daughter products, as well as technetium-99 and carbon-14. The term short-lived refers to half-lives of less than 50 years; the term long-lived refers to half-lives from 50 to over 50,000 years.

High-level liquid waste at Hanford and Savannah River sites is highly alkaline (pH in the range of 12 to 14) and is composed of supernate, saltcake, and sludge, with each containing heterogeneous mixtures of chemical constituents. The alkaline waste is stored primarily in carbon steel-lined concrete tanks with either single-shell or double-shell designs. Some of these tanks have known or suspected leaks, with liquids leaking into the surrounding soil or the annulus space between the double-shell walls. In contrast, high-level liquid waste at Idaho National Laboratory is highly acidic (pH approximately 0), consisting of viscous liquid and sludge stored in single-shell, stainless steel tanks. At all three sites, tank capacities vary from 5,000 to 1,300,000 gallons; and most tanks are below grade, covered by three to ten feet of soil. Access to tanks is limited to small openings on top of the tanks.

The baseline plans for management of the tank waste involve waste characterization, retrieval, pretreatment and separation, immobilization, and tank closures. Pumping liquid to dissolve and/or entrain saltcake and sludge is the baseline technique for waste retrieval. Various separation and sludge washing steps are performed to treat and separate the retrieved waste into high-level waste and low-level waste, with the objectives of reducing the high-level waste volume for immobilization and maximizing the waste loadings in each immobilized waste form. The baseline plan for immobilizing high-level waste is vitrification in borosilicate glass, while immobilization of low-level waste is to be in vitrified borosilicate glass and also in grout. The current DOE estimated cost for the baseline high-level waste management exceeds $50 billion to be spent over several decades (2).

At the beginning of the decade, EM requested that the National Academy of Sciences National Research Council (NRC) help to develop the long-term research agenda for high-level waste. The NRC recommendations are documented in *Research Needs for High-Level Waste Stored in Tanks and Bins at U.S. Department of Energy Sites: Environmental Management Science Program* (3). EMSP staff have used the report to help guide development of its science research program, including the science projects described in this volume. These needs were further summarized into the four topic areas of the FY2001 EMSP solicitation, which are excerpted below:

- Long-term issues related to tank closure, e.g., innovative methods for in situ characterization of the high-level waste remaining in the tanks after retrieval to facilitate tank closure
- High-efficiency, high-throughput separation methods that would reduce high-level waste program costs over the next few decades
- Robust, high loading, immobilization methods and materials that could provide enhancements or alternatives to current immobilization strategies
- Innovative methods to achieve real-time, and, when practical, in situ characterization data for high-level waste and process streams that would be useful for all phases of the waste management program

The Deactivation and Decommission Problem and Science Research Agenda

This volume has two chapters (13 and 14) focusing on applying science research for decontamination solutions needed by the EM Deactivation and Decommissioning operations.

Approximately 5,000 of the 20,000 facilities (buildings, structures, and equipment) used for research, manufacturing, and testing of nuclear weapons have been designated as surplus, no longer needed for the current mission of stockpile stewardship and management. Many of these facilities are contaminated with radioactive and hazardous materials, requiring decontamination and/or removal of contaminated equipment and structures to meet the requirements of the desired end state for the facility. Desired end states include: demolition of the facility to make the area available for appropriate land uses (e.g., restricted access, open space, recreational, industrial, residential, and agriculture); and release of the facility itself for restricted or nonrestricted use.

Baseline decontamination techniques are labor intensive and prone to worker exposure to radiation, chemical, and safety hazards. The techniques often require hands-on contact by workers operating powerful equipment (e.g., grinders, cutters, torches) while wearing protective clothing. DOE estimates that completion of deactivation and decommissioning of these surplus facilities will cost some $30 billion over 50 years or more. Alternative science-based solutions over baseline techniques may save about half of the $30 billion, as estimated by EM and cited in the NRC report mentioned below (*4*).

The science research agenda is discussed in the 2001 NRC report *Research Opportunities for Deactivating and Decommissioning Department of Energy Facilities* (*4*). This report states that the facilities formerly used for the following processes would be especially challenging for deactivation and decommissioning:

- Radiochemical processing of irradiated nuclear fuel and target assemblies (for production of plutonium, tritium, and other nuclear materials)
- Uranium enrichment by gaseous diffusion
- Plutonium processing
- Tritium processing

The report's recommendations include research in several areas, such as decontamination of equipment and facilities, for which the two needs below are cited:

- Basic research toward fundamental understanding of the chemical and physical interactions of important contaminants with the primary materials of interest in deactivation and decommissioning projects, including concrete, stainless steel, paints, and strippable coatings
- Basic research on biotechnological means to remove or remediate contaminants of interest from surfaces and within porous materials

Scope and Organization of This Volume

This book is composed of 17 chapters, in addition to this introduction. The chapters are grouped into four sections, in accordance with the symposium sessions in which the corresponding presentations were made. The section titles listed below are the same as the symposium session titles.

Characterization, Monitoring, and Analysis Techniques

New or advanced analytical techniques are described in this section for chemical and physical characterization of high-level waste streams and immobilized glass, and for monitoring of tank corrosion and migration of waste in the subsurface. Also described is the development of basic instrumentation for long-term monitoring of chlorinated hydrocarbons in environmental media.

- *Chapter 2*: New fluorescence molecular sensors for determination of Cs^+ and Sr^{2+} in high-level waste streams. Three different types of fluorophores (i.e., anthracene, dansyl, and coumarin) covalently attached to calix[4]arene derivatives are studied as sensing elements.
- *Chapter 3*: Fluorescence and UV-Visible spectroscopic studies of the chemistry of uranium in glass melts to optimize the immobilization processes. Study of the fluorescence lifetime distribution of uranium as an in situ optical probe for its concentration in immobilized glass is also reported.
- *Chapter 4*: Surface-enhanced Raman scattering (SERS) spectroscopy of uranyl-humic complexes using a silver-doped sol-gel substrate. Determination of the complexation of natural humic substances with uranyl ions to gain better understanding of the migration of uranium in subsurface environments is investigated.
- *Chapter 5*: Electrochemical impedance spectroscopy (EIS) studies of the passive film formed on iron in contact with aqueous solutions. The film

growth and formation of dominant defects are explored to gain insight into the mechanisms of corrosion of storage tank walls.
- *Chapter 6*: Ultrasonic diffraction grating spectroscopy and reflection techniques for characterizing slurry properties. On-line techniques for measuring properties of liquids (velocity of sound, viscosity, and density) are explored for applications in waste slurry characterization during pipeline transport.
- *Chapter 7*: Evanescent wave cavity ring-down spectroscopy (EW-CRDS) for determination of trichloroethylene (TCE) and perchloroethylene (PCE). The design of a monolithic folded resonator with an analyte-enriching coating is revealed as potentially applicable for long-term environmental monitoring applications.

Separations Chemistry and Technology

The first three chapters in this section involve research in enhanced separation technologies to remove radionuclides from high-level waste, thus reducing the final volume of high-level waste for immobilization and disposal. Research in supercritical CO_2 extraction in Chapter 9 has potential for spent nuclear fuel applications as well. The latter two chapters involve new approaches for separating and removing hazardous metal ions from contaminated environments.

- *Chapter 8*: Separation of fission products based on room-temperature ionic liquids. Five nonflammable, nonvolatile ionic liquids are studied for solvent extraction of Sr^{2+} and Cs^+ in waste streams, and the use of sacrificial cation exchangers to enhance extraction processes is investigated.
- *Chapter 9*: Supercritical fluid extraction (SFE) of radionuclides (uranium dioxide, Cs^+, Sr^{2+}) from nuclear waste. CO_2-soluble chelating agents are investigated for use in the in situ chelation-SFE technique to extract target radionuclides from solid and liquid materials.
- *Chapter 10*: Fundamental chemistry of the Universal Extractant (UNEX) for the simultaneous separation of major radionuclides (cesium, strontium, actinides, and lanthanides) from radioactive wastes. Mechanisms of extracting multiple radionuclides in a single solvent extraction process and the influence of organic diluents on extraction properties are elucidated.
- *Chapter 11*: Macrocyclization during switch-binding of metal ions by a linear ligand. Equilibrium and kinetic studies are conducted to reveal the switch-binding process in which a rapidly reacting linear chelate undergoes

a cyclization structure change upon coordination with metal ions (Cu^{2+}, Hg^{2+}, Ni^{2+}, and Zn^{2+}) to form a stable macrocyclic complex.
- *Chapter 12*: Organofunctional sol-gel materials for toxic metal separation. Granular silica gels grafted or encapsulated with organic ligands are shown to have high selectivity and capacity and fast kinetics for removing target metal ions (Cu^{2+}, Hg^{2+}, and Sr^{2+}) from aqueous solutions, and are regenerable.

Facility Inspection, Decontamination, and Decommissioning; Materials Science

More effective decontamination techniques are investigated, with one involving studies of dynamics of airborne particle generation from laser plasma and one involving supercritical CO_2 extraction of actinides from porous solids employing in situ silicon-functionized ligands. Additionally, a new class of inorganic ion exchangers with high selectivity and capacity for Sr is shown to be directly convertible into a durable ceramic waste form.

- *Chapter 13*: Investigation of nanoparticle formation during surface decontamination and characterization by pulsed laser. The threshold energy needed to remove particles from concrete, stainless steel, and alumina is investigated; and the relationship between particle production and laser fluence correlates well to a log-log proportional plot.
- *Chapter 14*: Supercritical CO_2-soluble ligands for extracting actinide ions from porous solids. Various silicon-substituted alkylenediphosphonic acids are synthesized and show improved solubility and desirable extraction properties in supercritical CO_2 for Am^{3+}.
- *Chapter 15*: Investigation of new metal niobate and silicotitanate ion exchangers for Cs and Sr removal and their direct thermal conversion to perovskites as potential ceramic waste forms. A novel class of niobate-based ion exchangers, SOMS, is found to have high selectivity for Sr^{2+} and is readily converted to perovskites at 500–600 °C that exhibit low Sr leachability.

Modeling and Waste Treatment Chemistries

Two chapters are dedicated to the study of chemical behavior of technetium in high-level waste environments, because one of its isotopes, ^{99}TC, has a long half-life (213,000 years) and its mobility in the environment has long-term environmental and human health implications. The last chapter presents a study of the low solubility of aluminosilicate compounds and their phase transformation as causing clogging of pipes and transfer lines.

- *Chapter 16*: Solubility of $TcO_2 \cdot xH_2O$ in the presence of gluconate in aqueous solutions. The study suggests that Tc(IV) solution species may be stabilized by a series of polymerized gluconate species under alkaline conditions—understanding of Tc complexes and speciation in high-level waste is important for devising their removal processes before waste immobilization.
- *Chapter 17*: Identification of non-pertechnetate species (NPS) in high-level nuclear waste tanks at the Hanford reservation. The study reveals that the X-ray spectra of NPS of actual tank waste fit well with the spectra of the *fac*-$Tc(CO)_3$ complexes, which could conceivably be formed from TcO_4^- in alkaline tank waste environments.
- *Chapter 18*: Precipitation of aluminum containing species in tank wastes. The study shows that aluminosilicate compounds undergo phase transition, from amorphous precipitates to a zeolite phase to denser phases of sodalite and cancrinite, under operating conditions similar to high-level waste evaporators—corresponding to evaporator clogging problems.

Multidisciplinary Research Collaboration

A key aspect of all research reported herein, as well as other research projects conducted under EMSP sponsorship, is close collaboration among national laboratories, universities, and other private institutions, for interdisciplinary science research and development. The multiple science research areas covered in this volume include: actinide chemistry, analytical chemistry and instrumentation, engineering science, geochemistry, inorganic and organic chemistry, materials science, and separations chemistry. An interdisciplinary team of scientists helped each project to succeed in providing science-based solutions with direct relevance to solve the EM high-priority, long-term cleanup problems.

EMSP now resides in the Environmental Remediation Sciences Division (ERSD) in the Office of Biological and Environmental Research (BER) of the DOE Office of Science. It has been integrated with the former Natural and Accelerated Bioremediation Research (NABIR) Program to form the Environmental Remediation Sciences Program.

Further Reading

Information on additional aspects of research projects reported here as well as on all other EMSP-supported projects can be found in a searchable project database (5). The website also contains program documents, EMSP solicitations, and National Research Council reports, including the two cited above. Information on relevant research programs within the ERSD can be found on its website (6).

References

1. *Subsurface Contamination Remediation: Accomplishments of the Environmental Management Science Program;* Berkey, E., Zachry, T., Eds.; ACS Symposium Series 904; American Chemical Society: Washington, DC, 2005.
2. *Department of Energy FY 2006 Congressional Budget Request*; DOE/ME-0050, Vol. 5; U.S. Department of Energy: Washington, DC, 2005. Available on the DOE website (http://www.mbe.doe.gov/budget/06budget/Start.pdf).
3. National Research Council. *Research Needs for High-Level Waste Stored in Tanks and Bins at U.S. Department of Energy Sites: Environmental Management Science Program;* National Academy Press: Washington, DC, 2001. Available on the National Academy Press website (http://www.nap.edu/books/0309075653/html).
4. National Research Council. *Research Opportunities for Deactivating and Decommissioning Department of Energy Facilities;* National Academy Press: Washington, DC, 2001. Available on the National Academy Press website (http://www.nap.edu/catalog/10184.html).
5. See the Environmental Management Science Program website (http://emsp.em.doe.gov/index.htm).
6. See the Environmental Remediation Sciences Division website (http://www.sc.doe.gov/ober/ERSD_top.html).

Characterization, Monitoring, and Analysis Techniques

Chapter 2

Fluorophores as Chemosensors Based on Calix[4]arenes and Three Different Fluorescence Reporters

Gudrun Goretzki, Peter V. Bonnesen, Reza Dabestani, and Gilbert M. Brown[*]

Chemical Sciences Division, Oak Ridge National Laboratory, Oak Ridge, TN 37831–6119

The preparation and metal ion complexation reactions of three fluorescent ligands based on a calix[4]arene scaffold and crown ether or azacrown ether binding sites are reported. Three different fluorescent reporter groups were investigated: dansyl, anthracene, and coumarin. The N-dansyl azacrown-5 derivative in the cone conformation shows no change in fluorescence on complexation, but an open chain analog displayed highly selective quenching with mercury. The N-methylanthracene derivative with an azacrown-5 binding site in the 1,3-alt conformation is a photo-induced electron transfer (PET)-sensor showing chelation enhanced fluorescence (CHEF) effects with Hg(II), Cu(II), and Sr(II). The coumarin calix[4]arene crown-6 derivative in the 1,3-alt conformation is able to sense cesium selectively over sodium as well as potassium.

Introduction

Fluorescent sensors are being developed that combine a molecular recognition element with an optical transduction element. Such sensors will be needed for real-time application in the characterization of nuclear waste and waste process streams. The U.S. Department of Energy (DOE) has large volumes of high-level radioactive waste stored in tanks. Proposed waste remediation efforts include separation of the small amount of highly radioactive fission products from the bulk waste, and efficient processes are currently under development. As processes are implemented to remove soluble radionuclides, of which cesium and strontium are among the highest contributors to the activity, sensors will be beneficial for the characterization of the tank waste itself as well as of process streams. In DOE's Hanford waste tanks, the ratio of total Cs to ^{137}Cs varies from 2 to 3 depending on the individual tank. In general the ratio of total Sr to ^{90}Sr is even larger (*1–3*). The separation methods used to remove the radioisotopes will be selective for an element in a particular oxidation state, not to a single isotope. Thus, separation methods to remove ^{137}Cs and ^{90}Sr will be selective for the total concentration of Cs^+ and Sr^{2+} cations. Thus, fluorometric sensors that can carry out a rapid, in situ analysis, preferably using fiber optics, of the total concentration of alkali and alkaline earth elements are desirable and would be of immediate benefit to the DOE.

The fluorescent method is significant due to its high sensitivity and direct visual perception even in highly dilute solutions and its potential for remote application utilizing fiber optics. The rapidly developing fields of supramolecular and molecular recognition chemistry are the foundation for chemical sensing because of the ability to design molecules that are both simultaneously highly sensitive and highly selective (*4, 5*). The sensing and detection mechanism and the nature of the sensing unit in concert with the recognition chemistry play a crucial role in this approach.

The design of a fluorescent chemosensor is simple in its logic, because it involves two independently selectable sub-units: the ionophore based on a supramolecular framework and a photoactive dye architecture for response and signal tuning. The two units are combined in such a way that the supramolecular interactions of the metal ion that causes the electronic perturbation of the ligating atoms of the ionophore will be transmitted to the fluorophore through electron- or charge transfer.

Calix[4]arenes have been widely used as a three-dimensional platform for selective metal ion recognition (*6–8*). In fact, several calixarene-based fluorescence sensors have already been designed (*9–13*). In this approach, appropriate fluorescence molecules are covalently attached to calix[4]arene derivatives possessing the 1,3- alternate or the cone conformation. High selectivity for Cs^+ complexation has been shown by 1,3-alt-calix[4]arene-crown-6-ether ligands with signaling by a fluorophore (*9–10*). Previous investigations have shown that calix[4]arenes in the cone conformation having amide, ester, or free carboxylic acid binding sites possess a high affinity for alkaline-earth cations including Sr(II) (*14*). We report here the synthesis and binding properties of several new fluorescence molecular sensors which are based on three different types of fluorophores—dansyl, anthracene, and coumarin. For the dansyl fluorophore, a fluorescence change should occur by changing the reduction potential of the nitrogen during analyte binding. Methylamino-anthracene is a classical representative of photo-induced electron transfer (PET) fluorophores due to the nitrogen free electron pair and the methylene spacer between the sensing and the recognition unit. In the case of the coumarin fluorophore, we synthesized an intrinsic fluorescence sensor with the expectation that a change in fluorescence intensity due to inhibited charge transfer during complexation of the metal ion at the oxygen donors would occur. The sensing properties of these molecules are evaluated and compared.

Dansyl-Fluorophore

Synthesis

The synthesis for the fluorescent calix[4]arene **4** began with the reduction of the 1,3-disubstituted cyanomethyl calix[4]arene **2** (*15*) to give the diamine **3** (*16*), which on condensation with dansylchloride afforded the receptor molecule **4** in 60% yield (Scheme 1).

A related cyclic molecule was synthesized in two steps in order to compare it with its open chain analog. 1,3-Bis(2-chloroethoxy(2-ethoxy)calixarene **5** was prepared via reaction of the calix[4]arene **1** with tosylated 2-chloroethoxy-ethanol. Cyclization with dansylamide in dimethylformamide (DMF) gives the azacrown-5 derivate **6** (Scheme 2).

Fluorescence Measurements

To investigate the fluorescence response of compound **4** towards contact with various metal ions, we chose an extraction experiment. Aqueous solutions of mercuric nitrate with concentrations ranging from 0.1–5 µM (pH = 4.0, HNO_3) were extracted with a 1.2 µM solution of **4** in 1,2-dichloroethane (DCE). The phases were separated, and the fluorescence of **4** in the organic phase was measured. The fluorescence spectra of compound **4** in DCE show excitation and emission bands at λ_{ex} = 340 and λ_{em} = 502 nm respectively. On extraction of mercury(II) the dansyl group fluorescence is quenched. The dependence of the relative emission intensity I/I_0 (where I_0 and I are emission intensities observed

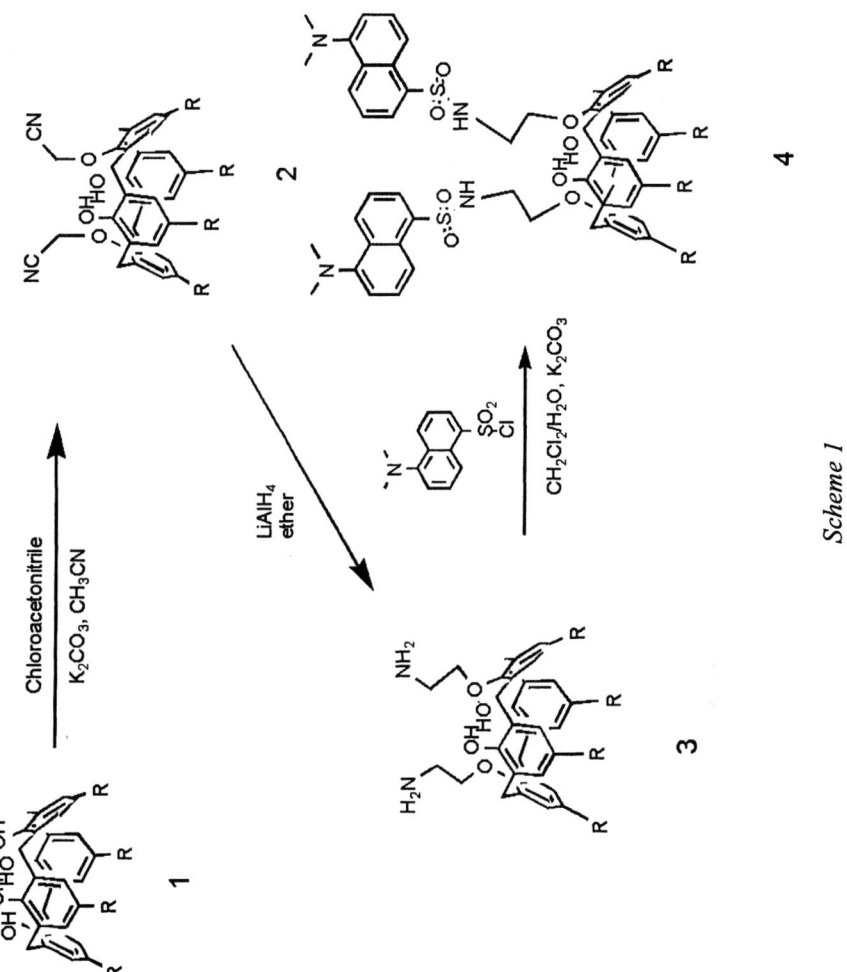

Scheme 1

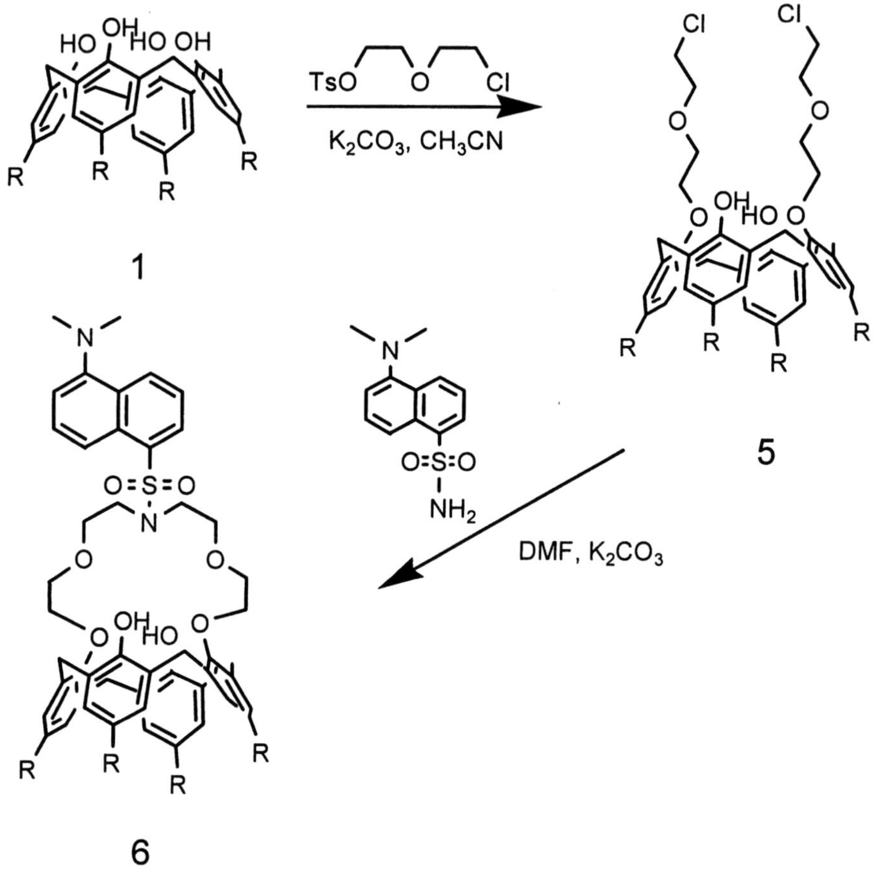

Scheme 2

R = *tert*-butyl

for **4** before and after mercuric ion extraction, respectively) on the initial Hg(II) concentration in the aqueous solution is shown in Figure 1.

Similar effects due to Hg(II) complexation have been reported (*17, 18*). Consistent with earlier suggested mechanisms, the quenching occurs by electron transfer from the excited dansyl moiety to the proximate mercuric ion (*19*). The emission intensity of **4** decreases with enhanced Hg(II) extraction into the organic phase. A maximum of 85% quenching of the initial ligand fluorescence was observed after extraction of 2.5 µM Hg(II) solution.

Similarly, the fluorescence quenching is enhanced as the pH of the aqueous phase increases, with the concentration of Hg(II) held constant. An aqueous mercuric nitrate solution (2.4 µM) with pH = 1–7 (HNO$_3$) was extracted with a 1.2 µM solution of **4** in DCE, and the fluorescence of **4** in the organic phase measured. The dependence of I/I_0 upon the aqueous phase pH is presented in Figure 2. Increasing the pH of the contacting aqueous phase beyond pH 8 in a system without metal ions yields a fluorescence enhancement due to deprotonation of the sulfonamide nitrogen. Accordingly, all experimental work was conducted at < pH 8.

In contrast, no change in the emission spectra of **4** was observed after extraction of aqueous Na$^+$, K$^+$, Cs$^+$, Sr^{2+}, Cu(II) or Pb(II) –nitrate solutions.

Unfortunately the dansyl functionalized azacrown-5 compound **6** shows no change in the emission spectra for a wide range of metal ions, although we have evidence for complexation via NMR experiments. This indicates that the acidic proton at the nitrogen plays the crucial role in this sensing mechanism, and consequently tertiary amines are not capable of performing signal transduction.

Anthracene-Fluorophore (PET/CHEF)

Synthesis

The synthesis of anthracene-functionalized calix[4]azacrown-5 **11** is outlined in Scheme 3. The crown-precursor **8** in the 1,3-alt. conformation was obtained by treatment of bis-benzyloxycalixarene with tosylated 2-chloroethoxyethanol in the presence of Cs$_2$CO$_3$. Cyclization with tosylamide and following deprotection of the amine with mercury-amalgam afforded a monoazacrown-5 derivative (*20*). Treatment with 9-chloromethylanthracene and triethylamine in THF led to fluorescent ionophore **11**.

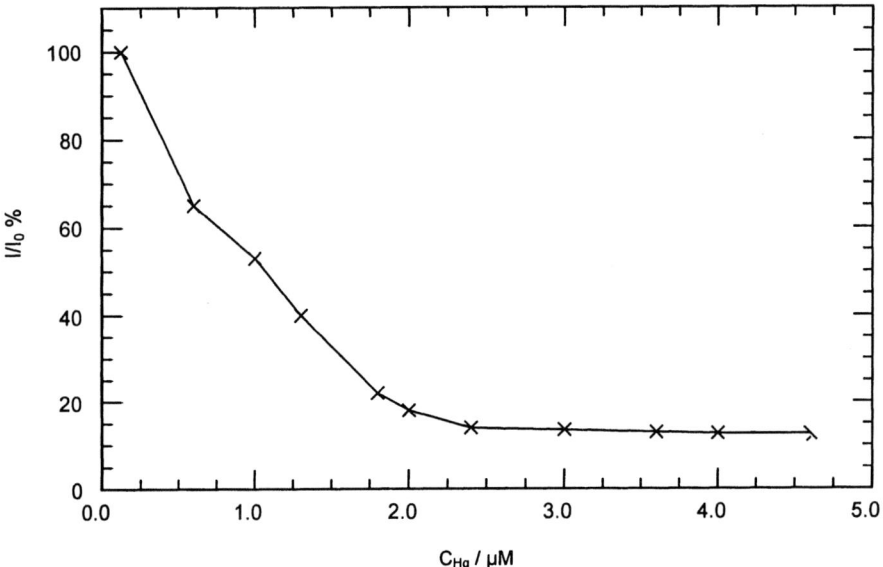

*Figure 1. Dependence on the relative emission intensity at 502 nm (λ_{ex} = 340 nm) on the initial aqueous phase concentration of Hg(II) (pH = 4.0, HNO$_3$) of 1.2 µM **4** in 1,2-DCE.*

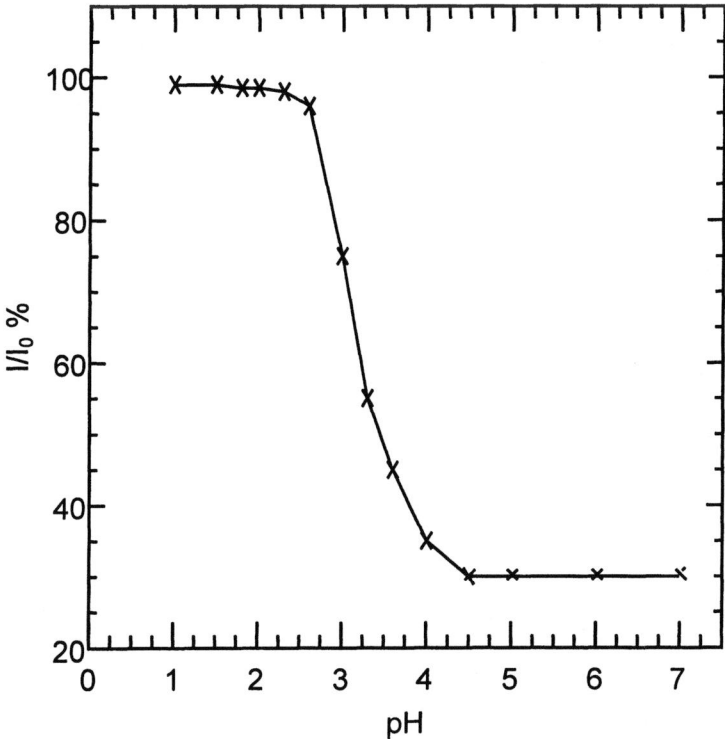

Figure 2. Dependence of the relative emission intensity at 502 nm (λ_{ex}= 340 nm) on the aqueous phase pH for extraction of 2.4 µM mercuric nitrate with 1.2 µM 4 in 1,2-DCE.

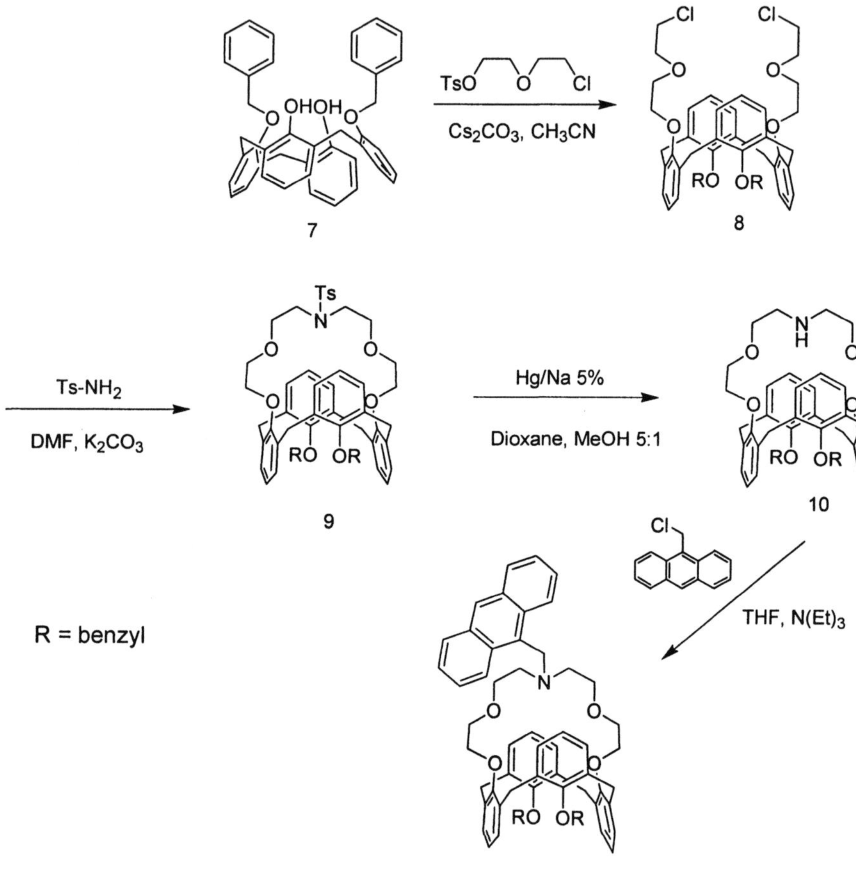

Scheme 3

Fluorescence Measurements

This sensor operates on the fluorescence photoinduced electron transfer (PET) principle. The fluorescence of the free ligand is quenched by electron transfer of the free nitrogen electron pair to the fluorophore. Upon complexation, the nitrogen lone pair no longer participates in PET, causing chelation enhanced fluorescence (CHEF) *(21-23)*. The quality of the PET and consequently the factor of enhancement upon complexation depend on the polarity of the solvent. We chose ethanol as the solvent and investigated the fluorescence change of **11** towards addition of varying concentrations of metal salts. The perchlorate salts of Hg(II), Sr(II), Cs(I), K(I), Ag(I), Na(I), and Cu(II) have been used to evaluate metal ion binding. Using these metal ions (100 μM), we found that **11** (10 μM) displayed large CHEF effects with Hg(II), Sr(II), and Cu(II) (Figure 3). Compound **11** also shows small CHEF effects with Ag(I), Cs(I), and Na(I). The CHEF effect induced by potassium was negligible, although there is some evidence that potassium is in fact bound inside the cavity of the macrocycle. It is likely that potassium as a hard metal ion only interacts with the oxygen donors and does not interact significantly enough with the nitrogen, and hence the sensing unit is unaffected.

Figure 4 shows the changes in the emission intensity of the anthracene calixarene, upon titration with strontium and mercury ions. The increase in the fluorescence intensity starts in both cases with approximately a 10-fold excess of the metal ion. This indicates low complexation association constants. The fluorescence intensity was saturated at a metal ion concentration that was an excess of 100 times the ligand concentration. The fluorescence changes upon titration with copper(II) shows a different behavior. We found a maximum enhancement at a concentration of 1×10^{-4} M copper. Upon further addition of copper, the fluorescence intensity decreases. This phenomenon is indicative of a second order quenching reaction of the excess paramagnetic copper(II) ions with the fluorophore.

Coumarin-Fluorophore (PCT)

It is desirable to have a fluorescence signaling mechanism that is independent of the polarity of the environment, and we chose to incorporate a coumarin derivative as the fluorophore. Coumarins show interesting photochemical and photophysical properties, are widely used in laser dye applications, and possess very good photochemical stability. The coumarin plays the role of a transducer by translating the energy change due to complexation of the metal inside the cavity into a measurable quenching of the

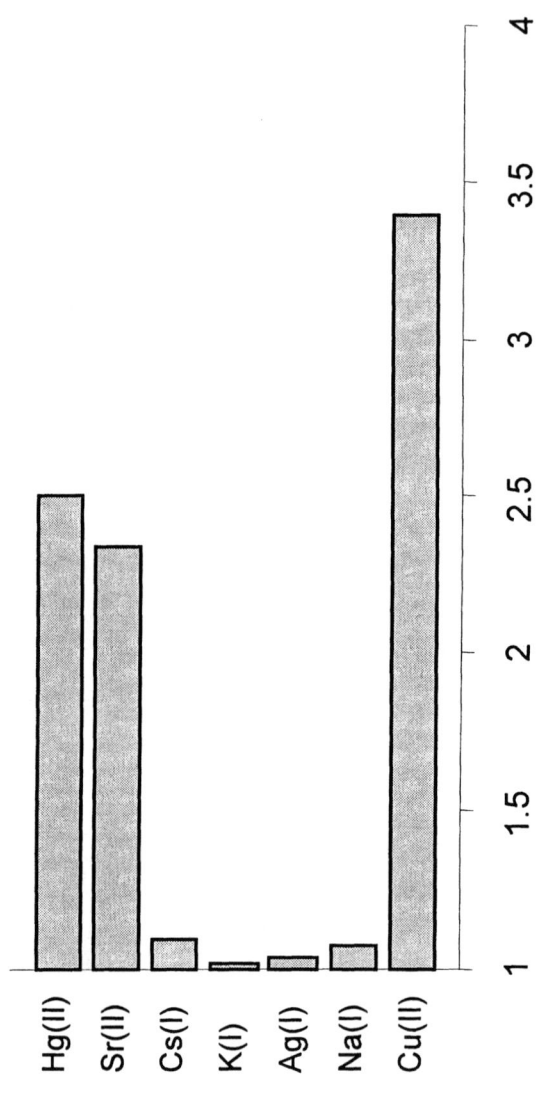

Figure 3. CHEF effects of compound 11 (1μM) with metal ions (100μM) in ethanol at 420 nm (λ_{ex} = 380 nm).

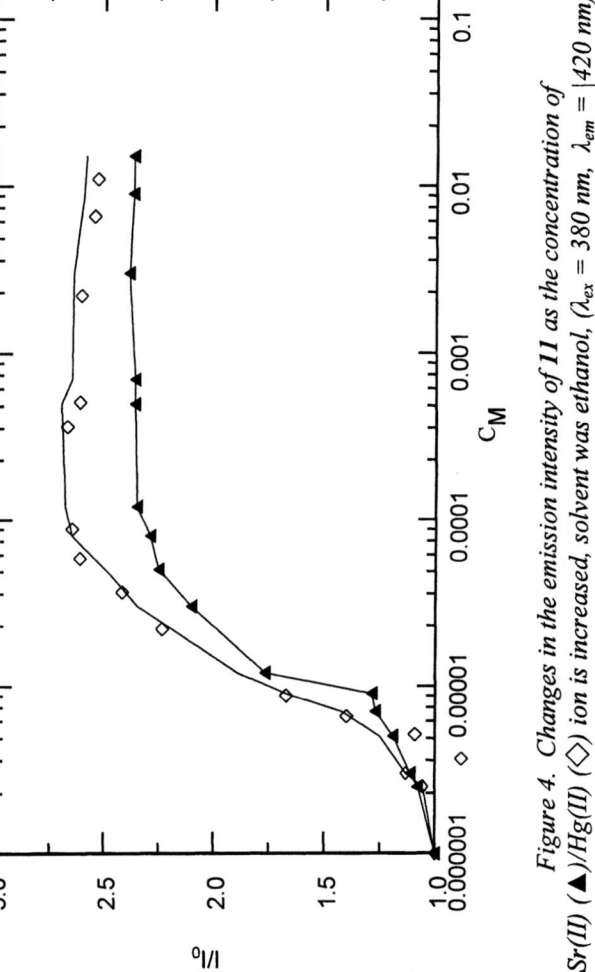

Figure 4. Changes in the emission intensity of 11 as the concentration of Sr(II) (▲)/Hg(II) (◇) ion is increased, solvent was ethanol, (λ_{ex} = 380 nm, λ_{em} = |420 nm).

fluorescence. Use of coumarin derivatives as the fluorophores changes the sensing mechanism to cation-controlled photoinduced internal charge transfer (PCT-sensor).

Synthesis

We synthesized a calix[4]crown-6 ether derivative **15** (see Scheme 4), which is well suited for the recognition of cesium ion (*24, 25*). It is also the aim to investigate the recognition of potassium, because the selectivity of this type of compound towards potassium ion against sodium ion was, in some cases, almost as good as the selectivity for cesium ion (*26*). We synthesized 4-methyl-6,7-dihyroxycoumarin **12** via the well established Pechmann synthesis (*27, 28*). Again reaction with tosylated 2-(2-chloroethoxy)ethanol yielded the bis-chloro compound **13**, which was easily converted by Finkelstein conditions to the bis-iodo compound **14**. Crowning of bis-benzyloxycalixarene with the coumarin crown precursor in the presence of cesium carbonate afforded the calixarene[4]coumarin- crown- 6 ether derivative **15**. The compound shows the ^{1}H-NMR characteristics of the 1,3-alternate conformation.

Fluorescence Measurements

Coumarin, without any substituents on the phenyl ring, is nonfluorescent. Introduction of an electron donating group on the 7-position leads to intense fluorescence due to charge transfer from the oxygen donor to the lactone carbonyl group (acceptor) (*29*). Interaction of a metal ion with the free electron pair is expected to decrease the electron donating character and consequently change the fluorescence. Previous work (*30*) with a related coumarin substituted calix[4]arene-crown-6-ether showed that coumarin fluorescence was quenched by metal ion complexation.

Fluorometric titration of **15** was carried out in ethanol at a ligand concentration of 0.6 µM. The fluorescence spectra show excitation and emission bands at λ_{ex} = 340 and λ_{em} = 411 nm, respectively. The equivalents of metal salt were varied from 0–20 times the concentration of the ligand. Evolution of the emission spectra upon addition of Cs^+ and K^+ are shown in Figures 5 and 6, respectively.

The fluorescence quenching due to Cs^+ complexation is not as effective as that by K^+. However, the maximum change is reached at a lower concentration. This correlates with earlier investigations showing that cesium forms very strong complexes with calixcrown-6-ether derivatives, but the metal is located deep inside the calixarene framework and interacts with the π-system of the phenyl units. Accordingly it is evident that the distance between the metal and the two

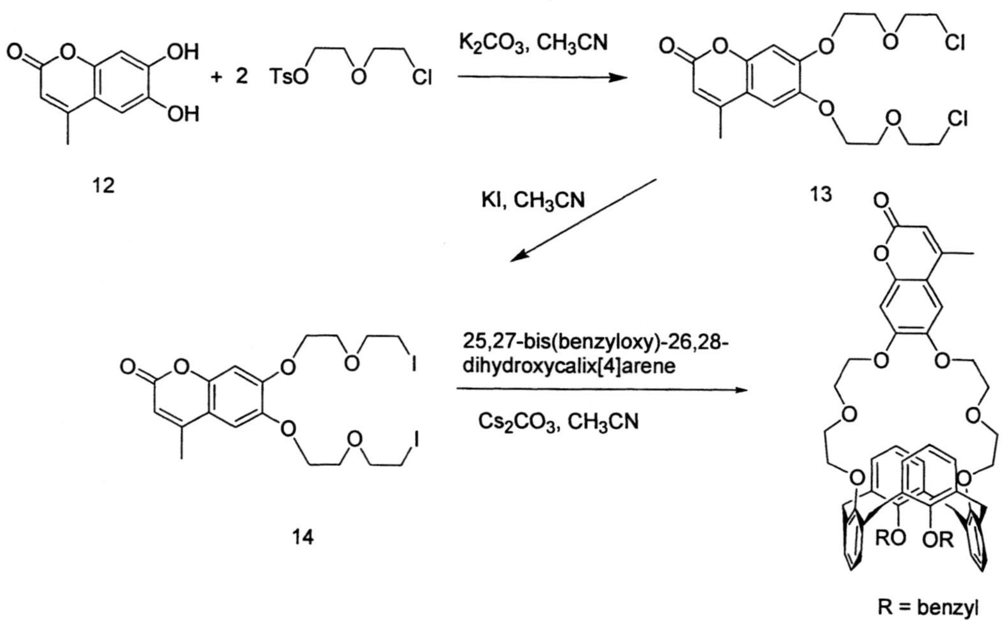

Scheme 4

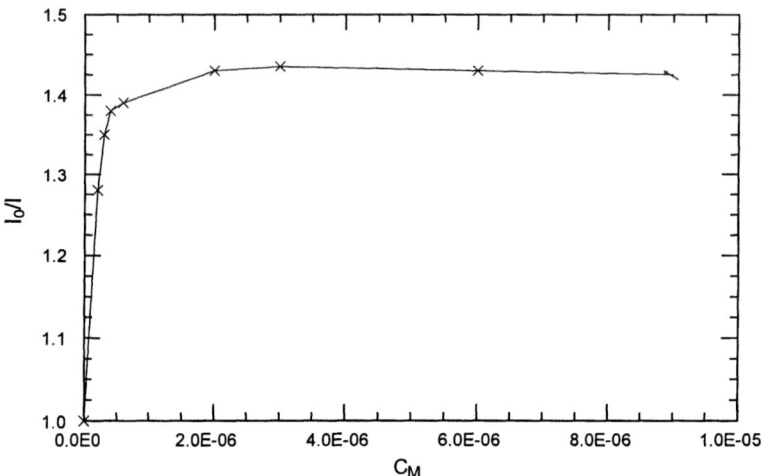

*Figure 5. Changes in the emission intensity of **15** (0.6 μM) as Cs ion is increased. Solvent was ethanol (λ_{ex} = 340 nm, λ_{em} = 411 nm, complex shows a blue shift of 6 nm from the free ligand).*

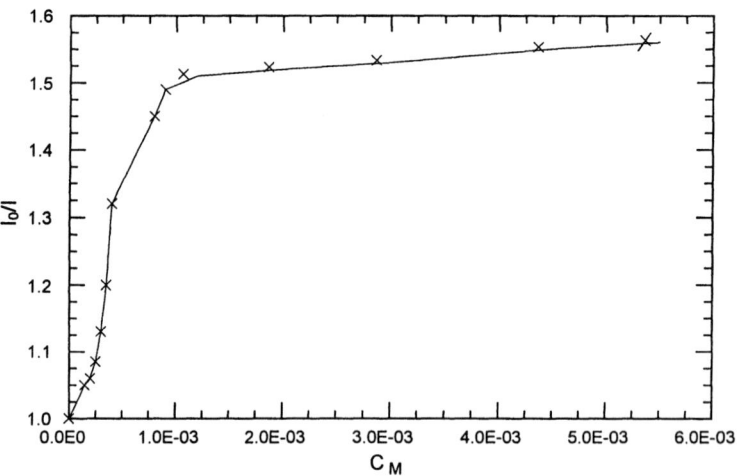

*Figure 6. Changes in the emission intensity of **15** (0.6 μM) as K ion is increased. Solvent was ethanol (λ_{ex} = 340 nm, λ_{em} = 411 nm, complex shows a blue shift of 6 nm).*

oxygen atoms of the coumarin is too large to influence the electron density and the internal charge transfer process sufficiently. In contrast, potassium forms looser complexes with smaller association constants, but is located close to the coumarin fluorophore (*26*) and provides a larger perturbation to the internal charge transfer.

Conclusion

In summary, we report the synthesis and binding study of three fluorescence ligands based on calix[4]arenes with different sensing units. Among the metal ions examined, the dansyl-functionalized compound **4** displayed highly selective quenching with mercury. The PET-sensor **11** shows CHEF effects with Hg(II), Cu(II), and Sr(II), and the coumarin calixarene crown compound **15** is able to sense cesium selective over sodium as well as potassium.

Experimental

General

All commercial solvents and all starting materials and basic reagents were used without further purification. All metal salts were dried at 110 °C for at least 48 h before use. Aqueous solutions were prepared using distilled deionized water.

The syntheses of compounds **1** (*31, 32*), **2** (*15*), **3** (*16*), **7** (*33*), and **12** (*27, 28*) were prepared as described in the literature, and compound **11** was prepared by a procedure analogous to that reported (*20*).

For the fluorescence titrations, the solutions were excited at their isosbestic point.

Synthesis

5,11,17,23-Tetrakis(1,1-dimethylethyl)-25,27-bis(2-dansylamidoethoxy)-26,28-dihydroxycalix[4]arene, **4**

To a vigorously stirred mixture of 300 mg (0.4 mmol) **3** in $CHCl_3$ and 125 mg (0.9 mmol) K_2CO_3 in H_2O was added 226 mg (0.85 mmol) dansylchloride in 10 ml $CHCl_3$ over a period of 20 min, and the reaction mixture was stirred at room temperature over night. The phases were separated, the organic layer washed with water, and after drying and evaporation the compound was obtained as yellow powder and recrystallized from ethanol to afford 173 mg (36%) of **4**.

Mp: 154–156 °C. ^{1}H NMR (400 MHz, CDCl$_3$, δ): 1.12 (s, 18H, t-butyl), 1.22 (s, 18H, t-butyl), 2.89 (s,16H, NCH_3 + NCH_2), 3.28 (d, J = 12.9 Hz, ArCH_2Ar), 3.78 (t, 4H, OCH_2), 4.20 (d, J = 12.9, ArCH_2Ar), 6.98 (2s, 8H, Ar*H*), 7.08 (d, 2H, dansyl), 7.30 (t, 2H, dansyl), 7.49 (t, 2H, dansyl), 7.85 (t, 2H, NH), 8.06 (d, 2H, dansyl), 8.35 (d, 2H, dansyl), 8.52 (d, 2H, dansyl), 8.62 (s, 2H, OH). ^{13}C NMR (100 MHz, CDCl$_3$, δ): 30.88 (q-*tert*.-butyl), 31.09, 31.51 (CH$_3$-tert.-butyl), 32.56 (ArCH$_2$Ar), 42.56 (NCH$_2$), 45.44 (NCH$_3$), 76.16 (OCH$_2$). MALDI (Matrix: 4-nitroaniline) m/z calcd. 1201.6, found 1201.8 (M$^+$), 1226.5 (M$^+$+Na), 1245 (M$^+$+K).

5,11,17,23-Tetrakis(1,1-dimethylethyl)-25,27-bis(5-chloro-3-oxapentyloxy)-26,28-dihydroxycalix[4]-arene, **5**

A mixture of 1.3 g (2 mmol) p-*tert*-butylcalix[4]arene **1**, 1.2 g (4 mmol) 2(2-chloroethoxy)ethanol *p*-toluenesulfonate and 0.56 g (4 mmol) K$_2$CO$_3$ carbonate in dry acetonitrile was refluxed for 3 days. The cooled solution was concentrated to dryness, diluted with dichloromethane, and extracted with 1N HCl, water, and brine. The organic layer was dried over MgSO$_4$, filtered, and the solvent removed. The resulting crude **5** was purified by column chromatography with ethylacetate/hexanes 1:3 v/v to yield 1.2 g (69%) as a white powder.

Mp: 96–98 °C, ^{1}H NMR (400 MHz, CDCl$_3$, δ): 0.93 (s, 18H, t-butyl), 1.27 (s, 18H, t-butyl), 3.28 (d, 4H, J = 13.06Hz, ArCH_2Ar), 3.73 (t, 4H, ClCH_2), 3.99 (m, 8H, OCH_2), 4.14 (t, 4H, OCH_2), 4.33 (d, 4H, J = 13.05 Hz, ArCH_2Ar), 6.78 (s, 4H, Ar*H*), 7.04 (s, 4H, Ar*H*), 7.15 (s, 2H, OH). ^{13}C NMR (100 MHz, CDCl$_3$, δ): 30.99, 31.69 (CH$_3$), 31.45 (ArCH$_2$Ar), 33.91, 33.82 (tert. Butyl, q), 42.87 (CH$_2$Cl), 70.02, 71.65, 71.93, 75.25 (OCH$_2$), 125.04, 125.52 (*Ar*-H), 141.37, 146.89, 149.67, 150.56 (*Ar*-q). MALDI (Matrix: 4-nitroaniline) m/z calcd. 864.03, found 863.33 (M$^+$), 885.93 (M$^+$+Na), 901.95 (M$^+$+K).

N-Dansyl-5,11,17,23-tetrakis(1,1-dimethylethyl)-26,28-dihydroxy-calix[4]arene azacrown-5, **6**

A solution of bis-(2-chloroethoxyethyl)oxy calix[4]arene **5** (0.5g, 0.58 mmol), dansylamide (0.152 g, 0.61 mmol), and 0.24 g (1.74 mmol) potassium carbonate in 30 ml of dry DMF was refluxed for 24 h. DMF was completely removed in vacuo, and 20 ml of 10% aqueous NaHCO$_3$ solution and

40 ml of CH$_2$Cl$_2$ were added. The organic layer was separated and washed with water, dried over MgSO$_4$, and then filtered. Evaporation of the solvent in vacuo gave a yellowish oil which was purified by column chromatography (R$_f$ = 0.54) using ethylacetate:hexanes (1:3) to provide 0.22 g (36.5%) of **6** as a slight greenish solid.

Mp: 122–125 °C; ^{1}H NMR (400 MHz, CDCl$_3$, δ): 0.93 (s, 18H, t-butyl), 1.26 (s, 18H, t-butyl), 2.86 (s, 6H, NC*H$_3$*), 3.26 (d, 4H, J = 12.9 Hz, ArC*H$_2$*Ar), 3.61 (t, 4H, NC*H$_2$*), 3.95 (m, 12H, OC*H$_2$*), 4.29 (d, 4H, J = 12.9, ArC*H$_2$*Ar), 6.77 (s, 4H, Ar*H*), 7.04 (s, 4H, Ar*H*), 7.15 (d, 1H, dansyl), 7.37 (s, 2H, OH), 7.50 (dt, 2H, dansyl), 8.11 (d, 2H, dansyl), 8.35 (d, 1H, dansyl), 8.49 (d, 2H, dansyl). ^{13}C NMR (100 MHz, CDCl$_3$, selected) δ 31.31 (ArCH$_2$Ar), 49.66 (NCH$_2$), 75.80, 71.17, 70.04 (OCH$_2$); MALDI (Matrix: 4-nitroaniline) m/z calcd. 1039.4, found 1040.25 (M$^+$), 1063.24 (M$^+$+Na), 1080.1 (M$^+$+K).

25,27-Bis(benzyloxy)-26,28-bis(5-chloro-3-oxapentyloxy)calix[4]arene, 1,3 alternate, 8

68% yield, ^{1}H NMR (400 MHz, CDCl$_3$, δ): 3.41–3.65 (m, 24H, ArC*H$_2$*Ar+OC*H$_2$*), 4.73 (s, 4H, OC*H$_2$* $_{benzyl}$), 6.31 (t, 2H, Ar*H*), 6.56 (d, 4H, Ar*H*), 6.66 (t, 2H, Ar*H*), 6.90–7.30 (m, 14H, Ar*H*$_{calix}$+Ar*H*$_{benzyl}$). ^{13}C NMR (100 MHz, CDCl$_3$, δ, selected): 36.72 (ArCH$_2$Ar), 42.48 (ClC*H$_2$*), 70.28, 70.47, 71.39, 72.13, 72.41 (OC*H$_2$*). MALDI (Matrix: 4-nitroaniline) m/z calcd. 818.3, found 820.13 (M$^+$).

N-Tosyl 25,27-Bis(benzyloxy)calix[4]arene azacrown-5, 9

72% yield, Mp: 198–201 °C; ^{1}H NMR (400 MHz, CDCl$_3$, δ): 3.27 (t, 4H, NC*H$_2$*), 3.36 (t, 4H, NCH$_2$C*H$_2$*O), 3.61–3.73 (m, 16H, ArC*H$_2$*Ar+OC*H$_2$*), 4.84 (s, 4H, OC*H$_2$* $_{benzyl}$), 6.35 (t, 2H, Ar*H*), 6.64 (d, 4H, Ar*H*), 6.83 (t, 2H, Ar*H*), 7.11 (d, 2H, Ar*H*$_{Tosyl}$), 7.14 (d, 4H, Ar*H*), 7.37 (m, 10H, Ar*H*$_{benzyl}$), 7.77 (d, 2H, Ar*H*$_{Tosyl}$). ^{13}C NMR (100 MHz, CDCl$_3$, δ selected): 21.5 (CH$_{3tosyl}$), 37.6 (ArCH$_2$Ar), 45.98 (NCH$_2$), 49.64 (NCH$_2$), 66.08, 70.08, 71.09 (OCH$_2$); MALDI (Matrix: 4-nitroaniline) m/z calcd. 916.2, found 919.8 (M$^+$), 943.5 (M$^+$+Na).

25,27-Bis(benzyloxy)calix[4]aren azacrown-5, 1,3-alternate, 10

36% yield, Mp: 173–175 °C; ^{1}H NMR (400 MHz, CDCl$_3$, δ): 2.81 (t, 4H, NC*H$_2$*), 3.52–3.65 (m, 12H, C*H$_2$*O), 3.69 (s, 8H, ArC*H$_2$*Ar), 4.82 (s, 4H, OC*H$_2$* $_{benzyl}$), 6.69–7.4 (m, 22H, Ar*H*). ^{13}C NMR (100 MHz, CDCl$_3$, δ selected):

38.2 (ArCH$_2$Ar), 44.32 (NCH$_2$), 48.65 (NCH$_2$), 67.24, 71.42, 71.2 (OCH$_2$). MALDI (Matrix: 4-nitroaniline) m/z calcd. 761.94, found 763.2 (M$^+$), 787.4 (M$^+$+Na).

N-(9-Anthrylmethyl)-25,27-bis-(benzyloxy)-calix[4]arene Azacrown-5, 1,3-Alternate, 11

52% yield, ^{1}H NMR (400 MHz, CDCl$_3$, δ): 2.69 (t, 4H, NC*H*$_2$), 2.82 (t, 4H, NCH$_2$C*H*$_2$O), 3.34–3.75 (m, 16H, C*H$_2$*O+ArC*H*$_2$Ar), 4.70 (s, 2H, AnC*H*$_2$N), 4.82 (s, 4H, OC*H$_2$* $_{benzyl}$), 6.9–7.4 (m, 22H, Ar*H*), 7.52 (dt, 4H, An*H*), 8.04 (d, 2H An*H*), 8.45 (s, 1H An*H*), 8.60 (d, 2H An*H*). ^{13}C NMR (100 MHz, CDCl$_3$, δ selected): 38.4 (ArCH$_2$Ar), 50.32 (NCH$_2$), 55.37 (NCH$_2$), 57.2 (NCH$_2$An), 69.24, 71.42, 71.2, 73.5 (OCH$_2$); MALDI (Matrix: 4-nitroaniline) m/z calcd. 952.18, found 954.3 (M$^+$), 977.8 (M$^+$+Na).

6,7-Bis-[2-(2-chloroethoxy)-ethoxy]-4-methylcoumarin, 13

A solution of 6 mmol (1.08 g) 6,7-dihydroxy-4-methylcoumarin, 13 mmol (3.22 g) (2-chloroethoxy)ethanol *p*-toluenesulfonate, and 1.7 g (12.4 mmol) K$_2$CO$_3$ in dry acetonitrile was refluxed for 3 days. The cooled solution was concentrated to dryness, diluted with dichloromethane, and extracted with 1N HCl, water, and brine. The organic layer was dried over MgSO$_4$, filtered, and the solvent removed. The tan oil was purified by column chromatography using ethylacetate/hexanes 4:1 v/v to yield 1.4 g (57%) as a yellowish powder.

^{1}H NMR (400 MHz, CDCl$_3$, δ): 2.37 (s, 3H, C*H$_3$*), 3.65 (t, 4H, C*H$_2$*Cl), 3.80–3.91 (m, 8H, OCH$_2$C*H$_2$*OC*H$_2$*CH$_2$Cl), 4.28 (t, 4H, Coum-OC*H$_2$*), 6.14 (s, 1H, Coum-ally-*H*), 6.89 (s, 1H, Coum-8-*H*), 7.25 (s, 1H, Coum-5-*H*).

6,7-Bis-[2-(2-iodoethoxy)-ethoxy]-4-methylcoumarine, 14

2.5 mmol (1 g) **13** and 7.5 mmol (1.275 g) KI were dissolved in acetone and heated 24 h under reflux, the solvent removed, and the residue dissolved in methylene chloride. The organic phase was washed with water, separated, dried, and the solvent removed. The crude product was used in the next step without further purification.

^{1}H NMR (400 MHz CDCl$_3$ δ): 2.37 (s, 3H, CH$_3$), 3.29 (t, 4H, CH$_2$I), 3.80–3.91(m, 8H, OCH$_2$C*H$_2$*OC*H$_2$*CH$_2$I), 4.28 (t, 4H, Coum-OCH$_2$), 6.14 (s, 1H, Coum-ally-H), 6.89 (s, 1H, Coum-8-H), 7.25 (s, 1H, Coum-5-H).

25,27-{4-methylcoumarin-6,7-diylbis[2-(2-oxoethoxy)ethoxy]}-26,28-bis-benzyloxy-calix[4]arene, 15

Using a procedure described earlier (*34*), 0.4 g (0.65 mmol) of 25,27-bisbenzyloxy-26,28-bishydroxy-calix[4]arene 7, was added to a mixture of **14** (0.37 mg, 0.65 mmol) and Cs_2CO_3 (1.06 g, 3.25 mmol) in 100 ml acetonitrile and refluxed under stirring for 3 days. After cooling to room temperature, the solvent was removed under reduced pressure. The residue was dissolved in CH_2Cl_2, and the organic layer was extracted with 1*N* HCl, brine, and water. After drying over anhydrous Na_2SO_4, the solution was concentrated, and the mixture purified by flash chromatography to give 220 mg (36%) of **15**.

^{1}H NMR (400 MHz, CDCl$_3$, δ): 2.43 (s, 3H, C*H₃*), 3.48–3.91 (m, 20H, OC*H₂*, ArC*H₂*Ar), 4.17–4.23 (dt, 4H, coumOC*H₂*), 6.19 (s, 1H, Coum-ally-*H*), 6.38 (t, 2H, Ar*H*), 6.66 (m, 6H, Ar*H*), 6.92 (s, 1H, Coum-8-*H*), 7.14 (s, 1H, Coum-5-*H*), 7.15 (m, 8H, Ar*H*$_{benzyl}$+ Ar*H*), 7.36 (m, 6H, Ar*H*$_{benzyl}$). ^{13}C NMR (100 MHz, CDCl$_3$ δ selected): 18.9 (CH3-coum), 37.46 (ArCH2), 69.45, 69.72, 70.11, 70.35, 70.47, 70.55, 70.90, 71.44, 71.94 (OCH$_2$). MALDI (Matrix: 4-nitroaniline) m/z calcd. 936.39 found 956.6 (M$^+$+Na).976.55 (M$^+$+K).

Acknowledgments

This research was funded by the Environmental Remediation Sciences Division, Office of Biological & Environmental Research, Environmental Management Science Program, DOE, under contract No. DE-AC05-00OR22725 with Oak Ridge National Laboratory (ORNL), managed and operated by UT Battelle, LLC, for the DOE. The participation of Gudrun Goretzki was made possible by appointment in the ORNL postgraduate Program administrated by the Oak Ridge Institute for Science and Education.

References

1. Kupfur, M. J.; Boldt, A. L.; Higley, B. A.; Hodgson, K. M.; Shelton, L. W.; Simpson, B. C.; Watrous, R. A.; LeClair, M. D.; Borsheim, G. L.; Winward, R. T.; Orme, R. M.; Colton, N. G.; Lambert, S. L.; Place, D. E.; Shulz, W. W. *Standard Inventories of Chemicals and Radionuclides in Hanford Site Tank Wastes;* HNF-SD-WM-TI-740, Rev 0B; Lockheed Martin Hanford Corporation: Richland, WA, 1999.

2. Kirkbride, R. A.; Allen, G. K.; Orme, R. M.; Wittman R. S. (NHC); Baldwin, J. H.; Crawford, T. W.; Jo, J. (LMHC); Fergestrom, L. J.; Hohl, T. M.; Penwell, D. L. (Cogema). *Tank Waste Remediation System Operation and Utilization Plan;* HNF-SD-WM-SP-012, Rev. 1, Vol. 1 and II. DOE-AC06-96RL13200; Numatic Hanford Corporation: Richland, WA, May 1999.
3. Hanlon, B. M. *Waste Tank Summary Report for Month Ending March 31, 2000*; HNF-EP-0182-144; CH2M Hill Hanford Group, Inc.: Richland, WA, May 2000.
4. *Fluorescent Chemosensors for Ion and Molecular Recognition*; Czarnik, A. W., Ed.; ACS Symposium Series 538; American Chemical Society: Washington, DC, 1992.
5. DeSilva, A. P.; Gunaratne, H. Q.; Gunnlaugsson, N. T. A.; Huxley, T. M.; McCoy, C. P.; Rademacher, J. T.; Rice, T. E. *Chem. Rev.* **1997**, *97*, 1515–1566.
6. Böhmer, V. *Angew. Chem., Int. Ed. Engl.* **1995**, *34*, 713–745.
7. Arnaudneu, F. A.; Collins, E. M.; Deasy, M.; Ferguson, G.; Harris, S. J.; Kaitner, B.; Lough, A. J.; McKervey, M. A.; Marques, E.; Ruhl, B. L.; Weill, M. J. S.; Seward, E. M. *J. Am. Chem. Soc.* **1989**, *111*, 8681–8691.
8. Arduini, A.; Pochini, A.; Reverberi, S.; Ungaro, R. *Tetrahedron* **1986**, *42*, 2089–2100.
9. Ji, H. F.; Brown, G. M.; Dabestani, R. *Chem. Comm.* **1999**, 609–610.
10. Ji, H. F.; Brown, G. M.; Dabestani, R. *J. Am. Chem. Soc.* **2000**, *122*, 9306–9307.
11. Kim, J. S.; Shon, O. J.; Rim, J. A.; Kim, S. K.; Yoon, J. *J. Org. Chem.* **2002**, *67*, 2348–2351.
12. Aoki, I.; Sakaki, T.; Shinkai, S. *J. Chem. Soc., Chem. Commun.* **1992**, 730–732.
13. Jin, T.; Ichikawa, K.; Koyama, T. *J. Chem. Soc., Chem. Commun.* **1992**, 499–501.
14 Casnati, A.; Baldini, L.; Pelizzi, N.; Rissanen, K.; Ugozzoli, F.; Ungaro, R. *J. Chem. Soc., Dalton Trans.* **2000**, 3411–3415.
15. Collins, E. M.; McKervey, M. A.; Madgin, E.; Moran, M.; Owens, M.; Ferguson, G.; Harris, S. J. *J. Chem. Soc., Perkin Trans. 1* **1991**, 3137–3142.
16. Szemes, F.; Hesek, D.; Chen, Z.; Dent, S. W.; Drew, M. G. B.; Goulden, A. J.; Graydon, A. R.; Grieve, A.; Mortimer, R. J.; Wear, T.; Weightman, J. S.; Beer, P. D. *Inorg. Chem.* **1996**, *35*, 5868–5879.
17. Talanova, G. G.; Elkarim, N. S. A.; Talanov, V.; Bartsch, R. A. *Anal. Chem.* **1999**, *71*, 3106–3109.
18. Sasaki, D. Y.; Padilla, B. E. *J. Chem. Soc., Chem. Commun.* **1998**, 1581–1582.
19. Fabrizzi, L.; Poggi, A. *Chem. Soc. Rev.* **1995**, *24*, 197–202.

20. Kim, J. S.; Shon, O. J.; Ko, J. W.; Cho, M. H.; Yu, I. Y.; Vincens, J. *J. Org. Chem.* **2000**, *65*, 2386–2392.
21. de Silva, A. P.; Sandanayake, K. R. A. S. *Tetrahedron Lett.* **1991**, *32*, 421–424.
22. de Silva, A. P.; Guanaratne, N. H. Q.; Sandanayake, K. R. A. S. *Tetrahedron Lett.* **1990**, *31,* 5193–5196.
23. Huston, M. E.; Haider, K. W.; Czarnik, A. W. *J. Am. Chem. Soc.* **1988**, *110*, 4460–4462.
24. Thuery, P.; Nierlich, M.; Bressot, C.; Lamare, V.; Dozol, J. F.; Asfari, Z.; Vicens, J. *J. Incl. Phenom.* **1996**, *23*, 305–312.
25. Thuery, P.; Nierlich, M.; Bryan, J. C.; Lamare, V.; Dozol, J. F.; Asfari, Z.; Vicens, J. *J. Chem. Soc., Dalton. Trans.* **1997**, 4191–4202.
26. Thuery, P.; Nierlich, M.; Lamare, V.; Dozol, J. F.; Asfari, Z.; Vicens, J. *Supramol. Chem.* **1997**, *8,* 319–332.
27. Bulut, M.; Erk, C. *Dyes and Pigments* **1996**, *30*, 89.
28. Bulut, M.; Erk, C. *Dyes and Pigments* **1996**, *32*, 61.
29. Takadate, A.; Masuda, T.; Murata, C.; Tanaka, T.; Irikura, M.; Goya, S. *Anal. Sci.* **1995**, *11*, 97–101.
30. Leray, I.; Asfari, Z.; Vicens, J.; Valeur, B. *J Chem Soc., Perkin Trans. 2*, **2002**, 1429–1434.
31. Gutsche, C. D. *Org. Synth.* **1989**, *68*, 234.
32. Gutsche, C. D.; Lin, L-G. *Tetrahedron* **1986***, 42*, 1633–1640.
33. Iwamoto, K.; Akaki, K.; Shinkai, S., *Tetrahedron* **1991**, *47*, 4325–4342.
34. Casnati, A.; Pochini, A.; Ungaro, R.; Ugozzoli, F.; Arnaud, F.; Fanni, S.; Schwing, M.-J.; Egberink, J. M.; de Jong, F.; Reinhoudt, D. N. *J. Am. Chem. Soc.* **1995**, *117*, 2767.

Chapter 3

Spectroscopic Properties and Redox Chemistry of Uranium in Borosilicate Glass

Zuojiang Li[1,2], Shannon M. Mahurin[1], and Sheng Dai[1,*]

[1]Chemical Sciences Division, Oak Ridge National Laboratory, Building 4500N, Mail Stop 6201, Oak Ridge, TN 37831
[2]Current address: 1541–F Honey Grove Drive, Richmond, VA 23229

Since borosilicate glass has been chosen as the primary matrix for the immobilization of radionuclide waste, study of the chemistry of actinides such as uranium in this glass network is essential for the design, construction, and optimization of advanced vitrification processes. In this work, the fluorescence and UV-vis properties of uranium doped in various glass matrices have been investigated. Results show that the fluorescence spectra as well as the lifetime distributions can be used to study the fundamental chemical properties of actinides in molten glasses such as the local structural heterogeneity. In addition, the fluorescence lifetime distribution of uranium can be used as an in situ optical sensor to determine its concentration in the immobilizing glass matrix. Study of the redox chemistry of uranium in the borosilicate glass suggests that the redox state of uranium is controlled by the basicity of the network as well as the imposed oxygen fugacity. The decomposition equations proposed in this paper can well explain the equilibria of uranium in molten glasses. These findings will have a broad impact on understanding the chemistry of uranium in glass melts and improving the current immobilization process.

Introduction

The immobilization and stabilization of high-level radionuclides is one of the largest and most costly challenges facing the U.S. Department of Energy (DOE) in its effort to clean up the historic and still growing nuclear wastes from various sources. In most cases, these radionuclides are immobilized in glass form and then permanently disposed in a geologic repository. Borosilicate glasses have been chosen as the primary matrices used in the immobilization of these nuclear wastes (*1-3*). This immobilization process begins by melting together mixtures of oxide powders and carbonates, usually consisting of defined quantities of SiO_2, B_2O_3, Al_2O_3, Na_2CO_3, CaO, MgO, ZnO, and Li_2CO_3 with high-level radionuclides. Molten glasses dissolve radionuclides and retain them in the inorganic frameworks. Any organic waste residues are decomposed and vaporized during the melting process. The efficiency of this waste immobilization process depends on two factors: (1) the solubility of the actinide oxides in the melts and (2) the structural stability of the final waste forms against leaching processes (*4-10*). The solubility of actinide oxides in mixed oxide media is strongly dependent on their redox states and the Lewis acid-base properties of the melts, while the structural stability of the final waste forms is fundamentally related to the local structures of the actinides in glasses and the structural modification of silica-based glasses induced by trapped actinides. Therefore, a fundamental understanding of the chemistry concerning actinides and their radioactive progeny in glass-forming melts is crucial to the efficient and long-term immobilization of such species.

The stability of the actinides against leaching from immobilized glass matrices is determined by the structure characters of the sites occupied by these molecules. Development of a structural model for actinide species in glass media requires the accurate determination of all local environments present in the glass. Fluorescence techniques are well suited for examining the site distribution and structural stability of actinides in glass matrices due to the sensitivity of the fluorescent species to their surroundings. Changes in the photophysical properties of the fluorescent probe, such as the fluorescence spectrum and the decay lifetime, can provide valuable information concerning the presence of structural variations in the glass. For example, site heterogeneities on a silica surface have been characterized by fluorescence decay kinetics of a probe molecule covalently bound to the silica (*11*).

Uranium glass exhibits strong fluorescence with good quantum yield and long emission lifetimes that are sensitive to both the local environment and mobility of the uranyl ion. It is an ideal probe for investigating local environments in glass matrices as a first step towards understanding actinide stability in these disposal media. In fact, previous studies have successfully demonstrated the effectiveness of steady-state and time-resolved fluorescence techniques for identifying variations in the surface structure of sol-gel glasses induced by a uranyl imprinting process (12).

When introduced into an oxide medium, a metal ion behaves as a Lewis acid while the anionic network behaves as a Lewis base. Thus, the basicity of the molten glass matrix can be envisaged as its ability to share electrons with an acidic solute such as the uranium ions. As in the case of a protonic system, a numeric scale such as *pH* is always expected for the comparison of acidity or basicity in two different solvent systems. Because the Lewis acid-base concept can only give a qualitative explanation of some chemical phenomena, the Lux-Flood concept (13–14) was proposed. This theory allows the establishment of oxyanion activity pO^{2-} for glass melts, which is analogous to *pH* in a protonic solvent. Unfortunately, the applicability of this concept is severely restricted to specific systems due to the difficulties in defining the thermodynamic oxide ion activities (pO^{2-}). Based on the nephelauxetic effects measured from UV spectra, Duffy et al. proposed an optical basicity by using ions such as Tl^+, Pb^{2+}, and Bi^{3+} as the probing ions (15–18). The oxygen atoms of the oxide network surrounding the metal ion donate the negative charge, which reduces the positive charge on the metal ion. The greater the negative charge borne by the oxygen atoms, the greater the electron donation. If the metal ion can signal the extent to which its positive charge falls, then there is the possibility of using it as a probe for the quantitative measurement of basicity of the corresponding glass media. The basicity scale thus obtained has been called optical basicity and is one of the most fundamental parameters that characterize glass matrices analogous to the way in which pH is used to describe properties of aqueous systems. The ratio of the nephelauxetic parameter of a probe ion for a particular oxide system to that for an ionic oxide (usually CaO) is defined as the optical basicity, Λ,

$$\Lambda = (\nu_f - \nu)/(\nu_f - \nu_{CaO}) \tag{1}$$

where ν_f is a specific electronic transition energy of the free, uncoordinated probe ion, and ν_{CaO} is the frequency for the probe ion in CaO. Accumulation of optical basicity data for oxidic systems has made it possible to assign Λ values to individual oxides: $\Lambda(AO_{a/2})$, $\Lambda(BO_{b/2})$, and so on. Accordingly, the optical basicity of a multi-component system can be calculated using

$$\Lambda = X_{AOa/2} \Lambda(AO_{a/2}) + X_{BOb/2} \Lambda(BO_{b/2}) + \ldots \tag{2}$$

where $X_{AOa/2}$ is the equivalent fractions of the individual oxides.

The concept of optical basicity makes it possible to compare the basicity of oxide media where the concepts of Lux-Flood and Bronsted-Lowry do not apply. In this work, the fluorescence emission spectra, decay profiles, and UV-vis spectra of several uranium-doped glasses were measured in an effort to evaluate the effect of compositional changes on the local environments experienced by the actinides. The spectroscopic results are compared and correlated to the structural differences in these samples. The redox chemistry of uranium in borosilicate glass is also discussed, and the decomposition equations presented are in excellent agreement with the experimental data.

Experimental

Materials

Four different kinds of uranium doped glasses were used in the experiment. The frits 165 and 202 used in the vitrification of radioactive waste materials were obtained from the Savannah River Site. The compositions of these two frits are listed in Table I. For samples prepared with frit 202, a specific amount of uranium trioxide (UO_3, 2.9 wt %) was mixed with the frit 202 and melted at 1400 °C for 10 h to produce a uranyl-doped frit 202 sample. In the case of frit 165, a series of samples with increasing amounts of uranium trioxide (0.2 to 20 wt %) were prepared at 1150 °C for 5 h in air. To investigate the influence of oxygen concentration on the redox state of uranium in glasses, a frit-165 based sample was prepared at 1150 °C in reducing atmosphere. The oxygen concentration in the tube furnace was controlled by placing carbon black beside the sample in the flow of high-purity nitrogen. The high temperature was achieved with a Linberg furnace with a maximum temperature of 1700 °C.

Table I. The Compositions of Two Frits from Savannah River Site and a Commercial Uranium Glass

Composition	Frit-165 (%)	Frit-202 (%)	Comm. Glass*
SiO_2	68.0	77.0	80.8
B_2O_3	10.0	8.0	12.0
Na_2O	13.0	6.0	4.2
Li_2O	7.0	7.0	
MgO	1.0	2.0	
ZrO_2	1.0		
Al_2O_3			2.0

* Another 1.0 wt % is uranium oxides.

Uranyl-doped sol-gel glasses were prepared using established literature procedures (19). Briefly, $UO_2(NO_3)_2 \cdot 6H_2O$ was dissolved in 2 N nitric acid to produce an 0.1 M aqueous uranyl solution. 1 mL of the $UO_2(NO_3)_2 \cdot 6H_2O$ aqueous solution (0.1M) was mixed with 1 mL of tetramethyl orthosilicate and 1 mL of methanol. The final mixture was cured at ambient temperature over a period of four weeks.

A borosilicate glass sample doped with uranium oxide was obtained from Yankee Glassblower, Inc. (Concord, Massachusetts). This green glass contains a very small amount of uranium oxide (1 wt %) and is generally used to bond dissimilar glasses in graded seals. The major components of this glass can be found in Table I.

Fluorescence Measurement

In a typical lifetime measurement experiment, a pulsed nitrogen laser (337 nm, pulse width = 5 ns) was transmitted to the luminescent uranium glasses by using an optical fiber. Fluorescence emission was collected by a second optical fiber at 180° with respect to the excitation direction and subsequently detected by an avalanche photodiode (APD, Perkin-Elmer). The excellent sensitivity of the APD detector and quantum yield of the uranium eliminated the need for focusing optics prior to the collection fiber. An interference filter (Thermo Oriel) with a transmission center at 550 nm and a 70-nm spectral bandpass was used to isolate the uranyl fluorescence and reject stray scattering from the excitation source. Time-resolved fluorescence decay traces were acquired by using a multichannel scaler (EG&G ORTEC) controlled by a personal computer. Emission spectra were obtained by connecting the collection fiber to the entrance port of a Spex 500M monochromator with a 150 groove/mm grating. Dispersed radiation was detected by a liquid nitrogen cooled charge-coupled device (CCD) detector (Spex System One) with a one-inch chip. No further corrections were made to the measured spectra, and no attempt was made to determine absolute emission intensity.

UV-Vis Near IR Spectra

The UV-Vis near infrared spectra of the frit-165 based uranium glasses were measured on a Varian Cary 5000 spectrophotometer and a homemade apparatus under the diffusive reflectance mode. The baseline was corrected with a frit-165 glass containing no uranium but prepared at the same conditions at 1150 °C for 5 h in air.

Results and Discussion

Fluorescence Properties of Uranium Doped Glass

Frit 165 shows very good solubility toward UO_3 when heated at 1150° C for 5 h. The uranium doped glasses appeared quite homogeneous, with colors ranging from light green to yellow when the UO_3 composition was increased from 0.2 to 20.0 wt %. Despite the uniform appearance of the glasses, structural changes were observed using fluorescence spectroscopy. Shown in Figure 1 are the fluorescence decay profiles of uranium trioxide immobilized in the frit 165. The time resolution of the experiment is 10 μs. To improve signal-to-noise ratios and reduce errors, 1200 decay traces were acquired and averaged for each sample. Noticeably, uranium showed strong fluorescence and a quite long decay time of ~1.6 ms for the glasses doped with a low concentration of uranium trioxide (0.20 wt %, 0.48 wt %). The fluorescence of uranium in the frit 165 decays rapidly with increasing UO_3 concentration due to self-quenching and the increased interactions among uranium species. Three samples doped with a large amount of uranium trioxide (> 10 wt %) exhibit rather similar decay profiles and a short decay time of ~ 0.06 ms, indicating the interactions among uranium atoms is so strong that the self-quenching of uranium fluorescence predominates in this situation. Because of the amorphous structure of glass, the uranium can occupy a variety of sites with a wide range of environments present to influence the fluorophore, thus leading to a complicated fluorescence decay curve (20, 21). It has been shown that these complex decay profiles can be more realistically modeled by using a continuous lifetime distribution, which can be mathematically represented by an integral equation of the form:

$$I(t) = \int_0^a e^{-t/\tau} s(\tau) d\tau \qquad (3)$$

where I(t) is the measured fluorescence intensity over time, and s(τ) is the lifetime distribution function. In this work, the continuous model analysis program, CONTIN (22), was used to perform the numerical Laplace inversion of the measured decay curves. Solutions to the inverse Laplace transform were constrained to positive distribution coefficients. No further assumptions concerning the lifetime distribution function are made in the CONTIN inversion of the experimentally measured fluorescence decay traces. The calculated lifetime distribution functions for the uranium-doped glasses are shown in Figure 2.

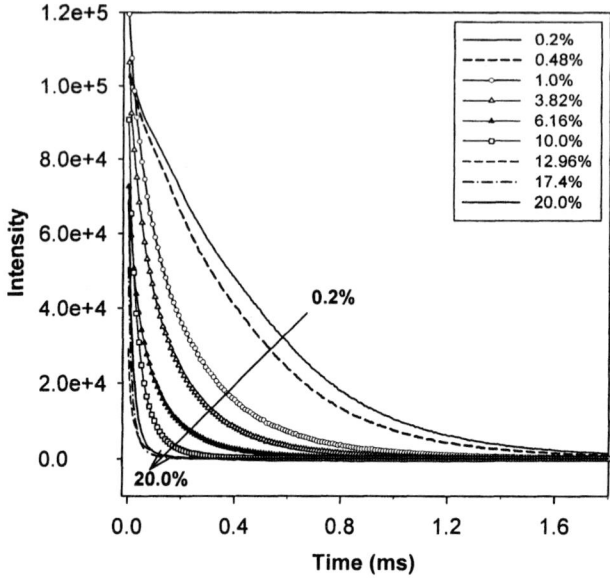

Figure 1. The fluorescence decay profiles of frit-165 based glasses doped with gradually increased amount of uranium trioxide.

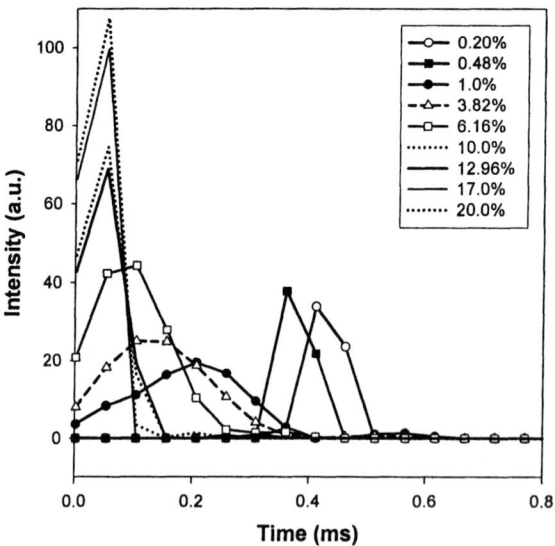

Figure 2. Lifetime distributions for uranium in frit-165 based glasses doped with gradually increased amount of uranium trioxide. For visual clarity and comparison, the distributions for samples doped with < 10.0 wt % UO_3 were magnified by a factor of 2.5.

The lifetime distributions for samples doped with 0.2 wt % and 0.48 wt % uranium trioxide show sharp and long-time peaks at 410 μs and 360 μs, respectively, indicating that the uranium is homogeneously immobilized in the glass matrix. In other words, the primary interactions in this situation should be ascribed to uranium and the surrounding oxides in the glass matrix. As the doping quantity increases, the bonding sites in which long lifetime fluorescence is emitted gradually disappear and the lifetime distribution peaks gradually broaden and shift to a short time position. When the doping amount is higher than 10 wt %, dramatic structural changes occur and the uranium species experience quite similar local environments in these glasses. The four samples doped with a large quantity of uranium trioxide exhibit rather similar lifetime distributions with a peak position at ~ 60 μs. In this situation, the interactions among uranium sites are so strong that self-quenching plays an important role in the fluorescence lifetime. The short lifetime and its broad distribution indicate that the doped uranium probably forms clusters in the glass matrix when its concentration is higher than 10 wt %. Previous study (4) showed that the maximum solubility of uranium oxide in the SRL-411 based borosilicate glass is about 10 wt % in reductive conditions. Investigations are currently in progress to further explore the relationship between the uranium solubility and the fluorescence lifetime. It should be noted that in highly doped glasses, the uranium trioxide behaves as a major component, which could significantly change the properties of the doping glass.

An interesting finding is that an approximately linear relationship was observed between the peak position in the lifetime distribution and the logarithmic value of uranium concentrations (see Figure 3) when the doping quantity is less than 10 wt %. This can be explained by the fact that interactions between uranium atoms and the glass matrix dominate in the slightly doped glass. Thus, the shift of lifetime distributions can be explained by a gradual change of the local environment experienced by uranium in the glass matrix as the uranium fraction increases. Therefore, uranium itself can be used as a probing atom to measure its concentration in borosilicate glass. In contrast, the fluorescence of uranium is rapidly absorbed by the adjacent uranium atoms in highly doped glasses. The linear relationship between the lifetime distribution and uranium concentration can be used for rapid, nondestructive determination of uranium quantities in the immobilized glass. The only requirement is that uranium species can form a homogeneous glass with the immobilizing matrix. Further systematic study and the validity of this method are still required in practical applications for other systems.

To compare the fluorescence properties in different glasses, the fluorescence decay profiles were measured for uranium in the sol-gel silica, a frit-202 based glass and a commercial uranium glass used to bond dissimilar glass. Shown in Figure 4 are the lifetime distributions of uranium in these glasses. No heating was involved in the preparation of the sol-gel sample. The uranyl was simply

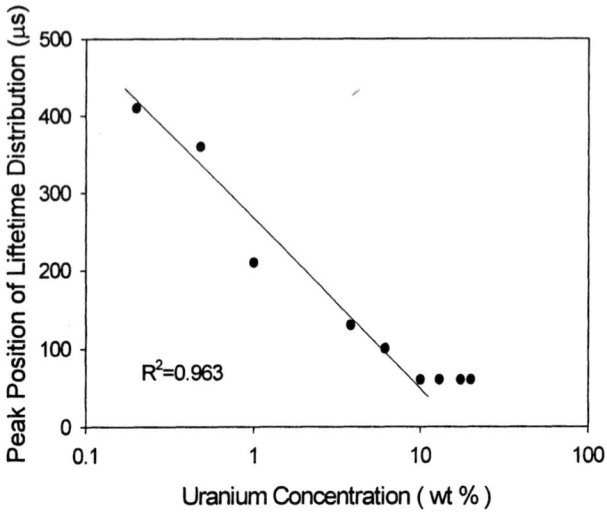

Figure 3. The linear relationship between the uranium concentration and the peak position on the calculated lifetime distribution plot.

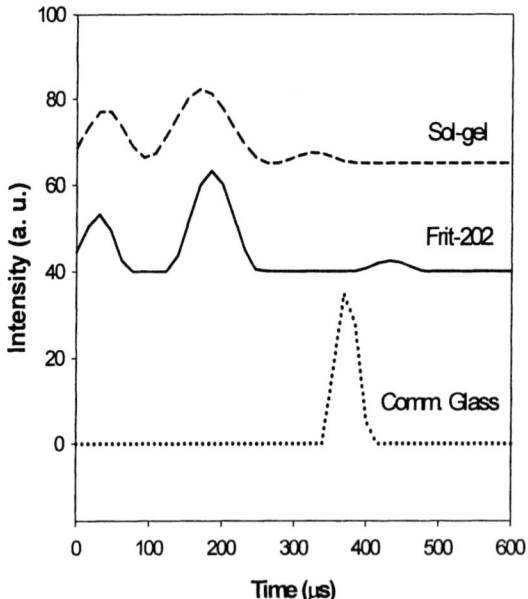

Figure 4. Lifetime distributions for uranyl in sol-gel glass, frit 202, and a commercial uranium glass. The distribution curves of Sol-gel glasses and frit-202 based glass are shifted upward for visual clarity.

entrapped in the silica network as the solvent evaporates, and hence, a large amount of heterogeneity in the local surroundings of the uranyl is anticipated. The lifetime distribution of the uranyl-doped sol-gel glass clearly exhibits three distinct peaks at 40 µs, 170 µs, and 325 µs. According to previous study of sol-gel glasses (12), the short lifetime peak can be associated with uranyl ions somewhat weakly bound at or near the silica surface, while the longer lifetime peaks are attributed to uranyl ions more strongly entrapped within the silica matrix. The relative peak can signify the relative occupancy of each environment.

The lifetime distribution of the frit-202 glass exhibited quite similar characteristics to that of the sol-gel glass. The two peaks at shorter lifetimes (40 µs and 170 µs) are nearly identical in peak position and relative size for these two samples. The primary difference in the lifetime distributions occurs in the position of the long lifetime peak. In the frit-202 glass, this peak is shifted to a longer time at 370 µs. Unlike frit 165, frit 202 does not form a homogeneous glass with uranium trioxide. A white, amorphous silicate material resulted when frit 202 was heated at 1400 °C, showing that a phase separation probably occurs and the structure is still quite heterogeneous. Frit 202 is designed to immobilize uranium in a waste stream combined with a number of other contaminants such as transition metal ions. In the absence of these secondary contaminants, it is not surprising to see three distinct peaks in the lifetime distribution plot. The fluorescence of frit-202 based glass probably comes from uranium on the glass surface, bound to the glass matrix and those well-dissolved in the glass network. The rather weak peak at 460 µs indicates that the solubility of uranium trioxide in this frit is rather small.

In contrast to frit 202 and the sol-gel glass, the commercial borosilicate glass doped with uranyl displays a significantly different distribution with only one sharp peak at 370 µs. The presence of a single peak suggests the existence of a homogeneous environment throughout the glass with little or no structural variations, which is consistent with the higher transparency and light-green appearance of this glass. It should be pointed out that the concentration of uranium in the commercial glass is 1 wt %, and its peak position at 370 µs is within the experimental error from that observed for the frit-165 glass doped with 1.0 wt % UO_3. Again, this result indicates that the lifetime distribution can be used to measure the uranium concentration in glass. Comparison of the lifetime distribution is a very useful method to analyze the structural homogeneity and the concentration of uranium in the corresponding uranium doped glass.

Presented in Figure 5 are the fluorescence spectra of uranium in the frit-165 based glasses measured at room temperature. The steady-state fluorescence emission spectra of these uranium-doped glasses also change noticeably with increasing UO_3 concentrations. For samples doped with a lower concentration of uranium trioxide (≤ 10 wt %), fluorescence spectra with broader peak widths were observed in addition to the slight red shift of the band maximum. Even though a detailed assignment of peaks is not available, it is

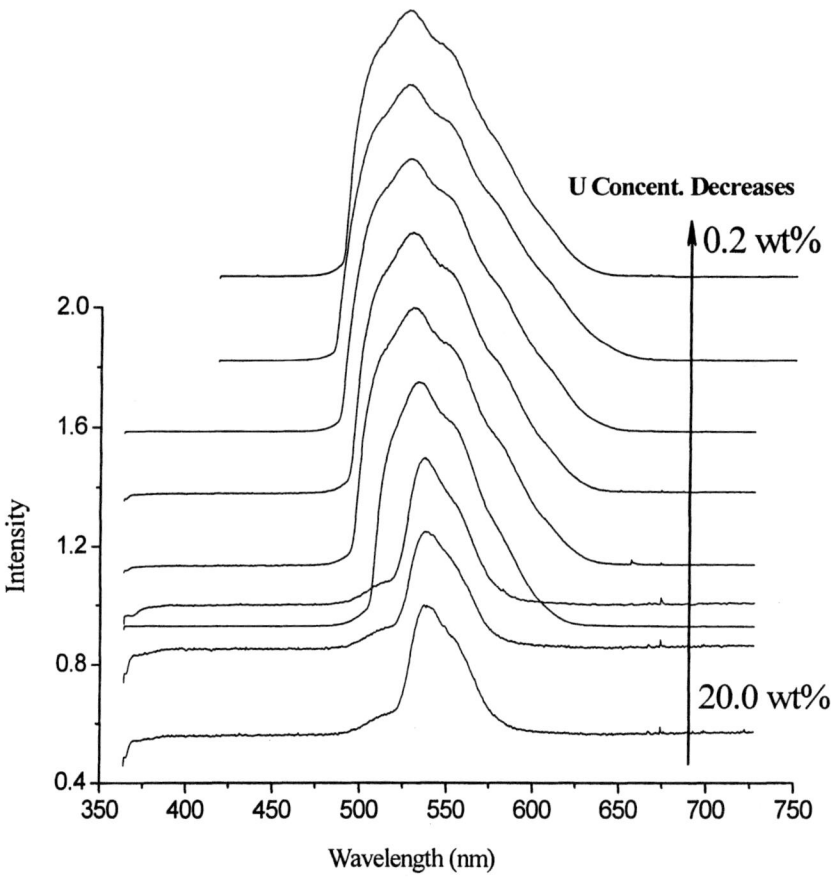

Figure 5. Steady-state fluorescence of uranium in frit-165 based glasses with gradually increased amount of uranium.

believed that the positions of these peaks are related to the basicity of the equatorial ligands, the length of the U-O bond in the uranyl group, and bonding interactions between the uranyl ions and the matrix oxide (*23*).

Significant differences are also observed in the fluorescence spectra for frit-165 glasses doped with large (> 10 wt %) and small (≤ 10 wt %) amounts of uranium trioxide. All samples exhibit a band maximum at about ~ 535 nm, but the secondary peak at ~550 nm disappears for samples doped with high uranium concentrations. In addition to the narrowed emission spectra, one can notice that the emission of short-wavelength fluorescence disappears or is significantly weakened for samples doped with over 10 wt % uranium trioxide. This again verifies that the mutual interactions among the uranium atoms dominate the fluorescent emissions in highly-doped glass, leading to self-absorption and the loss of high-energy emissions of short wavelengths. These observations are consistent with those of the lifetime distributions shown in Figures 3 and 4, proving the applicability of using fluorescence in the detection of structure homogeneity and uranium concentration in glass matrices. Uranium itself can essentially act as an in situ optical sensor for vitrification processes.

Redox Chemistry of Uranium Species in Molten Glass

Shown in Figure 6a is a representative UV-Vis-NIR absorption spectrum for the frit-165 based glass doped with 3.82 wt % uranium trioxide. The spectrophotometric features of uranium dissolved in borosilicate glasses are analogous to those reported previously. The prominent absorption peak at 415 nm and the shoulder at 490 nm is related to the U(VI) in glass (*4, 24*). The photometric absorption spectra of these samples have also been measured in the near infrared range. However, no characteristic absorption peaks were observed for U(V) at 900 nm, 1 400 nm, and 1580 nm, as well as for U(IV) at 640 nm, 1030 nm, and 1780 nm. It is evident that the uranium exists as U(VI) in the sample made in air. On the other hand, the sample doped with 1.0 wt % UO_3 and prepared in a reducing atmosphere in which the oxygen concentration was controlled by nitrogen plus carbon black, shows prominent absorption peaks at 1050 nm and 1 760 nm. These are characteristic absorption peaks associated with the U(IV). In addition, the prominent peaks of U(VI) at 415 nm and 490 nm were absent, showing that uranium exists as U(IV) in highly reduced conditions.

According to Schreiber, uranium can exist in glass in multiple valances such as U(VI), U(V), U(IV), or even U(III) in highly reductive conditions. The redox equilibria in the borosilicate glass were expressed by the following ionic equations (*4*):

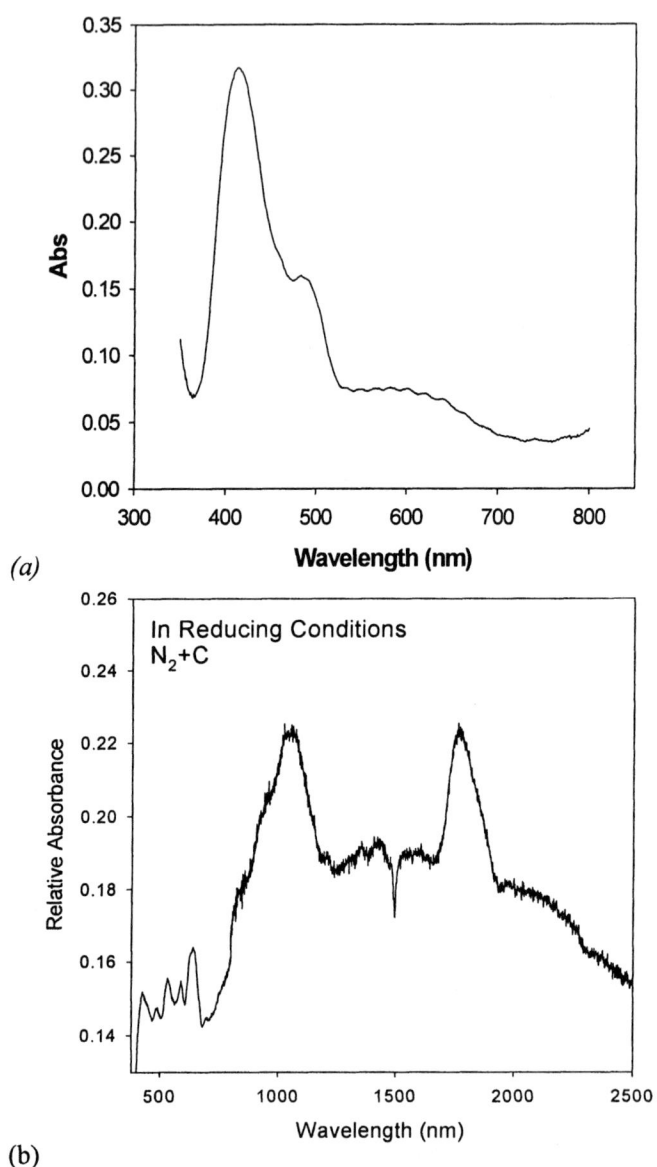

Figure 6. The UV-Vis-NIR spectrum of frit-165 glasses doped with (a) 3.82 wt % uranium trioxide in air at 1150 °C and (b) 1.0 wt % uranium trioxide in reducing atmosphere at 1150 °C.

$$4U^{6+} + 2O^{2-} \Leftrightarrow 4\ U^{5+} + O_2 \text{ (gas)} \tag{4}$$

$$4U^{5+} + 2O^{2-} \Leftrightarrow 4\ U^{4+} + O_2 \text{ (gas)} \tag{5}$$

So, the valances of uranium in the immobilized glass greatly depend on the imposed oxygen fugacity in the molten conditions. The different valance states can also impart distinct color to the glass. All the samples in this work were prepared in air with relatively high oxygen fugacity, which shifted the equilibrium to left in both eqs 4 and 5. Thus, these samples showed a color ranging from light green to light yellow. Combined with the UV-Vis absorption spectra in Figure 6, this indicates that uranium primarily exists as U(VI). No measurable amounts of U(V) and U(IV) were detected.

The redox equilibria of uranium in molten glass depend not only on the imposed oxygen fugacity but also on the glass composition. As pointed out earlier, the basicity of the glass medium has been used to rationalize the ability of a molten glass to stabilize specific oxidation states of certain ions, due to the different basicity of various oxides. Even for uranium doped glasses prepared at constant oxygen fugacity such as in carbon dioxide atmosphere, the U^{6+}-U^{5+} redox equilibrium is greatly affected by the melt compositions. For example, the U^{5+}/U^{6+} ratio increased significantly as the Na_2O fraction gradually decreased in the SiO_2-Na_2O glass system (25). Schreiber correlated this uranium equilibrium with the change of composition and explained that the uranium becomes systematically more oxidized as the melt becomes more basic, i.e., richer in sodium oxide. On the other hand, Duffy (26) suggested that the redox equilibria in silicate glasses can be better explained in terms of the change in optical basicity rather than the molar ratio of the glass compositions. The use of optical basicity can probably explain the nonlinear relationship observed by Schreiber et al. (25). By using an optical basicity of 0.48 for SiO_2 and 1.15 for Na_2O (27), the uranium redox equilibria data of previous study (25) are represented in Figure 7. One can notice that a linear relationship can be observed between the redox ratio of Log (U^{5+}/U^{6+}) and the overall optical basicity for each glass sample. Consequently, the optical basicity is a very useful method to explain the equilibrium of uranium in simple silicate glass. The validity of applying the optical basicity to more complicated glass media still requires further study.

Another significant result is the gradual increase in the U^{6+}/U^{5+} ratio with the increase in the basicity of the glass media at constant oxygen fugacity. Previous studies also proved that for many transition metal ions, the redox system becomes more oxidized with increasing glass basicity (25, 28, 29). As we know, the basicity of the glass media is controlled by the oxyanion activity (pO^{2-}). So, the

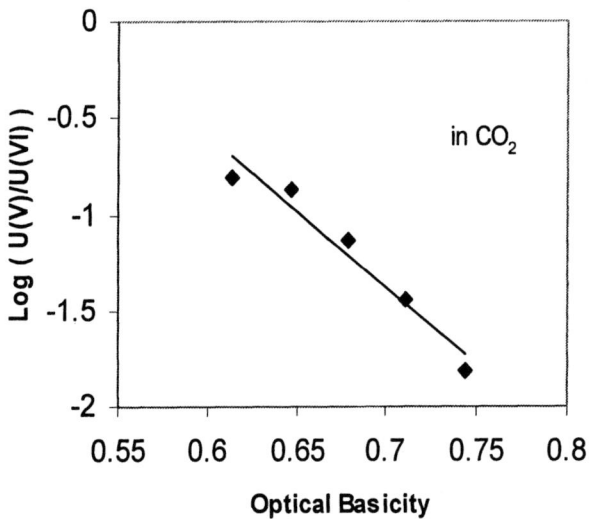

Figure 7. The redox equilibrium of uranium as a function of the optical basicity. Data from reference 25.

observations above concerning the equilibrium of uranium in borosilicate glass are contradictive to the reduction-oxidation equilibria expressed in eqs 4 and 5, according to which, the increase of the melt basicity (O^{2-}) will shift the equilibrium to a more reduced state. Therefore, the equilibria of uranium in molten glass can be better explained by the following equations:

$$UO_3 \Leftrightarrow UO_2^{2+} + O^{2-} \tag{6}$$

$$4UO_2^{2+} \Leftrightarrow 4UO^{3+} + 2O^{2-} + O_2 \tag{7}$$

$$4UO^{3+} \Leftrightarrow 4U^{4+} + 2O^{2-} + O_2 \tag{8}$$

In prior study, Schreiber et al. proved that the U(VI)-U(V)-U(IV) two-step redox equilibrium exists in the uranium doped borosilicate glasses (*4*). Their titration study of the relative amounts of U(V) and U(IV) showed that plots of $-log\ f_{O_2}$ against $log\{[U(V)/U(VI)]\}$ and $log\{[U(IV)]/[U(V)]\}$ gave two parallel straight lines with a slope of 4 at constant base composition and melt temperature. They explained the observations with the equilibria in eqs 4 and 5. However, the current results show that the uranium oxide undergoes decomposition reactions shown in eqs 6–8, instead of ionic reactions in eqs 4 and 5. In addition, the oxyanions act as products rather than reactants. The equilibria in eqs 6–8 can explain not only the slope of 4 between $-log\ f_{O_2}$ vs. $log\{[U(V)/U(VI)]\}$ and $log\{[U(IV)]/[U(V)]\}$, but also the fact that more oxidized valences are favored with increasing melt basicity.

Equations 6–8 are also very helpful in understanding the coordination chemistry of uranium in glasses and the stability of the immobilized uranium. They can be used to optimize the immobilization processes. These equations suggest that U^{6+} ions are covalently bonded in glass melts while U^{4+} is simply coordinated to oxygen ligands within the glass structure.

Previous study showed that the spectral positions and intensities of U(VI), U(V), and U(IV) are quite similar in borosilicate, aluminosilicate, and aluminophosphate glasses (*4*), indicating that the coordination sites of individual uranium redox species are similar to each other. It is believed that the characteristic absorption feature at ~ 415 nm in Figure 6 is related to the uranyl ions (UO_2^{2+}) within the glass matrix, which coincide with the equilibrium in eq 6. Consequently, the favored oxidized redox state of uranium in glasses with higher basicity is probably related to its tendency to form covalent complexes, UO_2^{2+}. The coordination site of U(V) in borosilicate glasses has an octahedral structure with tetragonal distortion. The U(V) complex is stable in glass but not in aqueous solution because of the disproportionation reactions. U(IV) is eightfold coordinated within glasses, and borate groups or silicate groups are closely associated with the U(IV) in this eightfold coordination (4). The redox state of uranium not only affects the coordination chemistry, but also the solubility of uranium in immobilizing glass matrices. It was found that U(VI) has a much higher solubility than U(IV) in the borosilicate glass at 1150 °C in air (4). It should be pointed out that the uranium solubility is only one of the many factors to be considered in the final choice of immobilization conditions, which also include the redox state of uranium in the glass matrix, the structure stability against leaching, the composition of the glass matrix, metal oxide additives, and the imposed oxygen fugacity.

The above fundamental chemistry of uranium provides essential insights into the immobilization of radionuclides in simple borosilicate glasses where no other multivalent ions are involved. In this case, the equilibria are controlled by the melt basicity and the imposed oxygen fugacity. However, the real glasses in immobilization processes are much more complicated because many multivalent ions such as iron, nickel, titanium, magnesium, and zirconium are present in the frit and sludge. These elements will significantly influence the redox equilibria and stability of uranium in the glass matrix, through mutual interactions as well as coordination complex formation in the melt. The mutual influences between metal ions can be expressed by the following simplified equilibrium:

$$mA^{(z+n)+} + nB^{x+} = mA^{z+} + nB^{(x+m)+} \qquad (9)$$

Duffy found that the logarithmic value of the above equilibrium quotient (*Log Q*) is related to the optical basicity through a simple linear expression for glass samples made in air at 1400 °C (*26*). Therefore, the redox state and the stability of uranium species in molten glass can be controlled by adding the proper amount of metal oxide additives in the immobilization process. Experiments are being conducted in our lab to investigate the effects of these transition metal ions on the redox chemistry of uranium. It is expected that by controlling the base compositions of the glass matrix, the melt temperature, and the imposed oxygen fugacity, uranium can be immobilized in the borosilicate glasses with high stability and leaching resistance.

Conclusion

The fluorescence properties of several different uranium-doped glasses were measured and the lifetime distributions were calculated by using a continuous model. The results show that fluorescence spectroscopy is a powerful method to measure the structure heterogeneity and local environment of uranium in borosilicate glasses. The fluorescence spectra and lifetime distributions are highly dependent on the sample base compositions and preparation conditions. In addition, the uranium can be used as an in situ optical probe for its concentration in immobilized glasses. UV-Vis spectroscopic studies suggest uranium exists as U(VI) in the glass samples made in air. Redox chemistry studies of uranium in borosilicate glasses indicate that the decomposition reactions in eqs 6–8 can more precisely explain the equilibria of uranium immobilized in borosilicate glasses. These findings make it possible to optimize the immobilization processes by controlling the base compositions of the glass matrix, the melt temperature, the imposed oxygen fugacity, and additives of transition metal oxides.

Acknowledgments

This work was conducted at the Oak Ridge National Laboratory (ORNL) and supported by the DOE Environmental Management Science Program, under contract No. DE-AC05-00OR22725 with UT-Battelle, LLC. This research was supported in part by an appointment for Z J. L. and S. M. M. to the ORNL Research Associates Program, administered jointly by ORNL and the Oak Ridge Institute for Science and Education. We would like to thank Dr. Ray Schumacher of DOE Savannah River Site for preparing glass frits for us.

References

1. Candall, J. L. *Sci. Basis Nucl. Waste Mgmt.* **1980**, *2*, 39.
2. Cunnane, J. C.; Allison, J. M. *Sci. Basis Nucl. Waste Mgmt.* **1993**, *XVII*, 3.
3. Roy, R. *Sci. Basis Nucl. Waste Mgmt.* **1996**, *XX*, 3.
4. Schreiber, H. D.; Balazs, G. B. *Phys. Chem. Glasses* **1982**, *23*, 139.
5. Schreiber, H. D.; Balazs, G. B.; Jamison, P. L.; and Shaffer, A. P. *Phys. Chem. Glasses* **1982**, *23*, 147.
6. Culea, E.; Milea, I.; and Iliescu, T. *J. Mat. Sci. Lett.* **1994**, *13*, 1171.
7. Culea, E.; Milea, I.; and Bratu, I. *J. Mol. Struct.* **1993**, *294*, 271.
8. Schreiber, H. D. *J. Less-Common Met.* **1983**, *91*, 129.
9. Schreiber, H. D.; Balazs, G. B.; Williams, B. *J. Am. Cer. Soc.* **1982**, *65*, 449.
10. Weber, W. J.; Ewing, R. C.; Angell, C. A.; Arnold, G. W.; Cormack, A. N.; Delaye, J. M.; Griscon, D. L.; Hobbs, L. W.; Navrotsky, A.; Price, D. L.; Stoneham, A. M.; Weinberg, M. C. *J. Mater. Res.* **1997**, *12*, 1946.
11. Wang, H.; Harris, J. M. *J. Phys. Chem.* **1995**, *99*, 16999.
12. Dai, S.; Shin, Y. S.; Toth, L. M.; Barnes, C. E. *J. Phys. Chem. B* **1997**, *101*, 5521.
13. Lux, H.; *Z. Elektrochem* **1934**, *45*, 303.
14. Flood, H.; Forland, T. *Acta Chem. Scand.* **1947**, *1*, 592.
15. Duffy, J. A.; Ingram, M. D. *J. Am. Chem. Soc.* **1971**, *93*, 6448.

16. Dent-Glasser, L. S.; Duffy, J. A. *J. Chem. Soc. Dalton Trans.* **1987**, 2323.
17. Duffy, J. A. *J. Chem. Soc. Faraday Trans.* **1992**, *88*, 2397.
18. Blair, J. A.; Duffy, J. A. *Phys. Chem. Glasses* **1993**, *34*, 194.
19. Dai, S.; Compton, R. N.; Young, J. P.; Mamantov, G. *J. Am. Ceram. Soc.* **1992**, *75*, 2865.
20. Lopez, M.; Birch, D. J. S. *Analyst* **1996**, *121*, 905.
21. Pandey, K. K. *Indian J. Pure Appl. Phys.* **1991**, 29, 362.
22. Provencher, S. W. *Comput. Phys. Commun.* **1982**, *27*, 213.
23. Denning, R. G. *Struct. Bonding* **1992**, *79*, 215.
24. Schreiber, H. D.; Balazes, G. B.; Williams, B. J.; Andrews, S. M. *Sci. Basis Nucl. Waste Mgmt.* **1981**, *3*, 109.
25. Schreiber, H. D.; Kochanowski, B. K.; Schreiber, C. W.; Morgan, A. B.; Coolbaugh, M. T.; Dunlap, T. G. *J. Non-Cryst. Solids* **1994**, *177*, 340.
26. Duffy, J. A. *J. Non-Cryst. Solids* **1996**, *196*, 45.
27. Lebouteiller, A.; Courtine, P. *J. Solid State Chem.* **1998**, *137*, 94.
28. Paul, A.; Douglas, R. W. *Phys. Chem. Glasses* **1965**, *6*, 206.
29. Pyare, R.; Nath, P. *J. Non-Cryst. Solids* **1991**, *128*, 154.

Chapter 4

Surface-Enhanced Raman Scattering of Uranyl-Humic Complexes Using a Silver-Doped Sol-Gel Substrate

Lili Bao[1], Hui Yan[2], Shannon M. Mahurin[1], Baohua Gu[2], and Sheng Dai[1,*]

[1]Chemical Sciences and [2]Environmental Sciences Divisions, Oak Ridge National Laboratory, Building 4500N, Oak Ridge, TN 37831

A silver-doped, surface-enhanced Raman scattering (SERS) substrate was prepared by an acid-catalyzed sol-gel method. Silver nitrate was first doped into the sol-gel film followed by chemical reduction of silver ions with sodium borohydride to produce silver particles. This silver-doped sol-gel substrate exhibits strong enhancement of Raman scattering from adsorbed uranyl ions and soil humic materials. A uranyl ion peak at 713 cm^{-1} and two humic acid peaks at about 1597–1610 and 1313–1338 cm^{-1} were observed. In comparison with Raman spectra of uranyl ions (with no humic acid), the intensity of the symmetric stretching band of uranyl decreased and its frequency increased slightly after complexation with humic acid. The complexation weakened the interaction of uranyl ions with the silver surface. For humic acid, only two peaks of the backbone structure were observed before and after the complexation, suggesting that humic molecules were only weakly sorbed on the silver surface.

© 2006 American Chemical Society

Introduction

Aqueous uranyl ions are known to form complexes with a variety of inorganic and organic ligands. Many techniques have been applied to study these complexes, such as potentiometry, ion-selective electrodes, conductivity, nuclear magnetic resonance, proton magnetic resonance, solvent extraction, ion exchange, solubility, UV-visible absorption spectrometry, photochemistry, fluorimetry and Raman spectroscopy (*1*). Surface-enhanced Raman spectroscopy, based on enhancement of Raman signals for analyte molecules near a characteristically roughened metallic surface, is a very sensitive technique for the trace detection and analysis of environmental samples. Using silver colloidal particles as the substrate, the SERS spectra of uranyl-hydroxide ion and uranyl-carbonate complexes were observed (*2*).

Substrate is the key factor for SERS detection, and silver is the most commonly used. There are a wide variety of silver-based substrates such as electrochemically roughened electrodes (*3*), silver colloids (*4*), silver colloid monolayers self-assembled on polymer-coated substrates (*5*), silver particles vacuum deposited on nano- or microspheres (*6*), and silver-doped sol-gel films (*7–10*). Recently, a silver-doped sol-gel substrate was developed with an acid-catalyzed procedure in our group. The substrate was found to be active for the enhancement of the uranyl Raman signal with a detection limit of ~8.5×10^{-8} M (*11*). This detection limit is below the drinking water standard and allows for potential detection and identification of U and/or U species in the natural subsurface environment. For example, the complexation of natural humic substances with uranyl ions is one of the important reactions that can play an important role for the migration of uranium in geosphere (*12, 13*). However, to date, no studies have reported on the SERS measurement of the uranyl-humate complexes. In this work, the new silver-doped sol-gel substrate was utilized to study the uranyl-humate complexes. A peak around 713 cm^{-1} that was assigned to the symmetric stretching mode of the O=U=O, and two peaks of the backbone of humic acid around 1597–1610 and 1313–1338 cm^{-1} were observed. There was a decrease in the intensity and a slight increase of the frequency of the uranyl ion peak in the complex, suggesting that after combination with humic acid, the interaction of the uranyl ion with the silver substrate became slightly weaker. For humic acid, only two backbone bands were observed before and after the complexation, indicating that humic acid was only physically adsorbed on the silver substrate.

Experimental

Tetramethyl orthosilicate (99%), silver nitrate (99%), Tween 80, and sodium borohydride were obtained from Aldrich and used without further purification. Nitric acid was purchased from VWR. Uranyl nitrate was obtained from Alfa Aesar. Humic acid was extracted from the forest surface soil of the Field Research Center of the Natural and Accelerated Bioremediation Research Program in Oak Ridge, Tennessee. Uranyl-humate complexes were prepared by addition of humic acid solutions to uranyl stock solutions at various concentrations. Glass plates for substrate preparation were prepared by cutting fresh frosted microscope slides (Erie Scientific) into 1 cm × 1 cm pieces.

The silver-doped sol-gel substrate was prepared by an acid-catalyzed sol-gel procedure. The silver ion containing sol was prepared by mixing 1 mL tetramethyl orthosilicate (TMOS), 1 mL water, 60 μL nitric acid (1 M), 1 mL silver nitrate solution (2 M), and 0.1 g Tween 80. Ten μL of the mixture was then deposited onto a frosted glass slide and dried at room temperature for 2 h before a gel film was formed. The silver ions trapped within the sol-gel film were then reduced to silver particles by immersing the gel film in sodium borohydride solution (2 M) for 10 s. Finally, the film was rinsed with water and dried with compressed air before use.

For SERS characterization, a sample solution of 10 μL was applied to the SERS substrate, and Raman spectra were recorded after evaporation of the water with a Renishaw system 1000 Raman spectrometer equipped with an integral microscope (Leica DMLMS/N). Radiation of 632.8 nm from a 25-mW air-cooled He-Ne laser (Renishaw) was used as the excitation source. Raman scattering was collected with a 50 ×, 0.75NA dry objective in 180° configuration and focused into a Peltier cooled charge-coupled device (CCD) camera (400 × 600 pixels). With a holographic grating (1800 grooves·mm^{-1}) and a 50-μm slit, a spectral resolution of 1 cm^{-1} can be obtained. A silicon wafer with a Raman band at 520 cm^{-1} was used to calibrate the spectrometer, and the accuracy of the spectral measurement was estimated to be better than 1 cm^{-1}. Scanning electron microscope (SEM) investigations were performed using a Philips XL-30 field emission scanning electron microscope. UV-Vis absorption spectra were obtained with Cary 5000 UV-Vis-NIR spectrophotometer.

Results and Discussion

Characterization of Silver-Doped Sol-Gel Substrates

The silver-doped sol-gel film, which is gray in color, was first characterized by UV-Vis absorption spectroscopy in order to obtain a general measure of the size distribution of the silver particles. As shown in Figure 1, the UV-Vis displays a broad peak with a maximum absorption at 426 nm. The broad absorption peak suggests a wide, heterogeneous size distribution of the silver particles. The sol-gel substrate was further evaluated by SEM. Figure 2a shows

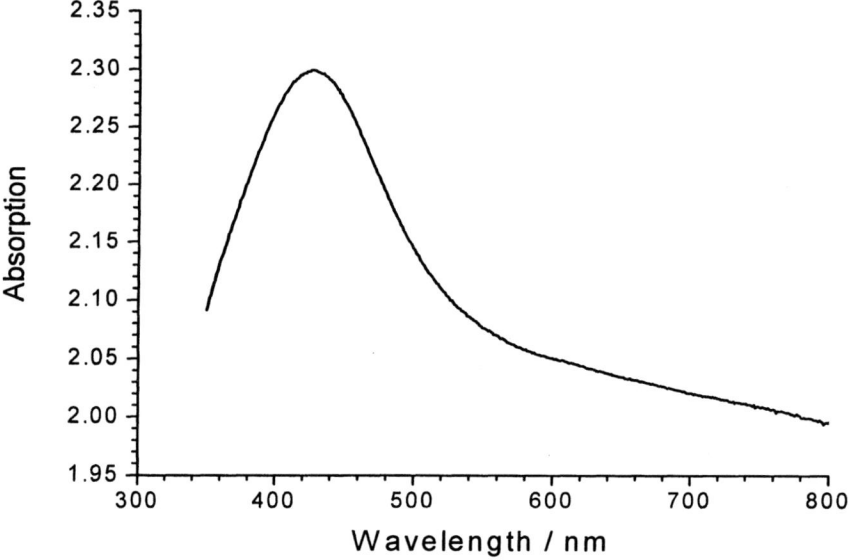

Figure 1. UV- Vis absorption spectrum of a silver-doped sol-gel substrate.

a SEM image where the bright particles correspond to silver and most of the particles are approximately 120 nm. A higher magnification SEM image shown in Figure 2b reveals a wide distribution of silver particles that range in size from 60–200 nm with the highest frequency of particles falling in the range of 80–120 nm. In general, it is the 80–100 nm silver particles that are primarily responsible for the enhancement of Raman signals. Clusters of silver particles such as those at the top left corner in Figure 2b were found to be among the most effective in enhancing the Raman signal compared to other spots in the substrate. These large clusters may be also responsible for the broad absorption peak in UV-Vis measurement. This increased sensitivity might be attributed to an increased analyte concentration in small pockets formed at the boundaries of neighboring silver particles (*14, 15*). Furthermore, the uranyl molecules that are present in these boundary regions experience a greater electromagnetic enhancement due to the proximity of multiple silver particles, which leads to greater SERS signals.

SERS Spectrum of Uranyl Ions

SERS spectrum of uranyl ions adsorbed on the silver-doped sol-gel film is shown in Figure 3a. The peak around 710 cm^{-1} is assigned to the O=U=O symmetric stretching vibration. To eliminate the nitrate ion as the source of the peak at 710 cm^{-1}, a concentrated sodium nitrate solution (2×10^{-2} M) was applied to the silver-doped sol-gel film, and a Raman spectrum was acquired. For sodium nitrate, no Raman signal was observed in this region, which confirms the assignment of this peak to the symmetric stretch of uranyl ions. Though there clearly is a substantial shift of the Raman band compared to uranyl ions in solution, which is located at 870 cm^{-1} (*16*), this type of shift in the symmetric stretching vibration of adsorbed uranyl ions is well documented. For example, Maya et al. examined the sorption of uranium (VI) carbonate on hydrous titania, zirconia, and silica gels, and noted that the O=U=O stretching frequency dropped to between 816–780 cm^{-1} (*16*). In addition, Tsushima et al. reported significant shifts in the Raman spectrum of uranyl adsorbed onto silver colloids to be ~ 798–751 cm^{-1} as a function of the pH (*2*). The lower frequency of the symmetric stretch indicates that there is a very strong interaction of uranyl ions with the silver surface. According to Maya (*16*) and Tsushima (*2*), this interaction suggests a considerable degree of electron density transfer from the silver substrate to uranyl ions with extensive donation from the surface groups on the silver to the equatorial plane of the adsorbed uranyl. This new substrate has shown high sensitivity or SERS enhancement for uranyl ions, and a detection limit of 8.5×10^{-8} M was obtained in our previous work (*11*).

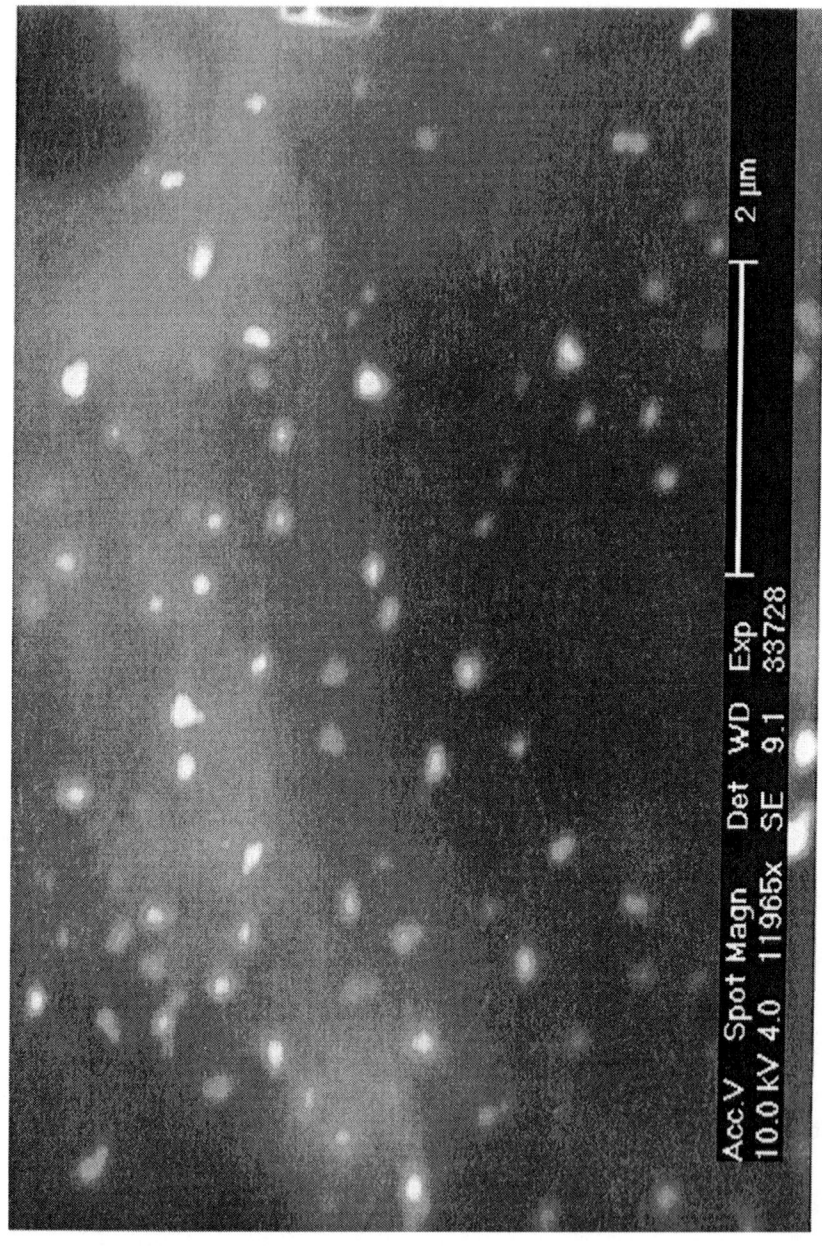

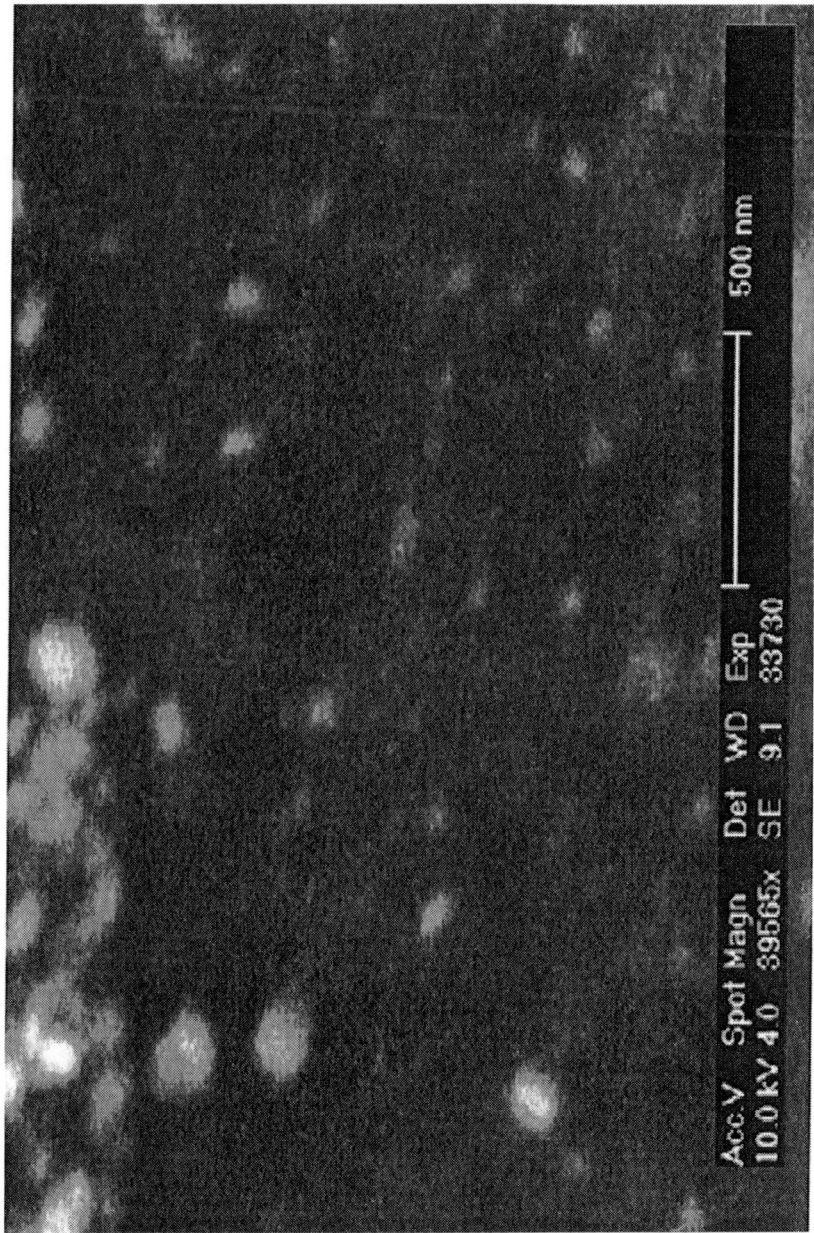

Figure 2. SEM micrograph of a silver-doped sol-gel substrate.

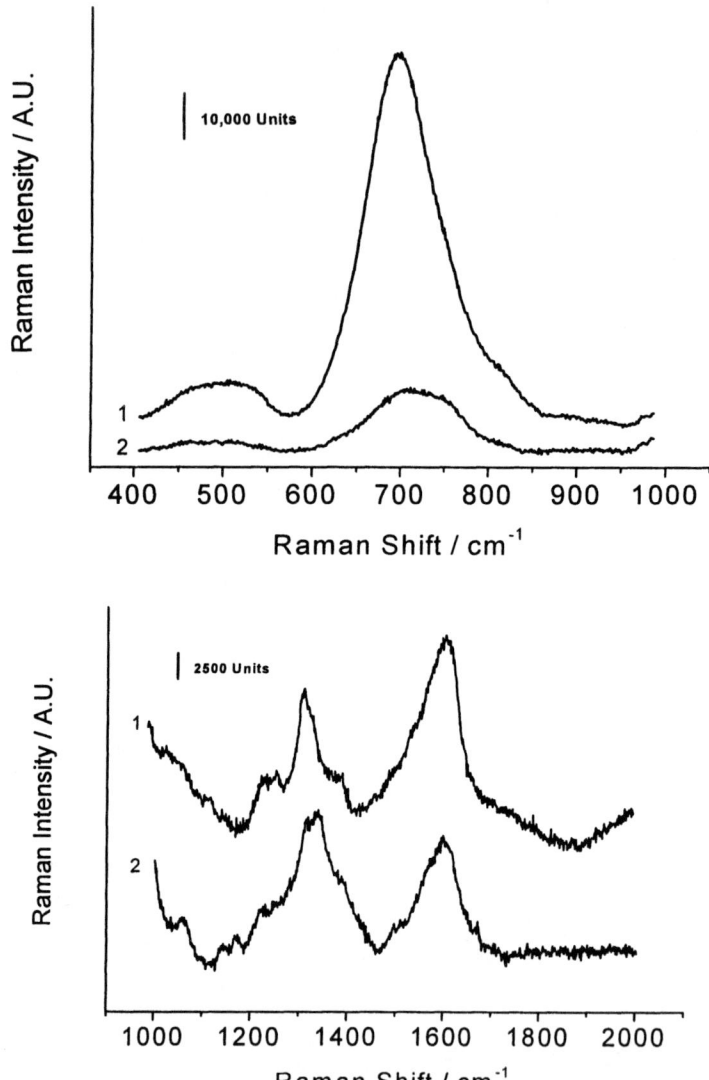

Figure 3. SERS spectra of uranyl, humic acid, or their complexes using the silver-doped sol-gel substrate. a1: 5×10^{-5} M uranyl, a2: 5×10^{-5} M uranyl-humic acid complex, b1: 10^{-5} M humic acid, b2: 5×10^{-5} M uranyl-humic acid complex.

SERS Spectrum of Humic Acid

SERS spectrum of humic acid is shown in Figure 3b. The band around 1601 cm^{-1} is assigned to the zone-center E_{2g} mode of graphite and is so called the G band, while the band around 1308 cm^{-1} is assigned to in-plane c-c photons at the M point of the Brillouin zone and is so called the D band ([17–20]). The D band reveals the degree of disorder of the carbon networks ([20]). The SERS spectrum reveals the graphite-like characteristics. It should be noted that Raman spectroscopy in the presence of silver characterizes humic substances in their neutralized form instead of revealing the information in acidic form, as the carboxylic groups are dissociated when the sample is adsorbed on silver surface ([20]). It was also found that the adsorption of the humic acid is not uniform in the substrate as the SERS peak position varies from 1600–1609 and 1308–1321 cm^{-1} respectively. The bands observed in this work can be verified by the previously reported results. In the work of Yang et al., SERS spectrum of peat humic acid was studied ([20]). The two corresponding bands were observed at 1594 and 1322 cm^{-1} by using a gold electrode while they were at 1610 and 1316 by using a HNO$_3$-ethched copper foil. More SERS bands were observed when gold electrode was used, while only two bands were obtained in the silver-doped sol-gel substrate. These observations were partially attributed to: (1) humic acids were only weakly or physically sorbed on the silver sol-gel substrate and (2) the dissociation constants of carboxylic groups are higher when adsorbed on silver surface than on the gold surface.

SERS Spectrum of Uranyl-Humate Complex

SERS spectra of uranyl-humate complex adsorbed on the silver-doped sol-gel film were shown in Figure 3. The uranyl symmetric stretching band was observed at 711 cm^{-1}, and the G and D band of humic acid were at 1598 and 1335 cm^{-1}. In the complex, not only the intensity of the uranyl band at 710 cm^{-1} decreased, but also the band position shifted. For uranyl ions, the band frequency was lower than 710 cm^{-1} and the average value obtained was 695 cm^{-1}. In the complex, however, the average value was 713 cm^{-1}. The decrease in the intensity and the slight increase of the frequency of the uranyl ion peak in the complex indicates that after complexation with humic acid, the interaction of the uranyl ion with the silver substrate becomes weaker. The vibrational bands of humic acid in the uranyl-humic complex were observed at 1597–1610 and 1313–1338 cm^{-1}, similar to those observed with humic acid alone (Figure 3b).

Conclusions

A silver-doped sol-gel SERS substrate was prepared using an acid-catalyzed procedure in which silver nitrate doped into the sol-gel film was reduced by sodium borohydride to produce silver particles. This substrate was used for enhanced Raman spectroscopy and for the detection of uranyl, humic acid, and their complexes in aqueous solution. In the complexes, a uranyl ion peak at 713 cm^{-1} and two humic acid peaks around 1597–1610 and 1313–1338 cm^{-1} were observed. In comparison with Raman spectra of uranyl ions (with no humic acid), the intensity of the symmetric stretching band of uranyl decreased and its frequency increased slightly after complexation with humic acid. The complexation weakened the interaction of uranyl ions with the silver surface. For humic acid, only two peaks of the backbone structure were observed before and after the complexation, suggesting that humic molecules were only weakly sorbed on the silver surface.

Acknowledgments

This work was conducted at the Oak Ridge National Laboratory and supported by the Environmental Management Science Program, U.S. Department of Energy, under contract No. DE-AC05-00OR22725 with UT-Battelle, LLC. This research was supported in part by the appointments for L. L. B. and S. M. M. to the ORNL Research Associates Program, administered jointly by ORNL and the Oak Ridge Institute for Science and Education. We would like to thank C. Liang and Z. Zhang for the SEM analysis of SERS substrates.

References

1. Nguyen-Trung, C.; Begun, G. M.; Palmer, D. A. *Inorg. Chem.* **1992**, *31*, 5280–5287.
2. Tsushima, S.; Nagasaki, S.; Tanaka, S.; Suzuki, A. *J. Phys. Chem. B* **1998**, *102*, 9029–9032.
3. Fleischmann, M.; Hendra, P. J.; McQuillan, A. J. *Chem. Phys. Lett.* **1974**, *26*, 163–166.
4. Laserna, J. J.; Torres, E. L.; Winefordner, J. D. *Anal. Chim. Acta* **1987**, *200*, 469–480.
5. Freeman, R. G.; Grabar, K. C.; Allison, K. J.; Bright, R. M.; Davis, J. A; Guthrie A. P.; Hommer, M. B.; Jackson, M. A.; Smith, P. C.; Walter, D. G; Natan, M. J. *Science* **1995**, *267*, 1629–1632.
6. Moody, R. L.; Vo-Dinh, T., Fletcher, W. H. *Appl. Spectrosc.* **1987**, *41*, 966–970.
7. Lee, Y. H.; Dai, S.; Young, J. P. *J. Raman Spectrosc.* **1997**, *28*, 635–639.

8. Volkan, M.; Stokes, D. L.; Vo-Dinh, T. *J. Raman Spectrosc.* **1999**, *30*, 1057–1065.
9. Murphy, T.; Schmidt, H.; Kronfeldt, H. D. *Appl. Phys. B* **1999**, *69*, 147–150.
10. Li, Y. S.; Lin, X.; Cao, Y. H. *Vib. Spectrosc.* **1999**, *20*, 95–101.
11. Bao, L.; Mahurin, S. M.; Dai, S. *Anal. Chem.*, **2003**, *75*, 6614-6620.
12. Kim, J. I. *Radiochim. Acta* **1991**, *52/53*, 71–81.
13. Czerwinski, K. R.; Buckau, G.; Scherbaum, F.; Kim, J. I. *Radiochim. Acta* **1994**, *65*, 111–119.
14. Bosnick, K. A.; Jiang, J.; Brus, L. E. *J. Phys. Chem. B* **2002**, *106*, 8096–8099.
15. Michaels, A. M.; Nirmal, M.; Brus, L. E. *J. Am. Chem. Soc.* **1999**, *121*, 9932–9939.
16. Maya, L. *Radiochim. Acta* **1982**, *31*, 147–151.
17. Mennella, V.; Monaco, G.; Colangeli, L.; Bussoletti, E. *Carbon* **1995**, *33*, 115–121.
18. Vidano, R. P.; Fischbach, P. B.; Willis, L. J.; Loehr, T. M. *Solid State Commun.* **1981**, *39*, 341–344.
19. Sinha, K.; Menendez, J. *Phys. Rev. B* **1990**, *41*, 10845–10847.
20. Yang, Y.; Chase, H. A. *Spectrosc. Lett.* **1998**, *31*, 821–848.

Chapter 5

Analysis of Electrochemical Impedance Data for Iron in Borate Buffer Solutions

Jun Liu[1], Brian M. Marx[2,4], and Digby D. Macdonald[3,*]

[1]Materials LifeCycle Solutions, R&D, ATMI, Inc., 7 Commerce Drive, Danbury, CT 06810
[2]MSE Department, Boise State University, Boise, ID 83725–2075
[3]Center for Electrochemical Science and Technology, The Pennsylvania State University, University Park, PA 16802
[4]Current address: 1702 Puddintown Road, State College, PA 16801

> The passive state of iron in borate buffer solutions containing EDTA (ethylenediaminetetraacetic acid, disodium salt, $C_{10}H_{14}N_2Na_2O_8$) ranging in pH from 8.15 to 12.87 at ambient temperature, has been explored using electrochemical impedance spectroscopy (EIS) and steady-state polarization methods. EDTA prevents the formation of the outer layer of the passive film, thereby permitting interrogation of the barrier layer alone. It has been found that, in the passive state, the impedance is only weakly-dependent on the solution pH and film formation voltage, but at high pH (pH > 12) and especially at high voltages, the impedance becomes very pH- and voltage-dependent. Under the steady-state conditions achieved in this work, passive iron is shown to satisfy the conditions of linearity, causality, stability, and finiteness, as required by Linear Systems Theory, on the basis of Kramers-Kronig (K-K) transformation of experimental impedance data. The experimental data have been interpreted in terms of the Point Defect Model (PDM), the predictions of which for an n-type, semi-conducting passive film have been experimentally observed. An impedance model for passive iron, based on the PDM, has been developed and optimized on the impedance data to yield values for fundamental parameters (transfer coefficients and standard rate constants) for the interfacial reactions occurring in the barrier oxide layer. We conclude that the dominant defect(s) in the barrier layer of the passive film on iron must be the oxygen vacancy or cation interstitial, with the latter being favored by the values of the kinetic parameters.

Introduction

Although the passivity of iron in contact with aqueous solutions has been extensively studied (*1–12*), consensus has not yet been reached as to the detailed structure and composition of the film or as to the defect structure. Thus, most workers have concluded that the passive film is a bi-layer structure, comprising "Fe_3O_4" adjacent to the metal and an outer layer of a Fe(III) species that has been variously described as "Fe_2O_3" or "FeOOH," most frequently upon the basis of ex situ X-ray or electron diffraction studies. As noted previously, part of the problem in studying the barrier layer stems from the fact that it is shielded from the experimental probe by the outer layer (*13, 14*).

A general consensus holds that the passive film formed on iron in borate buffer solutions has a bi-layer structure comprising a defective Fe_3O_4 inner layer (the barrier layer), which grows directly into the metal, and a γ-Fe_2O_3 outer layer that forms by precipitation (*12*). The outer layer effectively screens the barrier layer from optical investigation and also significantly modifies the electrochemical response of the surface. However, general agreement exists that the passivity of iron is primarily due to the barrier layer. Accordingly, this work is based on the hypothesis that EDTA (ethylenediaminetetraacetic acid, disodium salt, $C_{10}H_{14}N_2Na_2O_8$), which chelates the iron cations that are ejected from the barrier layer, should be effective in preventing the formation of the outer layer. If so, then the barrier layer alone may be interrogated alone to provide a much more accurate account of the passive state. Previous research (*13*) clearly shows that EDTA in borate buffer solutions effectively prevents the formation of the outer layer. The investigation of EIS for passive iron in this work focuses only on passive films formed on iron in borate buffer solutions with the addition of EDTA.

Experimental

All EIS experiments were carried out in a three-electrode polytetrafluoroethylene (PTFE) electrochemical cell, shown schematically in Figure 1a.

The working electrode was a pure iron wire (Alfa Aesar 99.99%) with a diameter of 1 mm and a surface area of 0.008 cm^2, which was embedded in two-component epoxy resin and mounted in a polyvinyl chloride (PVC) rod. The working electrode of pure iron was abraded with LECO SiC paper from 600 to 1200 grit, and then polished with 0.05 μm Al_2O_3/cotton cloth, which generated a

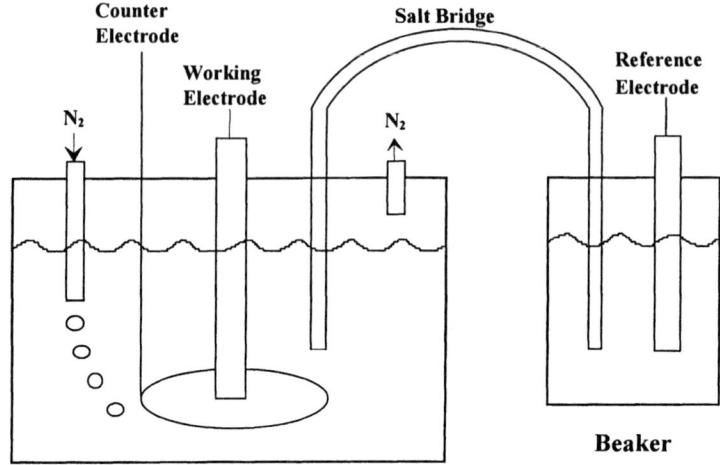

(a)

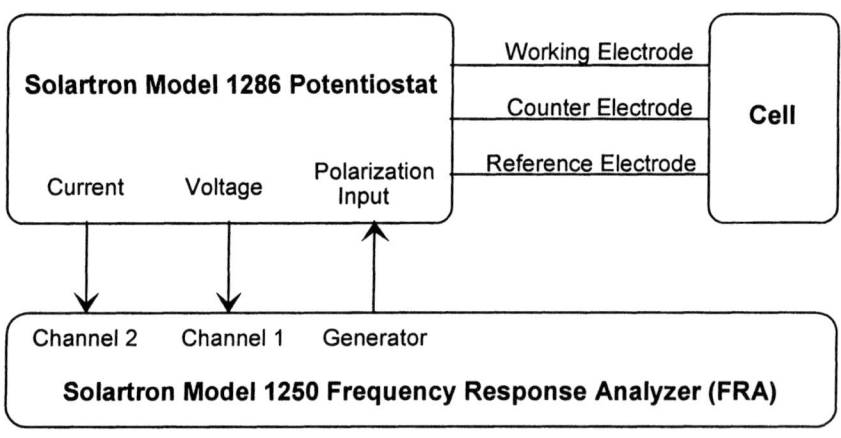

(b)

Figure 1. Electrochemical cell (a) and experimental apparatus (b) for electrochemical impedance spectroscopic measurements for the passive film on iron.

smooth and bright surface, as observed microscopically. The counter electrode was a platinum foil of high purity (99.998%) of large area. All potentials were measured against a saturated calomel reference electrode (SCE) located in a separate compartment that was connected to the main cell through a salt bridge and terminated close to the iron surface with a Luggin probe.

Six borate buffer solutions, having different pH values, were prepared from Aldrich A.C.S. reagents and conductivity grade water (Milli-Q system, 18.2 MΩ cm^{-1}), and 0.01 M EDTA was added to each. Table I displays the compositions and the pH values measured using a Fisher Scientific pH meter. Prior to and during all experiments, the solutions were purged with zero-grade nitrogen gas to reduce the oxygen concentration in solution. All experiments were performed at ambient temperature (~22 °C).

Table I. Compositions and pH Values of Borate Buffer Solutions Containing 0.01 M EDTA

Solution Name	Solution Compositions	Measured pH
BB-EDTA-8	0.3 M H$_3$BO$_3$ + 0.075 M Na$_2$B$_4$O$_7$ + 0.01 M EDTA	8.15
BB-EDTA-9	0.05 M H$_3$BO$_3$ + 0.075 M Na$_2$B$_4$O$_7$ + 0.01 M EDTA	8.94
BB-EDTA-10	0.075 M NaOH + 0.075 M Na$_2$B$_4$O$_7$ + 0.01 M EDTA	10.05
BB-EDTA-11	0.15 M NaOH + 0.075 M Na$_2$B$_4$O$_7$ + 0.01 M EDTA	11.27
BB-EDTA-12	0.22 M NaOH + 0.075 M Na$_2$B$_4$O$_7$ + 0.01 M EDTA	12.20
BB-EDTA-13	0.6 M NaOH + 0.075 M Na$_2$B$_4$O$_7$ + 0.01 M EDTA	12.87

The impedance measurement apparatus is presented in Figure 1b. A Schlumberger/Solartron 1286 electrochemical interface controlled and measured the voltage of the working electrode vs the reference electrode and monitored the current through the counter electrode. An alternating voltage excitation (10 mV peak-to-peak, unless otherwise noted) generated by the frequency response analyzer (FRA) was added to the stationary polarization voltage set by the potentiostat, thereby providing the perturbation to the electrochemical system. The voltage and resulting current signals were then transferred to a Schlumberger/Solartron 1250 FRA, which generated and recorded the electrochemical impedance data. Electrochemical impedance software, ZPlot (Version 1.4) and ZView (Version 1.4), were employed to control the measurement of the impedance.

Results and Discussion

Effect of Frequency Sweep

One of the most valuable advantages of EIS stems from its ability to detect relaxations at an electrode/solution interface over a broad frequency range in only a single experiment. This feature gives a high yield of useful information from an experiment, and EIS is more efficient than other, traditional electrochemical techniques in transferring information to the observer. For example, in, cyclic voltammetry, only the peak-to-peak voltages and peak currents are generally used, with the information contained in the off-peak regions of the voltammograms being normally discarded. Theoretically, a perfect yield obtained from an experiment should contain information within the frequency range from zero to infinity; in practice, however, information is only available in the range $f_{min} \le f \le f_{max}$, which means that data in the ranges of $0 < f < f_{min}$ and $f_{max} < f < \infty$ are not accessible by electrical perturbation. In this work, a wide Hertzian frequency (f) range covering seven decades, from 10 kHz to 1 mHz, has been explored.

Figure 2 displays a Nyquist plot of EIS data for passive iron in borate buffer solution with 0.01 M EDTA (pH 8.94) at an applied film formation voltage of 0.4 V vs SCE. The passive film was formed at 0.4 V vs SCE for 24 h, which was sufficiently long to achieve a steady state, as indicated by a time-independent passive current and other properties, as described below. Prior to the experiment, the iron working electrode was cathodically polarized at –1 V vs SCE for 5 min to remove the air-formed oxide film. The frequency was then stepped from 10^{-3} Hz to 10^4 Hz ("up") and then back from 10^4 Hz to 10^{-3} Hz ("down", immediately after the "up" sweep), with 20 evenly spaced data points being measured within each frequency decade.

Data for the decreasing frequency direction are shown in Figure 2. For both "up" and "down" frequency sweeps, it is obvious that, in the low frequency range, say from 10^{-2} Hz to 10^{-3} Hz (20 data points are recorded in this range), the Nyquist plots are not coincident between the "up" and "down" frequency sweep directions. Considering that the passive film formed after 24 h is in a steady state without any external perturbation (14), the deviation between the "up" and "down" frequency sweep experiments could arise from the measurement-induced instability of the electrochemical system at low frequencies, which requires a long time period to generate the perturbation and average the response (since time is inversely proportional to frequency). Such a poor correlation was consistently observed under all experimental conditions for all of our impedance measurements on passive iron. Consequently, the impedance data in the frequency range from 10^{-2} Hz to 10^{-3} Hz have been disregarded in investigating the passivity of iron.

Conversely, the correlation between the "up" and "down" frequency sweeps can also function as an indication of the stability (or instability) of the system. A steady-state interface should generate coincident data with regard to the reverse frequency sweep directions. Figure 3 presents the Nyquist plot and Bode plot for the passive film formed on iron in borate buffer solution with 0.01M EDTA (pH 8.15) polarized at 0.4 V vs SCE for 24 h, measured within the frequency range from 10^{-2} Hz to 10^{4} Hz with the frequency being swept "up" and "down." The Nyquist plot, the impedance modulus Bode plot, and the phase angle Bode plot are all coincident for the "up" and "down" sweeps of frequency over the chosen frequency range, which strongly suggests that the passive film formed on iron under these experimental conditions is indeed in a steady state and is well-suited for investigation using steady-state electrochemical techniques, such as EIS.

Effect of Perturbation Voltage

When making EIS experiments, a steady-state polarization voltage is specified at which the experiment will be performed: for instance, the voltage within the wide passive range that is employed for passive film growth on the surface of the pure iron electrode. A small-amplitude sinusoidal voltage of the form

$$V(t) = V_0 \sin(\omega t) \tag{1}$$

in which t is the time, $\omega = 2\pi f$ is the radial frequency, f is the Hertzian frequency, and V_0 is the amplitude, is then super-imposed on the steady-state polarization voltage. The sinusoidal voltage generates a perturbation in the electrochemical interface system, and, provided that V_0 is sufficiently small that the system is linear, the resulting perturbation current response is detected as

$$I(t) = I_0 \sin(\omega t + \theta) \tag{2}$$

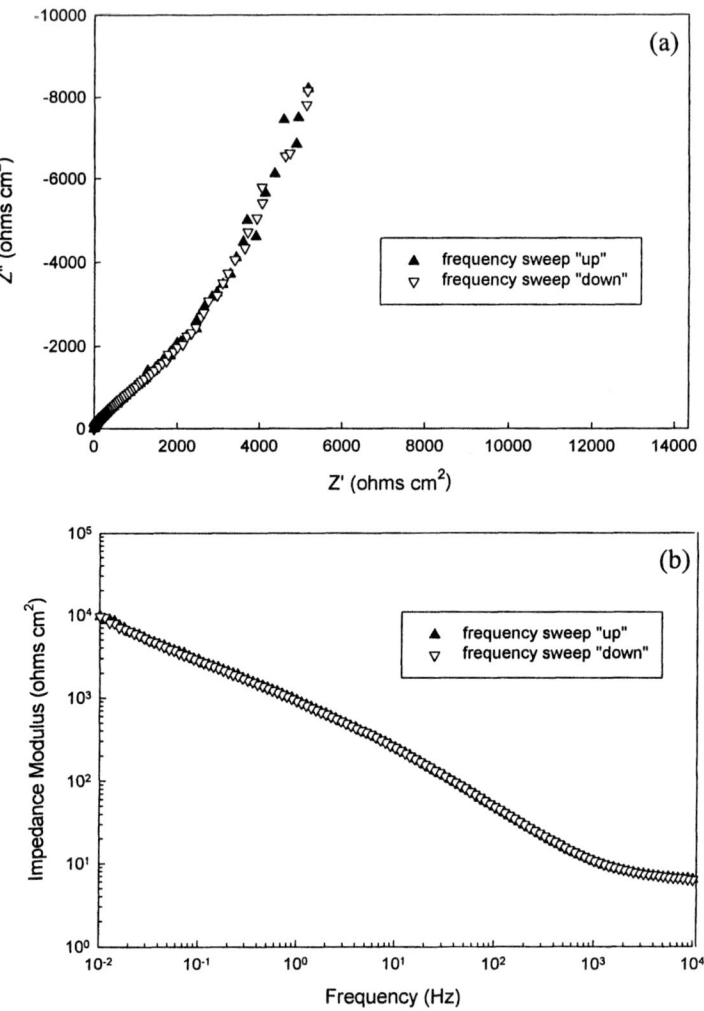

Figure 2. Nyquist plot for EIS data for the passive film on iron in borate buffer solution with 0.01 M EDTA (pH 8.94) at an applied film formation voltage of 0.4 V vs SCE formed for 24 h. Frequency range is from 10^4 Hz to 10^{-3} Hz. "Up" means sweeping the frequency in the low-to-high direction; "down" in the high-to-low direction.

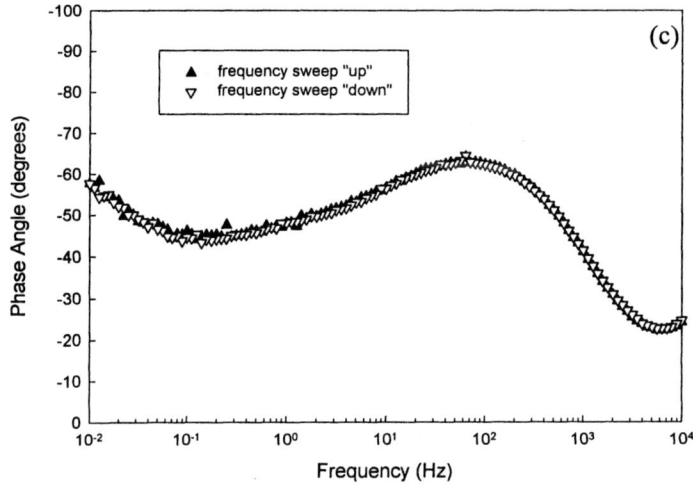

Figure 2. Continued.

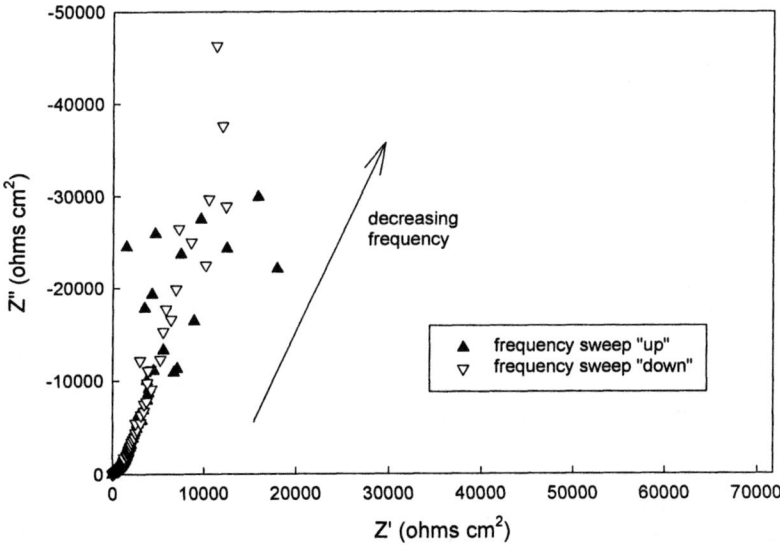

Figure 3. Nyquist plot (a) and Bode plots (b, c) for EIS data for the passive film on iron in borate buffer solution with 0.01 M EDTA (pH 8.15) at an applied film formation voltage of 0.4 V vs SCE formed for 24 h. Frequency range is from 10^4 Hz to 10^{-2} Hz. "Up" means sweeping the frequency in the low-to-high direction; "down" in the high-to-low direction. Perturbation voltage amplitude is 10 mV.

where I_0 is the current amplitude and θ is the phase angle. Note that both the sinusoidal perturbation voltage and the sinusoidal current response are a function of the excitation frequency over a wide frequency range.

The EIS data shown in Figure 3 were generated with a perturbation voltage amplitude of 10 mV (peak-to-peak). When the perturbation voltage amplitude for the same electrochemical system is increased to 20 mV, as shown in Figure 4, the correlation between the "up" frequency sweep and the "down" frequency sweep becomes much poorer. This phenomenon should result from the higher voltage perturbation, which causes the electrochemical system to become nonlinear.

Effect of Formation Voltage

Our previous work (*13, 14*) concluded that iron has a wide passive range in borate buffer solutions containing EDTA, and that the passive film reaches a steady state after about 24 h of polarization. The electrochemical impedance measurements were carried out in this work immediately after a 24-h film formation period at each applied voltage in the range from 0.1 to 0.6 V vs SCE, in both the ascending and descending voltage directions. Accordingly, impedance data for passive iron, at each formation voltage, have been measured twice in this experiment, once in the "up" voltage direction and then again in the "down" voltage direction.

As shown in the Nyquist plots for passive films formed on iron in borate buffer/EDTA solutions at 0.2 V vs SCE (solution pH = 8.15), presented in Figure 5, the "up" and "down" voltage step directions yield almost the same electrochemical impedance; that is, the impedance is independent of the previous history of the system. Note that, for the sake of clarity, only data for a single formation voltage are displayed in Figure 5. This is actually observed in most experiments under various conditions for the passive film formed on iron, provided that the time of film formation is at least 24 h.. Such an observation basically represents another diagnostic of the stability or instability of the system under investigation. In principle, a steady-state electrode/solution interface should give rise to the same electrochemical impedance characteristics, no matter the voltage step direction. The electrochemical system probed in Figure 5 is clearly at steady state, considering the fact that those two measurements under reversed voltage step directions were performed 8 days apart. We should also note that the solution contains EDTA, which has been previously shown (*13, 14*) to effectively inhibit the formation of the outer layer of the passive film by chelating iron cations being ejected from the barrier layer. Accordingly, the time-dependent formation of the outer layer does not interfere with the impedance characteristics of the barrier layer.

It has been clearly shown that the steady-state current density (each point being measured after 24 h of polarization in the ascending and descending voltage directions) for passive iron in borate buffer solution with 0.01 M EDTA does not depend on the applied film formation voltage, which indicates that the principal defects are oxygen vacancies and/or iron interstitials. This is consistent with the predictions of the Point Defect Model (PDM) for n-type semiconducting barrier oxide films (*15, 16*). In order that the impedance and steady-state polarization characteristics of passive iron can be correlated, it is also important to investigate the electrochemical impedance behavior of passive iron under exactly the same experimental conditions.

Impedance spectra (Nyquist plots) for passive iron in pH 10.05 borate buffer solution containing 0.01 M EDTA (in which only the barrier layer exists on the surface) are shown in Figure 6 as a function of formation voltage ranging from 0.1 V vs SCE to 0.6 V vs SCE. Throughout the frequency range, the Nyquist plots are almost independent of the formation voltage, again in agreement with the prediction of the PDM (*15, 16*). According to the PDM, point defect transport is mainly due to migration under the influence of the electric field within the barrier layer, and the electric field strength is assumed to be invariant with respect to the applied film formation voltage. A constant electric field is one aspect of the PDM that sets it apart from other models, such as the High Field Model. The fact that the impedance spectra are essentially independent of the formation voltage is therefore strong evidence that the electric field strength is indeed buffered against changes in the voltage of formation of the passive film. The PDM posits that, at high electric field strengths under steady-state conditions, Esaki (band-to-band) tunneling can occur, in which an internal current flows from the conduction band to the valence band (or in the reverse direction, depending on the charge carrier) resulting in charge separation that opposes any increase or decrease in the applied field, thereby leading to a constant electric field within the barrier layer. Accordingly, the electrochemical impedance should be independent of (or at least weakly dependent on) the film formation voltage, as observed. Although some deviations are observed in the Nyquist plots at higher formation voltages, these probably arise from a change in the oxidation state of iron (Fe) ions from +2 to +3; the experimental data presented in Figure 6 nevertheless confirm the validity of the PDM with regard to the nature of the electric field strength. Finally, if the electric field increased continually as the formation potential increased, as postulated by the High Field Model, dielectric breakdown should eventually occur. However, dielectric breakdown is seldom, if ever observed during the electrochemical formation/destruction of thin oxide films on metals other than the valve metals (Al, Ta, Nb, Zr, Ti, etc.), except under voltage pulse conditions.

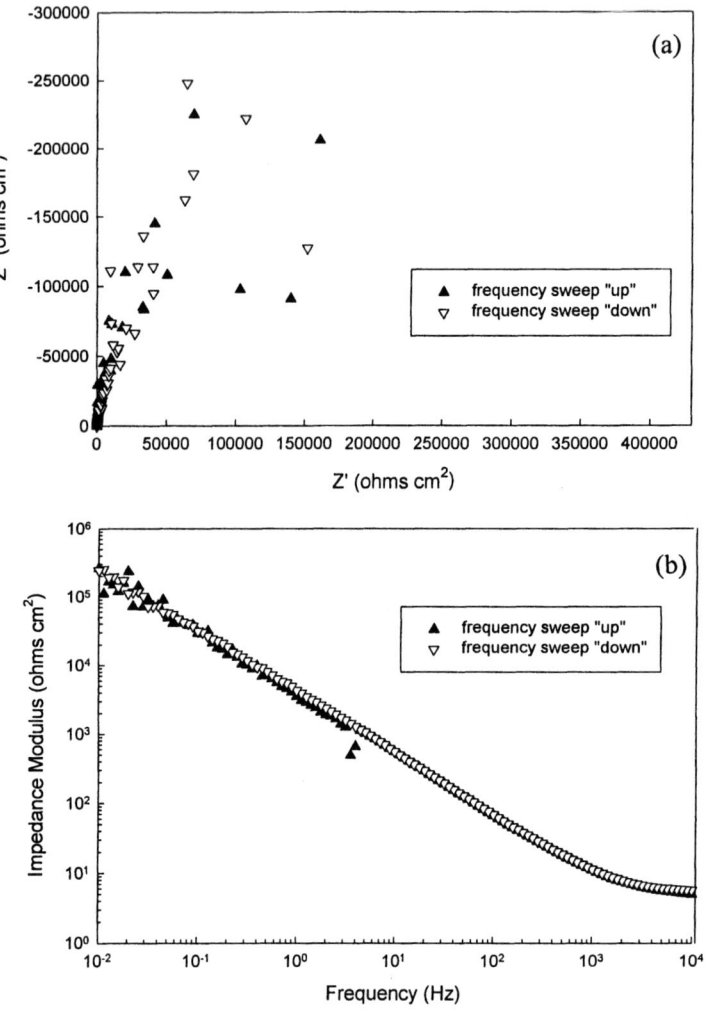

Figure 4. Nyquist plot (a) and Bode plots (b, c) for EIS data for the passive film on iron in borate buffer solution with 0.01 M EDTA (pH 8.15) at an applied film formation voltage of 0.4 V vs SCE formed for 24 h. Frequency range is from 10^4 Hz to 10^{-2} Hz. "Up" means sweeping the frequency in the low-to-high direction; "down" in the high-to-low direction. Perturbation voltage amplitude is 20 mV.

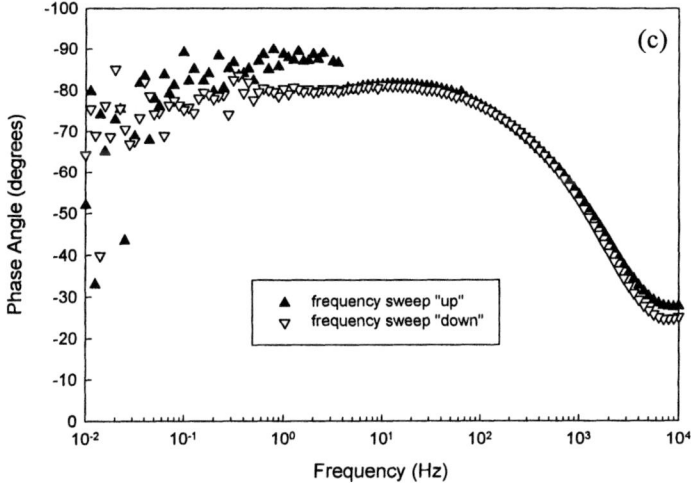

Figure 4. Continued.

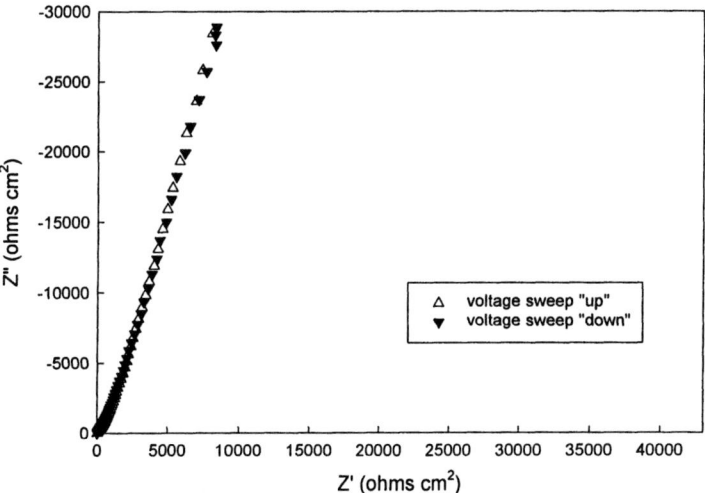

Figure 5. Nyquist plot for EIS data for the passive film formed on iron in borate buffer solution with 0.01 M EDTA (pH 8.15) at an applied film formation voltage of 0.2 V vs SCE. Frequency range is from 10^4 Hz to 10^{-2} Hz. Perturbation voltage amplitude is 10 mV. Passive film formation voltage has been applied at 0.1, 0.2, 0.3, 0.4, 0.5, 0.6, 0.5, 0.4, 0.3, 0.2, 0.1 V vs SCE continuously with a 24-h polarization period for each voltage. "Up" means changing the formation voltage in the negative-to-positive direction; "down" in the positive-to-negative direction. "Up" and "down" voltage measurements have a time interval of 8 days in this case. The spectra shown correspond to those at 0.2 V vs SCE on the up (open triangles) and down (closed triangles) voltage directions.

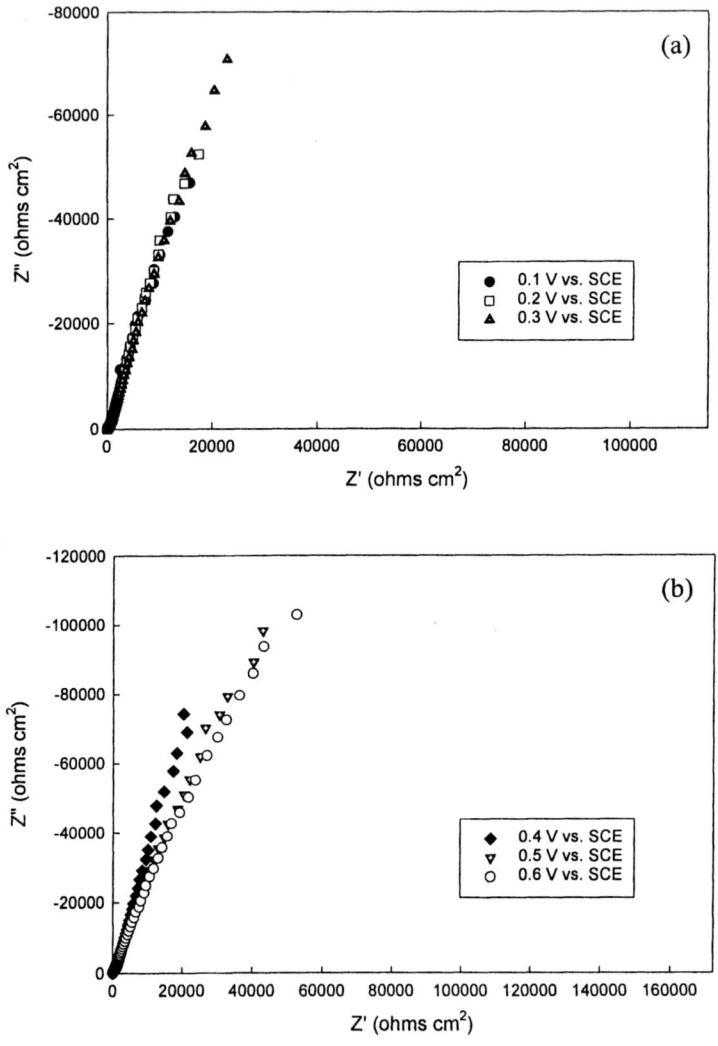

Figure 6. Nyquist plot for EIS data for the passive film formed on iron in borate buffer solution with 0.01 M EDTA (pH 10.05) as a function of formation voltage at a formation time of 24 h. Spectra were measured using an excitation voltage of 10 mV and an applied frequency ranging from 10^4 Hz to 10^{-2} Hz.

Effect of Solution pH

Six different pH-valued alkaline borate buffer/EDTA solutions were used in this work, the compositions of which are detailed in Table I. The electrochemical impedance data for passive iron in various pH-valued solutions are plotted in Figures 7 and 8.

The Nyquist and Bode plots for passive films formed on iron at an applied voltage of 0.6 V vs SCE in relatively low pH-valued, borate buffer solutions in the alkaline range, as shown in Figure 7. These data indicate that, over the pH range explored, the impedance is not strongly dependent upon pH, although a slight dependence is noted in the impedance modulus and in the phase angle. However, when the solution pH is higher than the range given in Figure 7, and especially when the film formation voltage increases to 0.6 V vs SCE, the impedance modulus and the phase angle are both significantly diminished, especially in the low frequency range (Figure 8).

The effect of solution pH on the polarization behavior of iron in borate buffer solution containing 0.01 M EDTA confirms that more highly alkaline solutions slightly inhibit the passive dissolution of iron (Figure 7b) and correspondingly improve passivity (*14*). On the other hand, a higher pH (> ca 12) might also greatly facilitate the oxygen evolution reaction, resulting in a relatively lower driving force/overpotential for the interfacial reaction. The significantly reduced impedance for passive iron shown in Figure 8 probably results from the oxygen evolution reaction or from the (partial) destruction of the barrier layer, due to enhanced dissolution rate resulting from the change in oxidation state of iron solutes entering the solution. Therefore, the impedance behaviors in relatively high alkaline solutions measured at high film formation voltages are actually generated from processes other than the passivity of iron. Such impedance data are not reliable for analyzing and characterizing the passive film formed on iron in borate buffer/EDTA solutions.

Kramers-Kronig Transforms

Because not every ratio of a perturbation and response can be defined as an impedance, the validity of experimental "impedance" data must be carefully established. It is important to ascertain whether there are experimental artifacts that could distort the actual impedance characteristics, casting doubt upon the validity of the experimental data. The Kramers-Kronig (K-K) transforms (*17–20*) have been developed as a principal method for verifying conformation of the system with the constraints of Linear Systems Theory (LST) and hence for validating impedance data.

Macdonald and Urquidi-Macdonald (*21, 22*) identified four general constraints for an electrochemical system to fulfill, in order for the system to

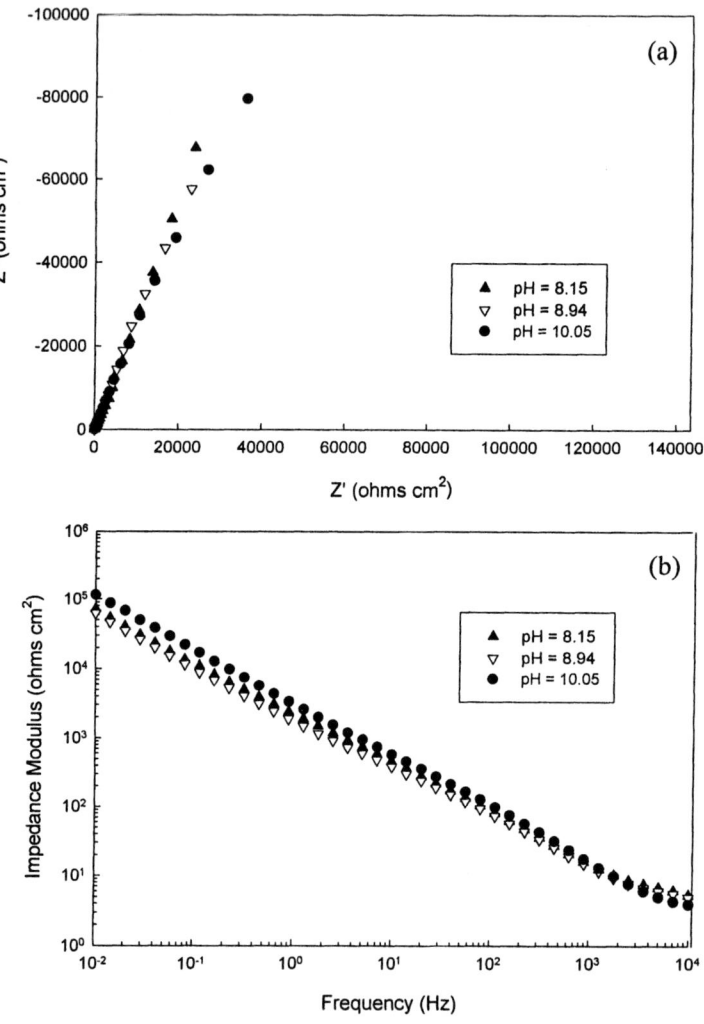

Figure 7. Electrochemical impedance spectra for the passive film formed on iron in borate buffer/EDTA solution (formation voltage = 0.6 V vs SCE) as a function of solution pH (pH = 8.15, 8.94, 10.05) at a formation time of 24 h. Spectra were measured using an excitation voltage of 10 mV and an applied frequency ranging from 10^4 Hz to 10^{-2} Hz.

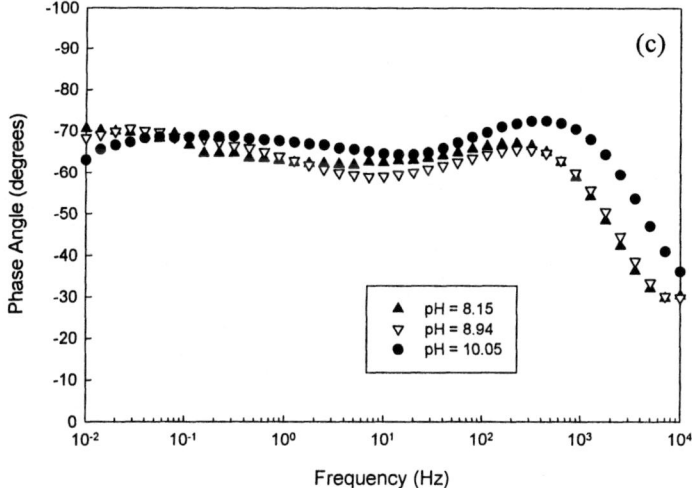

Figure 7. Continued.

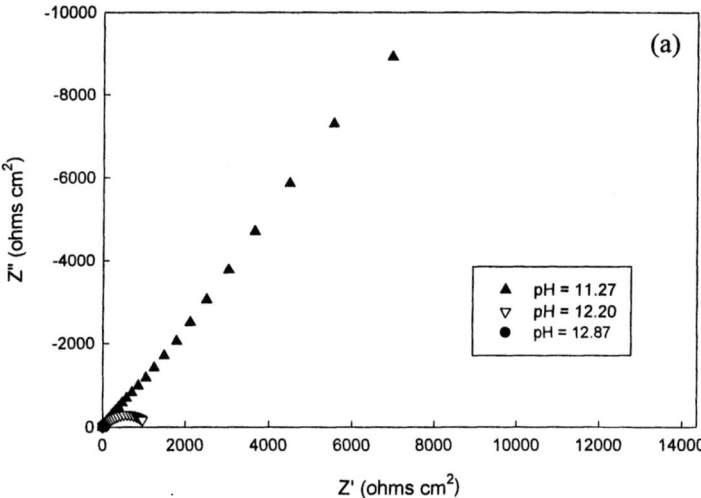

Figure 8. Electrochemical impedance spectra for the passive film formed on iron in borate buffer/EDTA solution (formation voltage = 0.6 V vs SCE) as a function of solution pH (pH = 11.27, 12.20, 12.87) at a formation time of 24 h. Spectra were measured using an excitation voltage of 10 mV and an applied frequency ranging from 10^4 Hz to 10^{-2} Hz. (Continued on next page.)

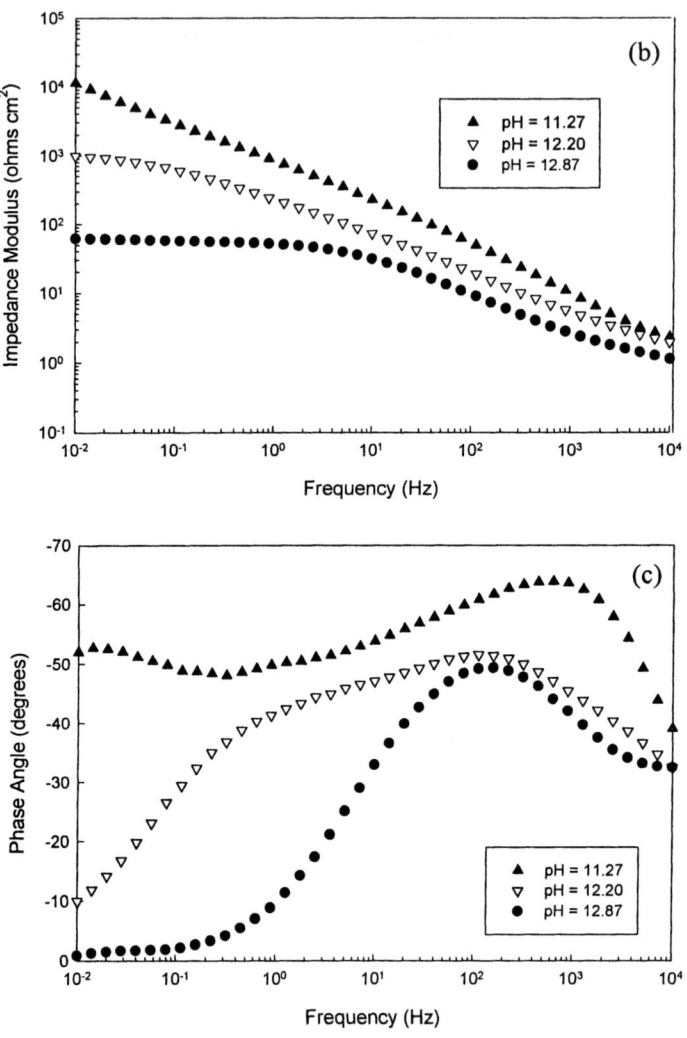

Figure 8. Continued.

comply with LST and hence for the impedance data to correctly transform via the K-K transforms:

1. Causality: The response of the electrochemical system must be due only to the perturbation applied, and should not contain significant components from other sources. There should be no response before time zero, with time zero being defined as the time at which the perturbation is first applied.
2. Stability: The electrochemical system must be stable, i.e., it should return to its original state after the perturbation is removed.
3. Linearity: The perturbation/response of the electrochemical system is described by a set of linear differential equations. In practice, this condition requires that the impedance should be independent of the perturbation amplitude (i.e., doubling the perturbation should double the response, but not change the impedance).
4. Finiteness: The impedance of the electrochemical system must be finite valued as $\omega \to 0$ and $\omega \to \infty$, and must be a continuous function at all intermediate frequencies.

Only those experimental data satisfying the above four conditions can be interpreted as "impedance" and used to describe the properties of the system in terms of LST. The K-K transforms are basically a set of mathematical relations transforming the real part of impedance into the imaginary part of impedance and vice versa, and can be stated as

$Z'' \to Z'$:

$$Z'(\omega) - Z'(\infty) = \left(\frac{2}{\pi}\right) \int_0^\infty \frac{x Z''(x) - \omega Z''(\omega)}{x^2 - \omega^2} dx \tag{3}$$

$$Z'(\omega) - Z'(0) = \left(\frac{2\omega}{\pi}\right) \int_0^\infty \left[\left(\frac{\omega}{x}\right) Z''(x) - Z''(\omega)\right] \cdot \frac{1}{x^2 - \omega^2} dx \tag{4}$$

$Z' \to Z''$:

$$Z''(\omega) = -\left(\frac{2\omega}{\pi}\right) \int_0^\infty \frac{Z'(x) - Z'(\omega)}{x^2 - \omega^2} dx \tag{5}$$

in which Z' and Z'' are the real and imaginary parts of the impedance, and ω and x are radial frequencies. Note that the K-K transforms are a purely mathematical result and do not reflect any physical property or condition of the system under investigation.

A MATLAB program developed by Urquidi-Macdonald and Macdonald was employed in this work to perform the K-K transforms. This program was used to test the consistency of the experimental data before applying any EIS modeling. Some typical results of this program are shown in Figure 9 for the passive film formed on iron in pH 12.20 borate buffer/EDTA solution at a formation voltage of 0.3 V vs SCE. The data calculated using the K-K transforms (from the experimental data) agree well with the experimental impedance data over most of the frequency range, although there are some deviations at low frequencies, especially for the imaginary part of the impedance. Such deviations arise from the "tails problem," resulting from the fact that the K-K transforms presented in eqs 3–5 require integration over an infinite bandwidth in frequency ($0 < x < \infty$), whereas the experimental data are restricted to a finite bandwidth (10^4 Hz to 10^{-2} Hz). Consequently, neglect of the regions $0 < f < f_{min}$ and $f_{max} < f < \infty$ introduces errors at both ends of the frequency range, with the error being most evident at the low frequency end in the real-to-imaginary axis transformation (lower plots). Clearly, the "tails problem" is a mathematical artifact, and the electrochemical system under EIS study clearly satisfies the conditions of linearity, causality, stability, and finiteness. Accordingly, the impedance data are therefore judged to be valid.

Impedance Model

The following section is dedicated to the development of an impedance model based on the PDM (*15, 16*), in order to perform mechanistic analysis of impedance data for passive iron.

Figure 10 summarizes the physicochemical processes occurring within the barrier layer formed on iron, in terms of the PDM, with all species being represented by Kroger-Vink notation. Reaction I describes the injection of a cation interstitial, $Fe_i^{\chi \bullet}$, from the metal into the barrier layer. The iron interstitial then migrates through the barrier layer and is ejected into the solution, as described by Reaction III. Reaction II results in the growth of the barrier layer into the bulk metal iron, and Reaction V leads to the destruction of the barrier layer by dissolution at the barrier layer/solution interface. Reaction IV describes the annihilation of oxygen vacancies at the film/solution interface, which are produced by Reaction II at the metal/film interface. Note that Reactions I, III, and IV are lattice conservative processes (i.e., they do not result in the movement of the barrier layer boundaries with respect to a laboratory frame of reference), while Reactions II and V are lattice nonconservative processes. A steady state must involve at least two nonconservative reactions, since only one nonconservative reaction would lead to monotonic growth or thinning of the passive film. Considering our previous findings that the dominant defects in the barrier layer of passive iron must be oxygen vacancies, or iron interstitials, or both, due to the n-type electronic

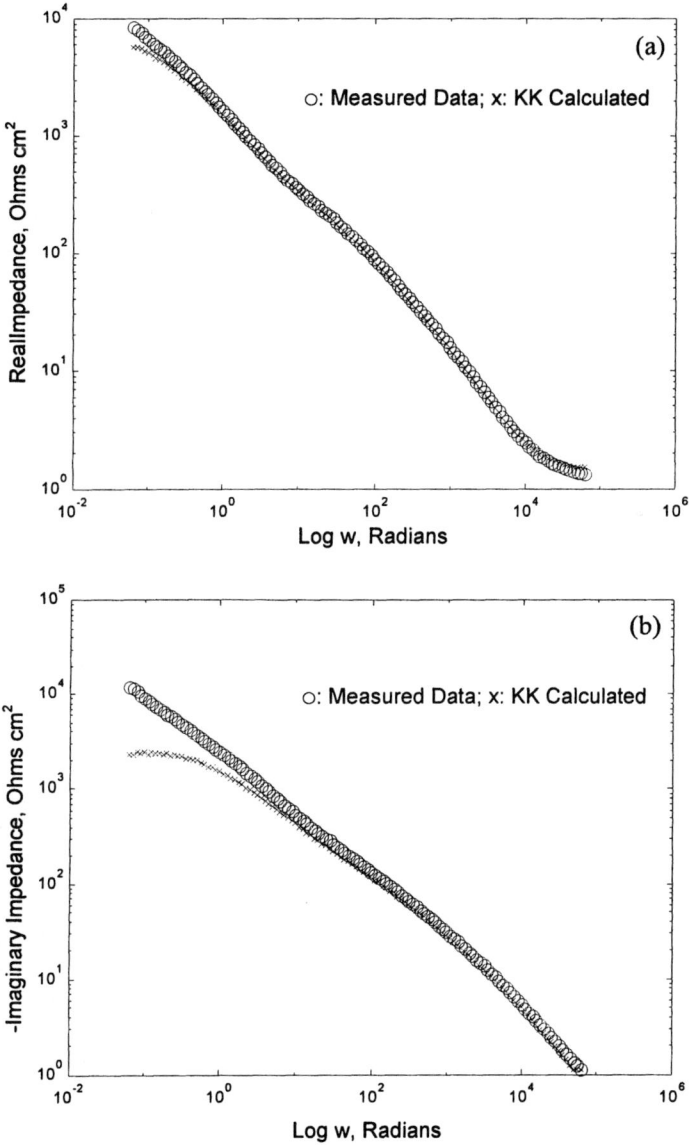

Figure 9. Comparison of EIS data calculated using K-K transforms and those measured from experiments for the real part (a) and the imaginary part (b) of impedance. Passive film on iron was formed in pH 12.20 borate buffer solution containing 0.01 M EDTA, with a film formation voltage of 0.3 V vs SCE and a polarization period of 24 h. Applied frequency ranged from 10^4 Hz to 10^{-2} Hz.

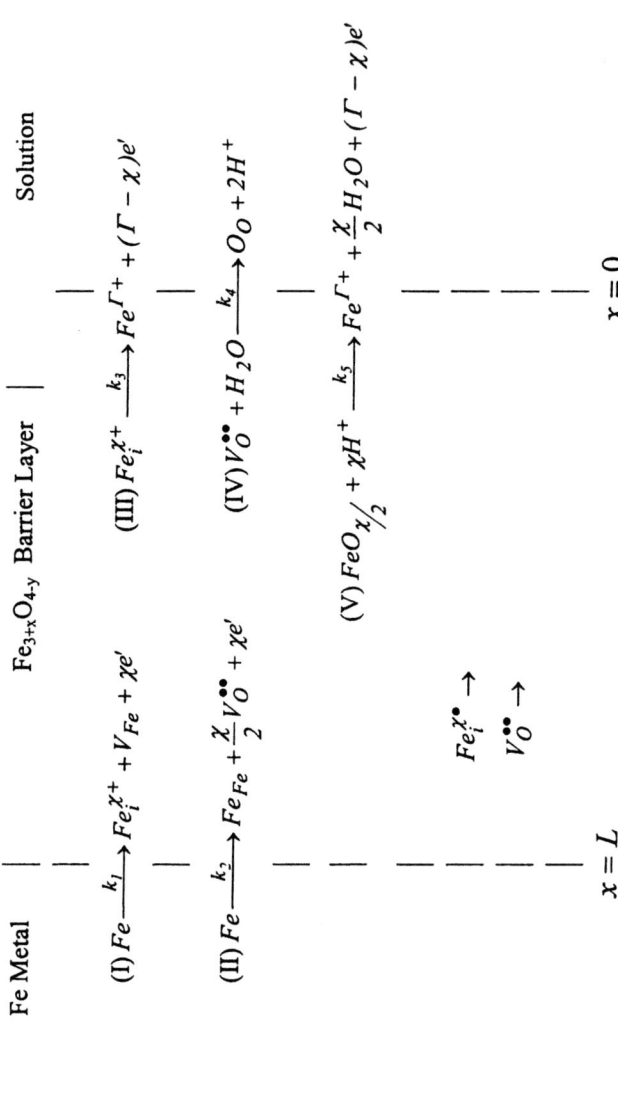

Figure 10. Schematic of physicochemical processes that occur within a defective barrier oxide ($Fe_{3+x}O_{4+y}$) layer on passive iron according to the Point Defect Model. $Fe \equiv$ iron atom; $V_O^{\bullet\bullet} \equiv$ interstitial iron cation; $v_{Fe} \equiv$ iron vacancy at the metal phase; $Fe_{Fe} \equiv$ iron cation in a normal cation position; $Fe^{\Gamma+} \equiv$ iron cation at the solution phase; $V_O^{\bullet\bullet} \equiv$ oxygen vacancy; and $O_O \equiv$ oxygen ion in anion site.

character of the passive film, the reactions regarding the consumption (at the metal/film interface) and production (at the film/solution interface) of iron cation vacancies are not included in the impedance model development. Cation vacancies would dope the barrier layer p-type.

The change in the barrier layer thickness with time involves two lattice nonconservative reactions, as follows:

$$\frac{dL}{dt} = -\frac{2\Omega}{\chi} J_o - \Omega k_5 C_{H^+}^n \qquad (6)$$

in which $\Omega = 14.9 cm^3 mol^{-1}$ is the volume per mole of the barrier layer, $\chi = 8/3$ is the average oxidation state of iron in the film, J_o is the flux for oxygen vacancies within the barrier layer, C_{H^+} is the hydrogen ion concentration in the solution at the film/solution interface, n is the kinetic order of the film dissolution reaction with respect to C_{H^+} at the film/solution interface, and k_5 is the rate constant for Reaction V. Table II displays rate constants for the five interfacial reactions; the definition of the standard rate constants will be detailed later in this section.

All electron-related interfacial reactions contribute to the total current density, as represented by

$$I = F[-\chi J_i^{m/f} - 2J_o^{m/f} - (\Gamma - \chi) J_i^{f/s} + (\Gamma - \chi) k_5 C_{H^+}^n] \qquad (7)$$

in which $F = 96487 C/mol$ is Faraday's constant, and $\Gamma = 3$ is the oxidation state of iron in solution. The fluxes are written as $J_i = -D_i(\partial C_i / \partial x) - \chi K D_i C_i$ for iron interstitials and $J_o = -D_o(\partial C_o / \partial x) - 2K D_o C_o$ for oxygen vacancies, where D and C are diffusivity and concentration, respectively, subscripts i and o are for iron interstitials and oxygen vacancies, $K = \varepsilon F / RT$, $R = 8.314 J mol^{-1} K^{-1}$ is the gas constant, T is the absolute temperature, and ε is the electric field strength, which is assumed to be constant according to the PDM. The continuity equations $\partial C / \partial t = -\nabla J$ then become

$$\frac{\partial C_i}{\partial t} = D_i \frac{\partial^2 C_i}{\partial x^2} + \chi D_i K \frac{\partial C_i}{\partial x} \qquad (8)$$

Table II. Rate Constants $k_i = k_i^0 e^{a_i V} e^{-b_i L}$ for Five Interfacial Reactions in Terms of the Point Defect Model

Reaction	a_i (V^{-1})	b_i (cm^{-1})	Unit of k_i^0
(I) $Fe \xrightarrow{k_1} Fe_i^{\chi+} + V_{Fe} + \chi e'$	$\alpha_1(1-\alpha)\chi\gamma$	$\alpha_1\chi K$	$\dfrac{mol}{cm^2 s}$
(II) $Fe \xrightarrow{k_2} Fe_{Fe} + \dfrac{\chi}{2} V_O^{\cdot\cdot} + \chi e'$	$\alpha_2(1-\alpha)\chi\gamma$	$\alpha_2\chi K$	$\dfrac{mol}{cm^2 s}$
(III) $Fe_i^{\chi+} \xrightarrow{k_3} Fe^{\Gamma+} + (\Gamma-\chi)e'$	$\alpha_3\alpha\Gamma\gamma$	0	$\dfrac{cm}{s}$
(IV) $V_O^{\cdot\cdot} + H_2O \xrightarrow{k_4} O_O + 2H^+$	$2\alpha_4\alpha\gamma$	0	$\dfrac{cm}{s}$
(V) $FeO_{\chi/2} + \chi H^+ \xrightarrow{k_5} Fe^{\Gamma+} + \dfrac{\chi}{2} H_2O + (\Gamma-\chi)e'$	$\alpha_5\alpha(\Gamma-\chi)\gamma$	0	$\dfrac{mol^{0.4}}{cm^{0.2} s}$

$$\frac{\partial C_o}{\partial t} = D_i \frac{\partial^2 C_o}{\partial x^2} + 2D_o K \frac{\partial C_o}{\partial x} \tag{9}$$

the initial conditions ($t = 0$) for which are $C_i(x) = C_i^0(x)$ and $C_o(x) = C_o^0(x)$. The boundary conditions are readily stated as: at $x = 0$ (the film/solution interface): $J_i(0) = -k_3^0 e^{a_3 V} C_i(0)$ (k_3^0 is in cm/s); $J_o(0) = -(\chi/2)k_4^0 e^{a_4 V} C_o(0)$ (k_4^0 is in cm/s); at $x = L$ (the metal/film interface): $J_i(L) = -k_1^0 e^{a_1 V} e^{-b_1 L}$ (k_1^0 is in $mol/cm^2 \cdot s$) and $J_o(L) = -(\chi/2)k_2^0 e^{a_2 V} e^{-b_2 L}$ (k_2^0 is in $mol/cm^2 \cdot s$).

Combined with the rate constants shown in Table II, the rate of film thickness change in eq 6 and current density in eq 7 then become

$$\frac{dL}{dt} = -\Omega k_2^0 e^{a_2 V} e^{-b_2 L} - \Omega k_5^0 e^{a_5 V} C_{H^+}^n \tag{10}$$

and

$$I = F[\chi k_1^0 e^{a_1 V} e^{-b_1 L} + \chi k_2^0 e^{a_2 V} e^{-b_2 L} + (\Gamma - \chi)k_3^0 e^{a_3 V} C_i(0) + (\Gamma - \chi)k_5^0 e^{a_5 V} C_{H^+}^n] \tag{11}$$

Note that the current density is a function of V, L, and $C_i(0)$. Accordingly, for any arbitrary changes δV, δL, and $\delta C_i(0)$

$$\delta I = \left(\frac{\partial I}{\partial V}\right)_{L, C_i(0)} \delta V + \left(\frac{\partial I}{\partial L}\right)_{V, C_i(0)} \delta L + \left(\frac{\partial I}{\partial C_i(0)}\right)_{V, L} \delta C_i(0) \tag{12}$$

For EIS, the variations are sinusoidal in nature, so that: $\delta V = \Delta V e^{j\omega t}$, $\delta L = \Delta L e^{j\omega t}$, and $\delta C_i(0) = \Delta C_i(0) e^{j\omega t}$; note that ΔX is the amplitude of the variation in X at a frequency of $\omega = 0$.

From eq 12, the Faradic admittance is defined as

$$Y_f = \frac{\delta I}{\delta V} = I^V + I^L \frac{\Delta L}{\Delta V} + I^{C_i(0)} \frac{\Delta C_i(0)}{\Delta V} \tag{13}$$

where $I^V = (\partial I/\partial V)_{L,C_i(0)}$; $I^L = (\partial I/\partial L)_{V,C_i(0)}$;   and $I^{C_i(0)} = (\partial I/\partial C_i(0))_{L,V}$. These differentials will be evaluated later from eq 18. Note that no account is taken of the transport of defects (oxygen vacancies or metal interstitials) through the barrier layer; instead, the impact of the defects is postulated to occur through the defect generation/annihilation reactions at the interfaces. Mathematically, this arises from writing the interfacial reactions as irreversible processes (Figure 10). We therefore refer to this model as being an "Interfacial Control Model" (ICM). A more general model has been developed (24), but results in more difficult optimization.

We now return to eq 10 and determine the response of dL/dt to δV, δL, and $\delta C_i(0)$ by taking the total differential. Thus, for the relaxation in film thickness:

$$\frac{d}{dt}(\delta L) = \Omega k_2^0 a_2 e^{a_2 V} e^{-b_2 L} \delta V - \Omega k_2^0 b_2 e^{a_2 V} e^{-b_2 L} \delta L - \Omega k_5^0 a_5 e^{a_5 V} C_{H^+}^n \delta V$$

Substituting $\frac{d}{dt}(\delta L) = j\omega \Delta L e^{j\omega t}$ into the previous equation leads to

$$j\omega \Delta L e^{j\omega t} = \Omega(k_2^0 a_2 e^{a_2 V} e^{-b_2 L} - k_5^0 a_5 e^{a_5 V} C_{H^+}^n) \Delta V e^{j\omega t} - \Omega k_2^0 b_2 e^{a_2 V} e^{-b_2 L} \Delta L e^{j\omega t}.$$

Solving this equation for $\Delta L/\Delta V$ yields eq 14 below, which describes the relaxation in film thickness with respect to the change in voltage.

$$\frac{\Delta L}{\Delta V} = \frac{\Omega(k_2^0 a_2 e^{a_2 V} - k_5^0 a_5 e^{a_5 V} C_{H^+}^n)}{\Omega k_2^0 b_2 e^{a_2 V} e^{-b_2 L} + j\omega} \quad \text{or} \quad \frac{\Delta L}{\Delta V} = \frac{\Phi_2}{1+j\omega} \quad (14)$$

By multiplying the denominator by the complex conjugate and rearranging terms, expressions for Φ_2 and τ_2 can be obtained and are shown in eqs 15 and 16, respectively.

$$\Phi_2 = \left(\frac{a_2}{b_2}\right) - \left(\frac{k_5^0}{k_2^0}\right)\left(\frac{a_5}{b_2}\right) e^{(a_5-a_2)V} e^{b_2 L} C_{H^+}^n \quad (15)$$

$$\tau_2 = \frac{1}{\Omega k_2^0 b_2 e^{a_2 V} e^{-b_2 L}} \quad (16)$$

For iron interstitials, the relaxation for $C_i(0)$ is determined through mass balance at the film/solution interface

$$\frac{dC_i(0)}{dt} = -k'_3 C_i(0) = -k'^0_3 e^{a_3 V} C_i(0) \text{ (Unit of } k'^0_3 \text{ is s}^{-1}).$$

The total differential is

$$\frac{d}{dt}(\delta C_i(0)) = -k'^0_3 [a_3 e^{a_3 V} C_i(0)\delta V + e^{a_3 V} \delta C_i(0)].$$

Substituting the sinusoidal variations for δV and $\delta C_i(0)$ one obtains

$$j\omega \Delta C_i(0)e^{j\omega t} = -k'^0_3 a_3 e^{a_3 V} C_i(0)\Delta V e^{j\omega t} - k'^0_3 e^{a_3 V} \Delta C_i(0) e^{j\omega t}$$

and

$$\Delta C_i(0)(j\omega + k'^0_3 e^{a_3 V}) = -k'^0_3 a_3 e^{a_3 V} C_i(0)\Delta V.$$

Therefore,

$$\frac{\Delta C_i(0)}{\Delta V} = \frac{-k'^0_3 a_3 e^{a_3 V} C_i(0)}{j\omega + k'^0_3 e^{a_3 V}} = -\frac{a_3 C_i(0)}{1 + j\omega \tau_3} \quad (17)$$

in which

$$\tau_3 = \frac{1}{k'^0_3 e^{a_3 V}} \quad (18)$$

The expression for the faradic admittance is finally derived from eq 17 as

$$Y_f = I^V + I^L \frac{\Phi_2}{1+j\omega\tau_2} - I^{C_i(0)} \frac{a_3 C_i(0)}{1+j\omega\tau_3} + j\omega C \quad (19)$$

in which the parallel geometric capacitance $C = \hat{\varepsilon}\varepsilon_0 / L$ (with dielectric constant $\hat{\varepsilon} = 30$ for passive iron (13) and vacuum permittivity $\varepsilon_0 = 8.85 \times 10^{-14} F/cm$) is taken into consideration, and

$$I^V = \left(\frac{\partial I}{\partial V}\right)_{L, C_i(0)} \quad (20)$$
$$= F[a_1 \chi k_1^0 e^{a_1 V} e^{-b_1 L} + a_2 \chi k_2^0 e^{a_2 V} e^{-b_2 L} + a_3(\Gamma - \chi) k_3^0 e^{a_3 V} C_i(0) + a_5(\Gamma - \chi) k_5^0 e^{a_5 V} C_{H^+}^n]$$

$$I^L = \left(\frac{\partial I}{\partial L}\right)_{V, C_i(0)} \quad (21)$$
$$= F[-b_1 \chi k_1^0 e^{a_1 V} e^{-b_1 L} - b_2 \chi k_2^0 e^{a_2 V} e^{-b_2 L}] = -\chi F[b_1 k_1^0 e^{a_1 V} e^{-b_1 L} + b_2 k_2^0 e^{a_2 V} e^{-b_2 L}]$$

and

$$I^{C_i(0)} = \left(\frac{\partial I}{\partial C_i(0)}\right)_{V,L} = (\Gamma - \chi)Fk_3^0 e^{a_3 V} \tag{22}$$

In the above equations, the parameters that appear on the right side [L and $C_i(0)$] are identified with the steady-state quantities. At steady state, eqs 8 and 9 become

$$\frac{\partial^2 C_i}{\partial x^2} + \chi K \frac{\partial C_i}{\partial x} = 0 \tag{23}$$

$$\frac{\partial^2 C_o}{\partial x^2} + 2K \frac{\partial C_o}{\partial x} = 0 \tag{24}$$

the solutions to which are

$$C_i(x) = A_i e^{-\chi K x} + B_i \tag{25}$$

$$C_o(x) = A_o e^{-2Kx} + B_o \tag{26}$$

The expressions for the fluxes therefore become

$$J_i = -\chi K D_i B_i \tag{27}$$

$$J_o = -2KD_o B_o \tag{28}$$

From the boundary conditions, the coefficients are obtained as

$$B_i = \frac{k_1^0 e^{a_1 V} e^{-b_1 L}}{\chi K D_i} \tag{29}$$

$$B_o = \frac{\left(\frac{\chi}{2}\right) k_2^0 e^{a_2 V} e^{-b_2 L}}{2KD_o} \tag{30}$$

Because the fluxes for a given species at the two interfaces are equal at steady state:

$$-k_3^0 e^{a_3 V} C_i(0) = -k_1^0 e^{a_1 V} e^{-b_1 L}$$

and

$$-\left(\frac{\chi}{2}\right) k_4^0 e^{a_4 V} C_o(0) = -\left(\frac{\chi}{2}\right) k_2^0 e^{a_2 V} e^{-b_2 L}.$$

The steady-state concentrations of iron interstitials and oxygen vacancies at the film/solution interface are then

$$C_i(0) = \left(\frac{k_1^0}{k_3^0}\right) e^{(a_1 - a_3)V} e^{-b_1 L} \qquad (31)$$

$$C_o(0) = \left(\frac{k_2^0}{k_4^0}\right) e^{(a_2 - a_4)V} e^{-b_2 L} \qquad (32)$$

Also in the case of steady state, eq 6 becomes

$$J_o(L) = -\left(\frac{\chi}{2}\right) k_5 C_{H^+}^n \qquad (33)$$

and from eqs 28 and 30:

$$-\left(\frac{\chi}{2}\right) k_5 C_{H^+}^n = -2 K D_o \left(\frac{\left(\frac{\chi}{2}\right) k_2^0 e^{a_2 V} e^{-b_2 L}}{2 K D_0} \right).$$

Therefore, the steady-state barrier layer thickness is

$$L = \left(\frac{a_2 - a_5}{b_2}\right) V - \frac{1}{b_2} \ln\left[\left(\frac{k_5^0}{k_2^0}\right) C_{H^+}^n\right] \qquad (34)$$

The parameters k_1^0, k_2^0, $k_3^{'0}$, k_4^0, and k_5^0 in the above equations and Table II are a function of solution pH and are not the standard rate constants for interfacial reactions, since the PDM defines $k_1^0 = k_1^{00} e^{-\alpha_1 \beta \chi p H} e^{-\alpha_1 \chi \gamma \phi_{f/s}^0}$, $k_2^0 = k_2^{00} e^{-\alpha_2 \beta \chi p H} e^{-\alpha_2 \chi \gamma \phi_{f/s}^0}$, $k_3^0 = k_3^{00} e^{-\alpha_3 \beta \chi p H} e^{-\alpha_3 \chi \gamma \phi_{f/s}^0}$,

$k_3'^0 = k_3'^{00} e^{-\alpha_3 \beta \gamma pH} e^{-\alpha_3 \Gamma \gamma \phi_{f/s}^0}$, $\quad k_4^0 = k_4^{00} e^{2\alpha_4 \beta \gamma pH} e^{2\alpha_3 \gamma \phi_{f/s}^0}$, and $k_5^0 = k_5^{00} e^{\alpha_5 (\Gamma - \chi) \beta \gamma pH} e^{\alpha_5 (\Gamma - \chi) \gamma \phi_{f/s}^0}$, in which α_1, α_2, α_3, α_4, and α_5 are the transfer coefficients, and k_1^{00}, k_2^{00}, k_3^{00}, $k_3'^{00}$, k_4^{00}, and k_5^{00} are the standard rate constants. The potential drop across the film/solution interface is $\phi_{f/s} = \alpha V + \beta pH + \phi_{f/s}^0$ and that across the metal/film interface is $\phi_{m/f} = (1-\alpha)V - \varepsilon L - \beta pH - \phi_{f/s}^0$, in which $\alpha = d\phi_{f/s}/dV$ is the dependence of the potential drop across the film/solution interface on V (i.e., the "polarizability" of the barrier layer/solution interface), $\beta = d\phi_{f/s}/dpH$ is the dependence of the potential drop across the film/solution interface on pH, $\phi_{f/s}^0$ is the value of $\phi_{f/s}$ at the standard state of $V = 0$ and $pH = 0$; $\gamma = F/RT$, $K = \varepsilon\gamma = \varepsilon F/RT$, $\varepsilon = 1.10 \times 10^6 V/cm$, and $\alpha = 0.728$ for passive iron (23).

DataFit software (version 7.1) was employed in this work to obtain the transfer coefficient α_i and standard rate constant k_i^{00} for the i-th elementary interfacial reaction via nonlinear optimization of the impedance model on the experimental impedance data. The values of β and $\phi_{f/s}^0$ were also generated for passive iron through this method. The optimization algorithm depends on finding the minimum residual sum of squares between the experimental data and the simulated data. Fundamental parameters for passive iron are presented in Table III, and typical comparisons between the experimental and simulated impedance data are displayed in Figure 11. Parameters for Reaction IV do not show up in the model, because no electrons are involved in the reaction.

The correlation between the experimental impedance data and the simulated data is judged to be very good for the Nyquist and modulus Bode plots (Figures 11a and b, respectively), in most cases. This validates the impedance model based on the PDM, through which a single set of parameters for passive iron (shown in Table III) has been derived via nonlinear optimization to obtain the simulated impedance data. However, the phase angle Bode plot displays some difference between the experimental data and the simulated data, especially in the higher frequency range (Figure 11c). A possible reason is that the phase angle, defined as $\theta = \tan^{-1}(-Z''/Z')$, is a very sensitive function of Z' and Z'', so that as these quantities become smaller with increasing frequency, slight deviations in Z' or Z'' could give rise to large fluctuations in θ, as observed.

Table III shows that the standard rate constant for the iron interstitial production, Reaction I in Figure 10, is higher than that for the oxygen vacancy generation reaction (Reaction II in Figure 10) by several orders of magnitude. This implies that the generation of iron interstitials at the metal/film interface is

Table III. Fundamental Parameters for the Barrier Oxide Layer Formed on Iron in Borate Buffer Solution at 22 °C

α_1	0.01 (from optimization)
α_2	0.24 (from optimization)
α_3	0.39 (from optimization)
α_5	0.30 (from optimization)
k_1^{00} ($mol \cdot cm^{-2} \cdot s^{-1}$)	3.8×10^{-12} (from optimization)
k_2^{00} ($mol \cdot cm^{-2} \cdot s^{-1}$)	1.1×10^{-15} (from optimization)
k'^{00}_3 (s^{-1})	2.4×10^{-6} (from optimization)
k_5^{00} ($mol^{0.4} \cdot cm^{-0.2} \cdot s^{-1}$)	3.3×10^{-8} (from optimization)
β	-0.0047 (from optimization)
$\phi^0_{f/s}$ (V vs SCE)	-0.29 (from optimization)
$\hat{\varepsilon}$	30 (*13*)
ε (V/cm)	1.10×10^6 (*23*)
α	0.728 (*23*)

much faster, and hence that the majority of the passive current density is due to cation interstitial production rather than film formation. If so, then the principal point defect in the barrier oxide layer on iron is the cation interstitial, $Fe_i^{\chi \bullet}$, with the value of χ most likely being 2. This contrasts the work of Oblonsky, et al. (*10*), which implies that the principal point defect is the cation vacancy, $V_{Fe}^{\chi'}$. If this latter assignment was correct, the barrier layer should display p-type electronic character, which is at odds with experiment.

Interestingly, the transfer coefficient for Reaction I, α_l, is close to 0, which indicates that the activated complex (transition state) for interstitial generation at the metal/barrier layer interface (Reaction I, Figure 10) is "*Fe*-like," rather than being similar to the product, $Fe_i^{\chi \bullet}$. Accordingly, the activated complex occurs at a position on the reaction coordinate that is minimally different from the origin (initial state) and hence is characterized by minimal redistribution of electron density.

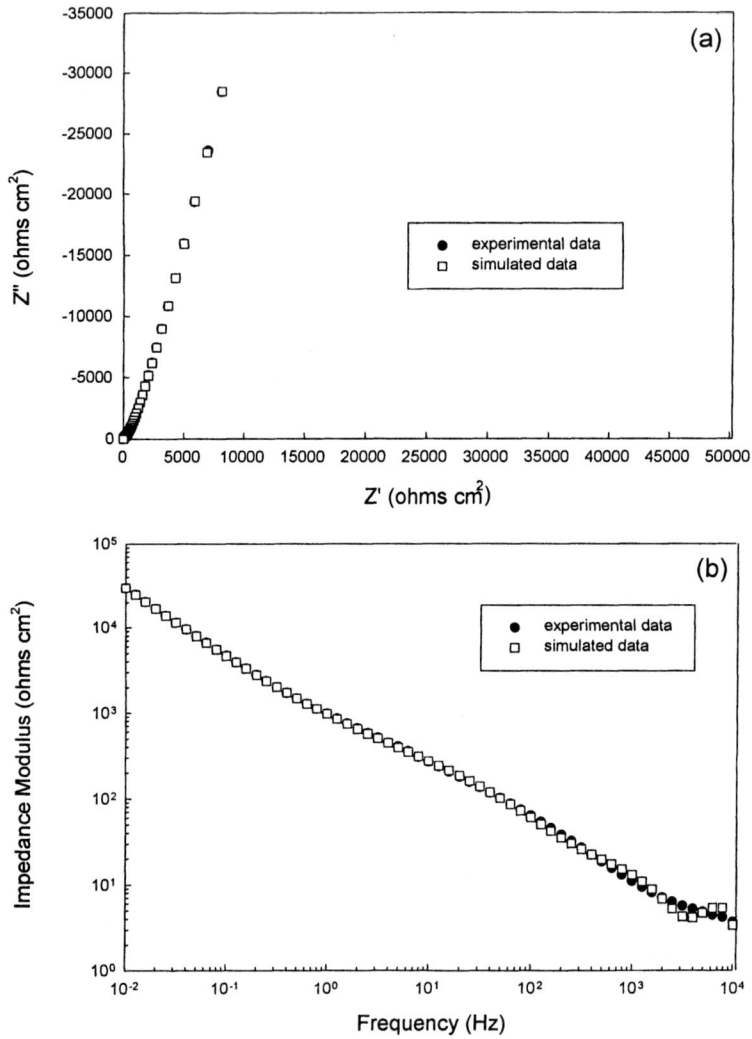

Figure 11. Nyquist plot (a) and Bode plots (b, c) of impedance data for the passive film formed on iron in borate buffer solution with 0.01 M EDTA (pH 8.15) at an applied film formation voltage of 0.2 V vs SCE. Closed circles represent experimental data, and open squares represent simulated data using nonlinear fitted parameters.

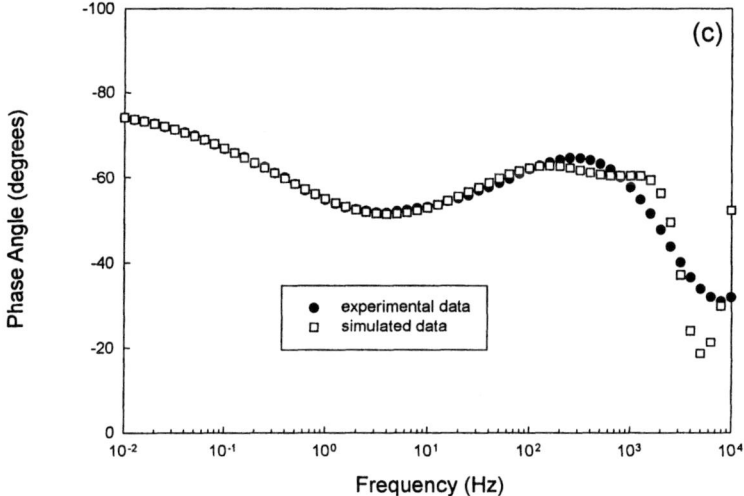

Figure 11. Continued.

The value of α (0.728), which was not determined in this work by optimization, but rather which was taken from a previous study (23), is quite large, but is consistent with the expectation that the value should be larger the thinner the film. This is, because the electric field strength is independent of applied voltage (in the steady-state), so that the voltage drop across the film is directly determined by the (linear) dependence of the film thickness on the applied voltage and the barrier layer thickness. The value indicates that, for each increment in the applied voltage, ΔV, 0.73 ΔV appears across the barrier layer/solution interface and 0.27 ΔV appears across the film and the metal/film interface.

The value of β is found to be -0.0047, which is very close to that found in previous work (23), suggesting that β is very small for the barrier layer formed on iron in borate buffer solutions. This value is consistent with the weak dependence of the impedance in the passive state on pH (Figure 7). Obviously, the value is not applicable at high pH (pH > 12), particularly at high applied voltages (Figure 8).

The parameters in Table III were used to calculate the steady-state properties, including film thickness and current density, of passive films formed on iron. The results are shown in Figure 12, and good agreement has been achieved between the simulated data and the experimental data from our previous research (14). This is taken as further evidence for the validity of the PDM.

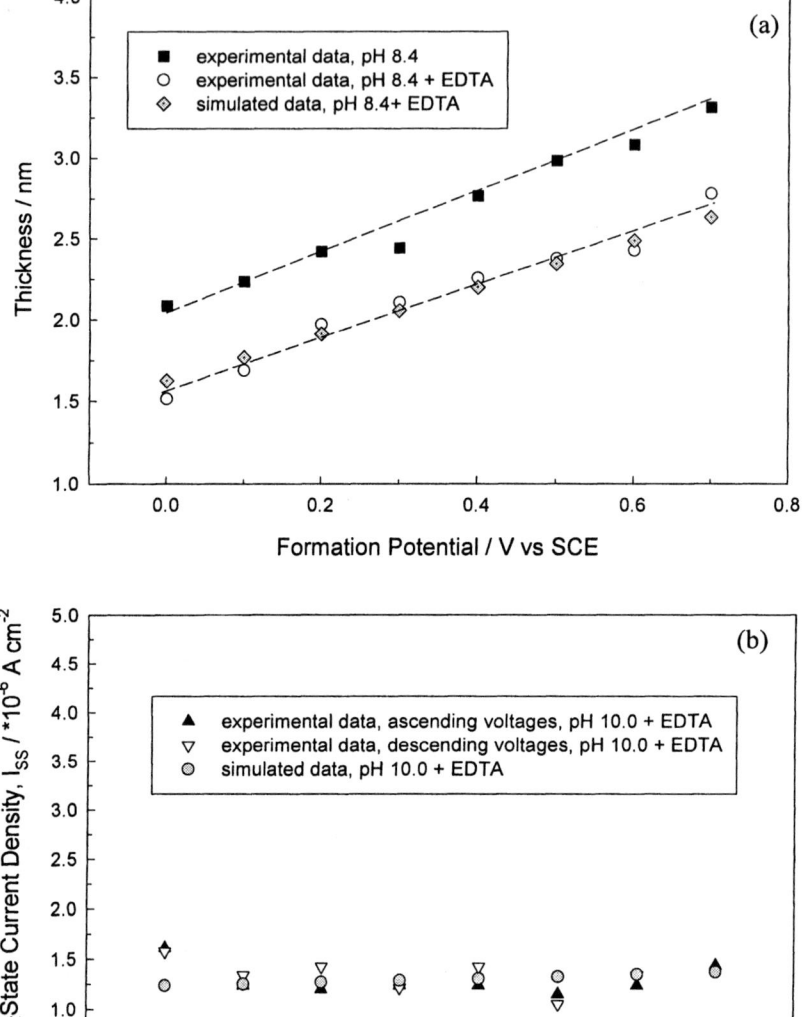

Figure 12. The steady-state film thickness (a) and current density (b) for the passive film on iron. Simulated data were calculated using fundamental parameters in the Point Defect Model.

Summary and Conclusions

EIS is a powerful, in situ technique for investigating steady-state electrochemical and corrosion systems. This technique employs a very small sinusoidal voltage perturbation to drive the electrode/solution interface not far from the steady state, and then detects interfacial relaxations over a wide frequency range.

There are three methods to test whether the experimental impedance data are correct and reliable:

1. The correlation between EIS data measured at "up" and "down" steps in frequency is indicative of the stability or instability of the electrode/solution interface under investigation. A steady-state electrochemical system should generate good agreement between two frequency step directions.
2. The impedance data at one film formation voltage can be measured twice, i.e., in both the ascending and descending voltage directions. A steady-state electrode/solution interface should yield the same electrochemical impedance characteristics, no matter the voltage step direction. "Up" means changing the passive film formation voltage in the negative-to-positive direction; "down" in the positive-to-negative direction.
3. The K-K transforms are employed to compare experimental data and calculated data, as a major test of the steady state, causality, and linearity constraints of Linear Systems Theory, as applied to electrochemical systems in EIS studies. An electrochemical system satisfying the conditions of linearity, causality, stability, and finiteness should give rise to good agreement between the measured impedance data and the data calculated by the K-K transforms throughout most of the frequency range.

It was found that the sinusoidal perturbation voltage should be of the order of 10 mV in order to satisfy the linearity constraint. In the passive state and in weakly alkaline solutions, the impedance is found to be essentially independent of pH. However, at high pH values, particularly at high voltages, the impedance becomes quite pH-dependent.

Throughout the frequency range, the Nyquist plots are shown to be almost independent of formation voltage, in agreement with predictions of the PDM. The point defect transport is mainly due to migration under the influence of the electric field within the passive film, and the electrical field strength is assumed to be invariant of the applied film formation potential in terms of the PDM. Therefore, the impedance should be independent of the film formation voltage.

The n-type electronic character of passive films on iron results from the dominant defects in the barrier oxide layer on passive iron being oxygen vacancies or iron interstitials, or both. An impedance model has been developed to derive a single set of fundamental parameters for the interfacial reactions occurring within the passive film formed on iron. These parameter values were then used to calculate the impedance over a wide range of conditions (frequency, voltage, and pH) and to estimate the steady state passive current density and barrier layer thickness as a function of voltage at ambient temperature (22 °C). The good agreement between experimental and calculated impedance data and calculated and experimental steady state passive current density and barrier layer thickness confirms the validity of the PDM for describing the passive state on iron.

Acknowledgments

The authors gratefully acknowledge the support of this work by the U.S. Department of Energy Environmental Management Science Program through Grant No. DE-FG07-97ER62515.

References

1. Nagayama, M.; Cohen, M. *J. Electrochem. Soc.* **1962**, *109*, 781.
2. Nagayama, M.; Cohen, M. *J. Electrochem. Soc.* **1963**, *110*, 670.
3. Sato, N.; Kudo, K. *Electochim. Acta.* **1971**, *16*, 447.
4. Azumi, K.; Ohtsuka, T.; Sato, N. *J. Electrochem. Soc.* **1986**, *133*, 1326.
5. Ord, J. L.; De Smet, D. J. *J. Electrochem. Soc.* **1976**, *123*, 1876.
6. Bojinov, M.; Fabricus, G.; Laitinen, T.; Makela, K.; Saario, T.; Sundholm, G. *Electochim. Acta.* **2000**, *45*, 2029.
7. Chen, C-T.; Cahan, B. D. *J. Electrochem. Soc.* **1982**, *129*, 17.
8. Ohtsuka, T.; Ohta, A. *Mater. Sci. Eng., A* **1995**, *A198*, 169.
9. Chanson, C.; Blanchard, P. *J. Electrochem. Soc.* **1989**, *136*, 3690.
10. Oblonsky, L. J.; Davenport, A. J.; Ryan, M. P.; Isaacs, H. S.; Newman, R.C. *J. Electrochem. Soc.* **1997**, *144*, 2398.
11. Jovancicevic, V.; Kainthla, R. C.; Tang, Z.; Yang, B.; Bockris, J. O'M. *Langmuir* **1987**, *3*, 388.
12. Büchler, M.; Schmuki, P.; Böhni, H.; Stenberg, T.; Mäntylä, K. *J. Electrochem. Soc.* **1998**, *145*, 378.
13. Sikora, E.; Macdonald, D. D. *J. Electrochem. Soc.* **2001**, *147*, 4087.
14. Liu, J.; Macdonald, D. D. *J. Electrochem. Soc.* **2001**, *148*, B425.
15. Macdonald, D. D. *Pure Appl. Chem.* **1999**, *71*, 951.
16. Chao, C. Y.; Lin, L. F.; Macdonald, D. D. *J. Electrochem. Soc.* **1982**, *129*, 1874.
17. Kramers, H. A. *Physiol. Zool.* **1929**, *30*, 522.
18. Kronig, R. de L. *J. Opt. Soc. Am.* **1926**, *12*, 547.
19. Tyagai, V. A.; Kolbasov, G. Y. *Elektrokhimiya* **1972**, *8*, 59.

20. Van Meirhaeghe, R. L.; Dutoit, E. C.; Cardon, F.; Gomes, W. P. *Electrochim. Acta* **1976**, *21*, 39.
21. Macdonald, D. D.; Urquidi-Macdonald, M. *J. Electrochem. Soc.* **1985**, *132*, 2316.
22. Urquidi-Macdonald, M.; Real, S.; Macdonald, D. D. *J. Electrochem. Soc.* **1986**, *133*, 2018.
23. Chao, C. Y.; Lin, L. F.; Macdonald, D. D. *J. Electrochem. Soc.* **1981**, *128*, 1187.
24. Macdonald, D. D.; Smedley, S. I. *Electrochim. Acta* **1990**, *35*, 1949.

Chapter 6

Investigating Ultrasonic Diffraction Grating Spectroscopy and Reflection Techniques for Characterizing Slurry Properties

M. S. Greenwood[1], A. Brodsky[2], L. Burgess[2], and L. J. Bond[3]

[1]Pacific Northwest National Laboratory, P.O. Box 999, Mail Stop K5–26, Richland, WA 99352
[2]Center for Process Analytical Chemistry, University of Washington, Seattle, WA 98195–1700
[3] Director, Center for Advanced Energy Studies, Idaho National Laboratory, Idaho Falls, ID 83415

An objective of ultrasonic diffraction grating spectroscopy (UDGS) is to measure the velocity of sound in a liquid. In addition, recent research shows the ability to determine the particle size of a slurry. The grating surface is in contact with the liquid or slurry and ultrasound reflected by the grating shows a peak at a critical frequency, from which the velocity is obtained for sugar water and NaCl solutions and information about particle size. The reflection of shear waves at a quartz-liquid interface is dependent upon the viscosity of the liquid. By using multiple reflections the small effect is amplified. A self-calibrating method for measuring the viscosity is described.

The objective of this project, sponsored by the U. S. Department of Energy (DOE) Environmental Management Science Program, is to develop noninvasive methods for measuring particle size, velocity of sound, and viscosity that can be implemented on-line and in real time. The concept of a pipeline sensor is shown in Figure 1, in which the ultrasonic diffraction grating is positioned in the wall of a pipeline. A shear wave transducer measures the viscosity and a longitudinal wave transducer measures the acoustic impedance (defined as the product of density and velocity of sound) of the liquid or slurry. Thus, the velocity of sound from the ultrasonic diffraction grating can be combined with the acoustic impedance to yield the density. Also, the reflection of the shear wave and the measurement of density will yield the viscosity. At DOE sites, there is a need to characterize slurries, as the radioactive wastes are retrieved from underground storage tanks and transported through pipelines, in order to prevent pipeline plugging. In many industries, there is also a need for liquid and slurry characterization for process control.

The ultrasonic diffraction grating technique and the reflection technique for measuring viscosity are quite different, but are linked by their importance for slurry characterization. Ultrasonic diffraction grating spectroscopy (UDGS) will be discussed first.

Two sets of experiment data for liquids using UDGS (*1, 2*) and measurements of viscosity using reflection techniques (*3*) have been reported. Reference 4 describes the density sensor that was installed in the pipeline between the radioactive waste tanks SY101 and SY102 on the Hanford Site and was used to obtain density data during the transfer in November 2002. Greenwood and Bamberger describe the next-generation density sensor that operates through a stainless steel pipeline wall (*5*).

The method of UDGS is analogous to that for optics using grating light reflection spectroscopy (GLRS) (*6–8*). The optical method has been successful in determining the particle size of a slurry in the range from about 2 to 200 nanometers and also for measuring the index of refraction. Because of the larger wavelength, ultrasonics can measure larger particle sizes.

Current research in UDGS has centered on experiments with slurries of polystyrene spheres to determine the effect of varying the particle size. The first set of measurements was carried out with a stainless steel grating at a critical frequency of 7.0 MHz. Notable effects were observed and these results have been reported (*9*). In order to increase the ultrasound transmitted into the slurry, experiments have been carried out with an aluminum grating at a critical frequency of 3.5 MHz for slurries of polystyrene spheres having diameters of 215 microns, 275 microns, 363 microns, and 463 microns, as a function of the weight percentage of the slurry. At 11 wt % the signal for these four slurries has widely separated amplitudes (10). A U.S. patent has been awarded for the techniques of ultrasonic diffraction grating spectroscopy (*11*). Additional data for the measurement of viscosity have been obtained using the 70° wedge (*12*). Current research in the measurement of viscosity has been carried out using a fused silica wedge having 45° angles and shows that the self-calibrating method gives good agreement to measurements obtained using a laboratory viscometer (*13*).

Part I. Ultrasonic Diffraction Grating Spectroscopy

Description of Method

The experimental data reported in references 1 and 2 were obtained using diffraction grating, formed by machining parallel triangular-shaped grooves on the flat surface of a stainless steel half-cylinder, as shown in Figure 2. The half-cylinders have a diameter of 5.08 cm, with a groove spacing of 200 microns on one and 300 microns on the other. The included angle of the triangular groove is 120°. The grating is placed in the immersion chamber (27.9-cm diameter) in a recess in the mounting plate to fix its position and angle relative to the send transducer, as shown in Figure 3, and the chamber filled with the desired liquid. The mounting plate can rotate in order to change the incident angle θ, and is then fastened at the desired (fixed) angle to the base of the chamber with set screws. The send and receive transducers are mounted in brackets that are placed in the aluminum housings. The send transducer housing is fastened to the base of the chamber. The chamber itself is mounted upon a turntable, which is motorized and computer-controlled to an angular placement with accuracy of 0.1°. The receive transducer is mounted to the (non-rotating) base of the turntable and, thus, the receive transducer does not move. The desired angle between the send and receive transducers is obtained by rotating the turntable with the immersion chamber (and send transducer) attached. In Figure 3 the send and receive transducers make equal angles with respect to the normal to the grating surface; the receive transducer is pointed at the *back* surface of the diffraction grating in this photograph. The system also permits the receive transducer to be pointed to the *front* surface of the diffraction grating.

For the data reported here, the transducers are fastened directly to the stainless steel grating, as will be discussed shortly. However, the basic concepts and measurement techniques are not changed.

Computer-Controlled Data Acquisition System

The data acquisition system consists of a commercially available pulser-receiver card and a digitizer card mounted in a personal computer. The pulser sends out a sinusoidal toneburst signal having an amplitude of about 330 volts peak-to-peak. The receiver has a maximum gain of 70 dB. The digitizer has 8 bits of resolution with a sampling rate of 100 MHz. A schematic diagram is shown in Figure 4. Custom software was written at Pacific Northwest National Laboratory (PNNL) to control the two cards and consists of five modules: (1) instrument setup, (2) data acquisition and display, (3) parameter measurement, (4) data storage, and (5) control of a motorized turntable to

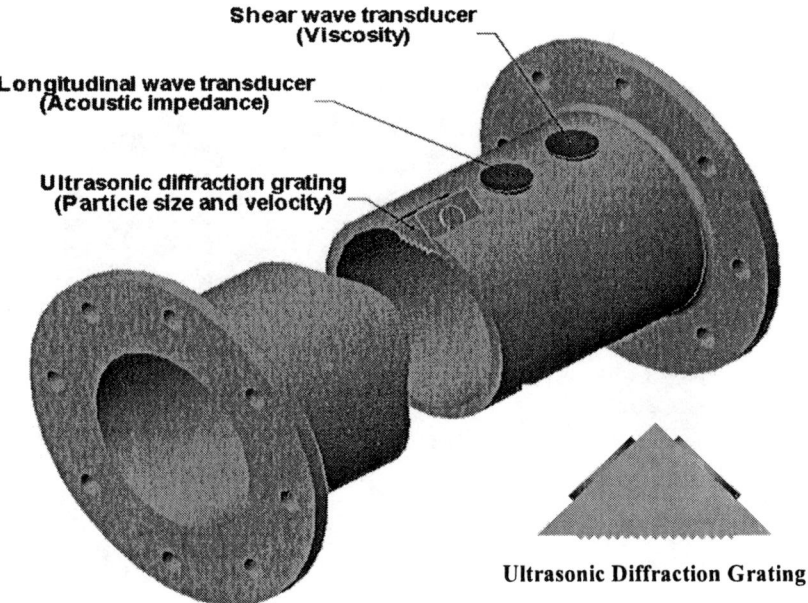

Figure 1. Pipeline concept.

Figure 2. Stainless steel ultrasonic diffraction grating.

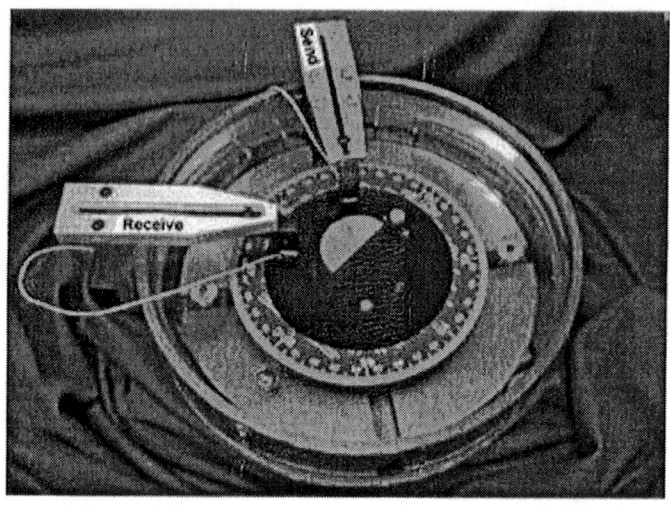

Figure 3. Immersion chamber showing grating and transducer configuration.

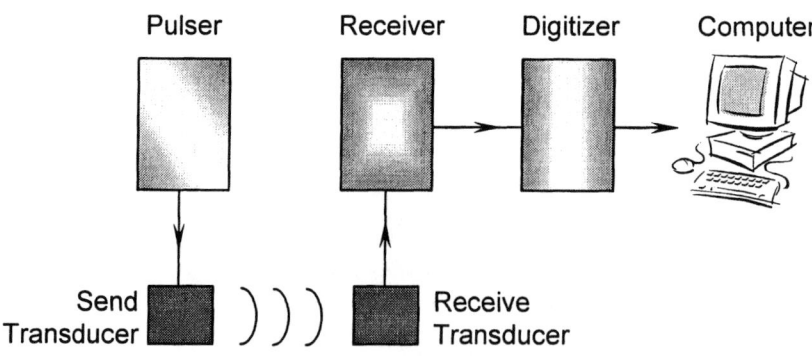

Figure 4. Schematic diagram of computer-controlled data acquisition system.

change angles between send and receive transducers. The software was written in 'C' and operates on a DOS-based platform on a PC.

Instrument setup consists of setting parameters for the pulser (frequency and number of cycles), receiver (gain), and digitizer (sample rate). Figure 5 shows the receive signal and cursors displayed on the computer monitor. The operator sets the software cursors: a time-of-flight gate labeled TOF, cursor G1, and cursor G2, and the threshold (TH) shown slightly above the time axis. The peak amplitude is found by examining the peak-to-peak amplitude between cursors G1 and G2 and selecting the largest value. The time of flight can also be obtained and is found by first locating the time point after the TOF gate where the waveform first exceeds the set threshold. Next, the software finds the preceding zero-voltage cross-over point, which becomes the resulting time-of-flight. Averaging is also used to factor out random noise.

Two types of measurements are possible with this data acquisition system: scan over frequency and scan over angle.

Scan over Frequency

In the setup of the data acquisition system, the receive transducer is positioned at a desired fixed angle, and the initial frequency, step in frequency Δf, and the final frequency are chosen. The system operates by sending a sinusoidal toneburst signal with the initial frequency. The maximum amplitude between the cursors G1 and G2 is determined and stored in a file. Then the frequency is increased by Δf and the maximum amplitude is determined again. This process is repeated until the final frequency is reached. The data file contains the amplitude for each frequency. It also contains the amplitude corrected for receiver gain, which is defined as the amplitude that would have occurred if there were no gain.

Scan over Angle

In this case, a constant toneburst frequency is chosen. The initial angle of the turntable, the change in angle $\Delta\varphi$, and the number of steps are chosen by the operator. The system operates by positioning the turntable at the initial angle and sending a toneburst signal of the constant frequency. The maximum amplitude between cursors G1 and G2 is determined and recorded to a file. Then the angle is increased by $\Delta\varphi$, and the process is repeated for the desired number of steps. The data file contains the amplitude for each angle. The amplitude corrected for receiver gain is also provided.

Measurements with the Blank

To account for the transducer response as a function of frequency, measurements were made with the blank, which has the same dimensions as the grating except that the flat surface of the half-cylinder is smooth.

Grating Equation and Critical Frequency Calculation

Figure 6 shows an ultrasonic beam of frequency f traveling through a solid, and striking the grating-liquid interface at an incident angle θ. As a result of constructive interference, a refracted beam in the liquid is produced at angle φ_m. The distance between adjacent grooves in the ultrasonic diffraction grating is d. (For simplicity, the grooves are shown schematically here as "slits.") In the solid the speed of sound is c_L and the wavelength is λ_1. In the liquid the speed of sound is c and the wavelength is λ. Constructive interference occurs when:

$$\frac{OC}{\lambda} - \frac{AB}{\lambda_1} = m$$
$$\frac{d f \sin \varphi_m}{c} - \frac{d f \sin \theta}{c_L} = m \quad (1)$$

where m is zero, or a positive or negative integer. When m = 0, Snell's Law is obtained:

$$\frac{\sin \varphi_0}{c} = \frac{\sin \theta}{c_L} \quad (2)$$

Using the results of eq 2, eq 1 becomes

$$\sin \varphi_m - \sin \varphi_0 = \frac{mc}{fd} \quad (3)$$

Eq 3 is the so-called grating equation and determines the angle φ_m. Note that as the frequency decreases, the angle φ_1 increases. At some frequency, called the critical frequency (5.65 MHz in this case), angle φ_1 is 90° and eq 3 becomes:

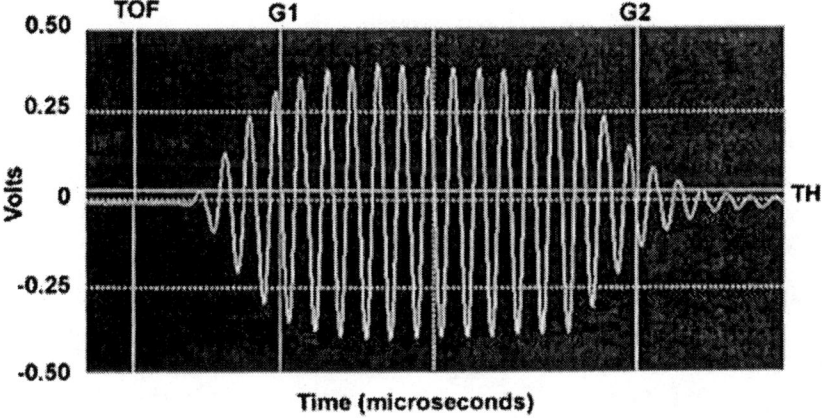

Figure 5. Toneburst signal obtained by receive transducer.

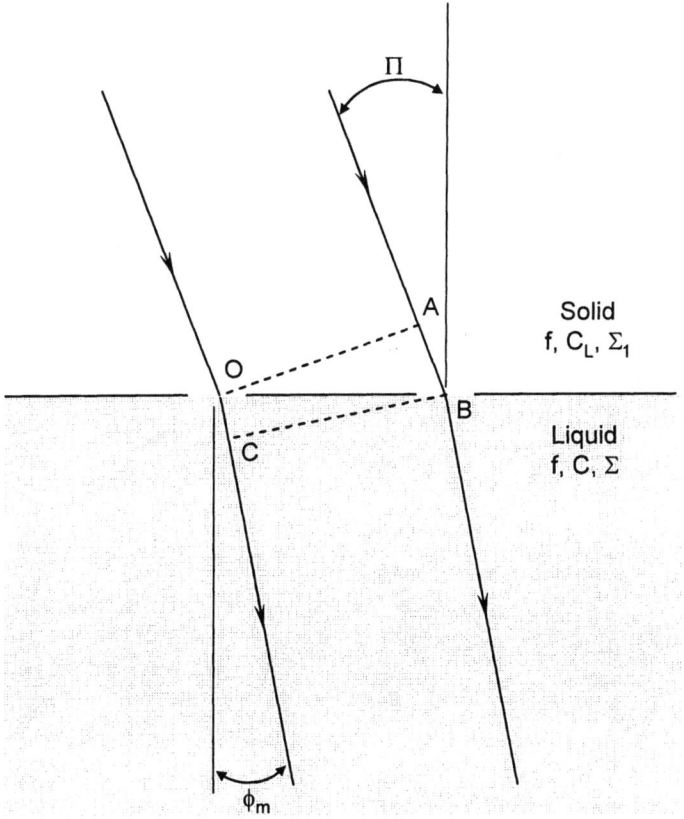

Figure 6. Definition of angles and path lengths.

$$F_{CR} = \frac{c}{d(1-\sin\varphi_0)} \qquad (4)$$

To understand how UDGS can be used to measure the speed of sound, consider a longitudinal wave striking the back of a 300-micron stainless steel grating at an incident angle of 30°, as shown in Figure 7a. The specularly reflected m = 0 longitudinal wave is measured by the receive transducer. In Figure 7b the incident and reflected waves are shown, as well as the reflected m = 0 and m = -1 shear waves, and the transmitted m = 0 and m = 1 longitudinal waves. The positions of the m = 0 waves are, of course, unchanged when the frequency changes. However, the diffracted m = 1 longitudinal wave in water does change with frequency. Figure 7b shows the position of the m = 1 transmitted longitudinal wave at 7 MHz, 6 MHz, and 5.65 MHz. As the frequency decreases, the angle increases, and at a frequency of 5.65 MHZ, the angle becomes 90°. At this frequency, called the critical frequency, the wave transforms from a traveling wave and becomes evanescent. This means that it is an exponentially decaying wave in the liquid or slurry. At a slightly smaller frequency, the evanescent wave disappears. To conserve energy, the energy is redistributed to all other waves, and an increase in amplitude of the specularly reflected signal is expected. This increase in amplitude has been observed (1, 2) and one objective in the current research is to obtain the critical frequencies with a smaller experimental error and also for two different liquids.

Figure 7b also shows the m = -1 reflected shear wave in the stainless steel. As the frequency decreases, this wave approaches -90° and also becomes evanescent at a frequency of 8.6 MHz. The energy redistribution at this frequency should also be observed in the receive transducer.

The possibility of using UDGS to measure particle size is due to the penetration of the evanescent wave into the slurry. As the wave interacts with the particles, some attenuation of the signal occurs. As a result, the signal of the reflected m = 0 wave, which is measured by the receive transducer, decreases in amplitude. Since the attenuation is dependent upon particle size, an algorithm can be developed to determine particle size. Collaborators at the University of Washington will develop such an algorithm, similar to that for the data from GLRS experiments (6–8). An important point is that the particle size measurement will be a *bulk* measurement.

Experimental Measurements

Figure 8 shows a photograph of the stainless steel grating, in which the transducers are epoxied directly to the stainless steel. The triangular grooves are spaced 243 microns apart, with 120° included angle. The longitudinal send and

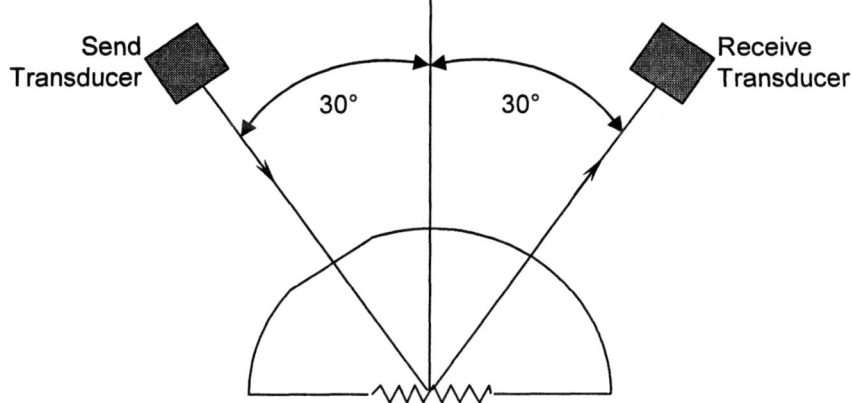

Figure 7a. Grating and transducer configuration for obtaining scan-over-frequency data.

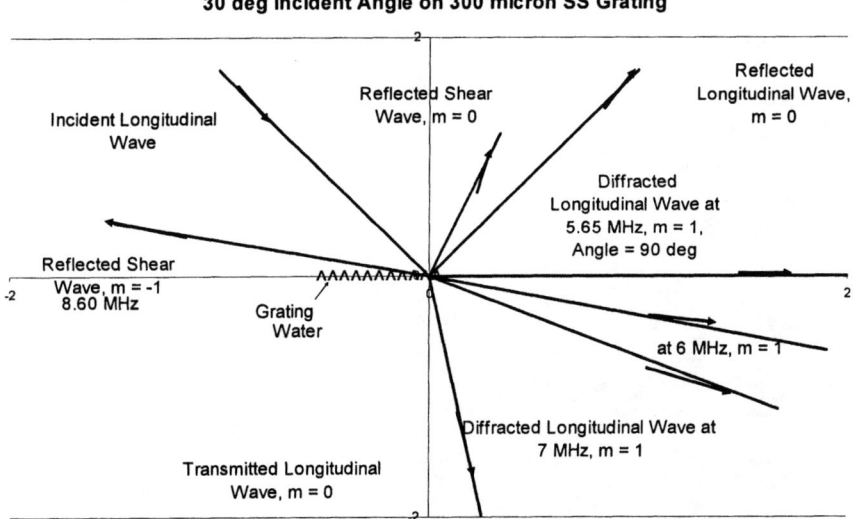

Figure 7b. Diagram showing reflected and transmitted diffracted waves.

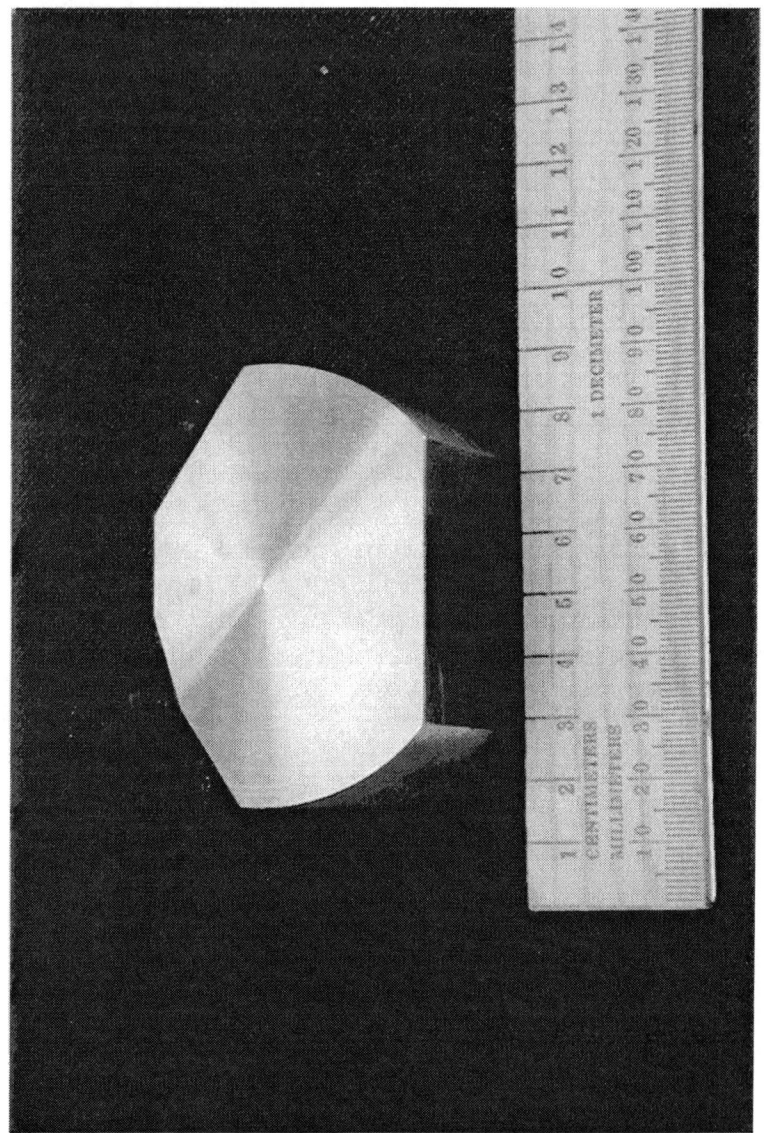

Figure 8. Photograph of 243 micron stainless steel grating.

receive transducers (0.95 cm diameter) make an angle of 33° with respect to a normal to the grating surface. A shear wave receive transducer is also positioned at 17°, but data are not presented here.

To account and correct for the transducer response, the stainless steel part was fabricated with a *smooth* front face, and it is called a "blank." The transducers are epoxied in place, and data for a scan-over-frequency are obtained for the blank. Then the grating is machined on the front face. At each frequency in the scan-over-frequency, the data for the grating are corrected by the following:

$$V_{\text{grating corrected for transducer response}} = V_{\text{grating}}/V_{\text{blank}} \quad (5)$$

Data were obtained for: (1) water, (2) sugar water (SW) solutions having weight percentages of 5, 10, 15, 17.5, 20, 22.5, 30, 35, 40, 43, 46, 50, 52, 54, and 56, and (3) NaCl solutions having weight percentages of 1.07, 2.75, 4.44, 6.31, 10.2, 12.3, 16.0, 19.0, and 22.2. Two or more trials were taken for each solution. The density of each one was measured using a pychnometer, and the velocity of sound was measured by using a time-of-flight method. The weight percentages of the NaCl solutions were chosen so that the velocity was very close in value to a given SW solution. For example, 15% SW has a velocity of sound of 1538 m/s, and 4.44% NaCl, 1535 m/s. 50% SW has a velocity of 1720 m/s, and 22.2% NaCl, 1744 m/s.

The viscosity of SW solutions increases with its concentration. For example, 15% SW has a viscosity of 1.6 cP, while 50% SW has a viscosity of 15.4 cP at 20 °C. The sodium chloride solutions have a nearly constant viscosity over the range of interest, with only a viscosity of 2.0 cP at a concentration of 26 wt %.

Data were obtained for a scan-over-frequency for all of the solutions. The objective is to compare the experimental value of the critical frequency with the theoretical value obtained from eq 4, using the measured velocity of sound. Figures 9–12 show the peak that occurs at the critical frequency. The frequency listed in each figure is the maximum value for that trial and is, of course, close in value to the critical frequency.

The Savitsky-Golay smoothing and differentiation procedure was used to analyze the data to obtain the critical frequency. This procedure smooths the data over a desired number of points (filter) and then fits a polynomial of desired order through these points. The value of the derivative is obtained by differentiating the polynomial. Our previous studies (*1, 2*) have shown that the critical frequency is located at the peak value of the grating data, corrected for transducer response in eq 5. Therefore, determining the frequency, at which the first derivative is zero, is the method used to determine the critical frequency.

The first-derivative method is difficult when the peak of interest is close to the m = -1 shear wave peak in the data. It is more precise to use the grating data

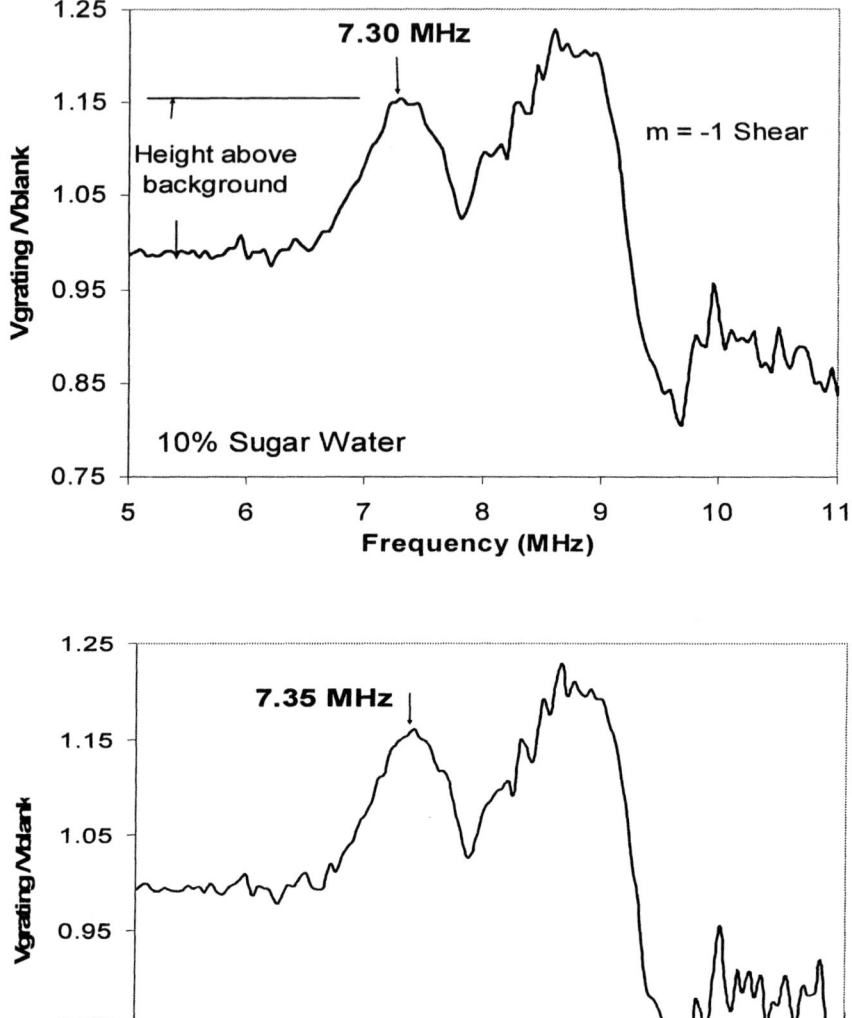

Figure 9. Scan-over-frequency data for 10% sugar water and 2.75% NaCl. The velocity of sound in 10% sugar water is 1515 m/s, and in 2.75% NaCl is 1513 m/s.

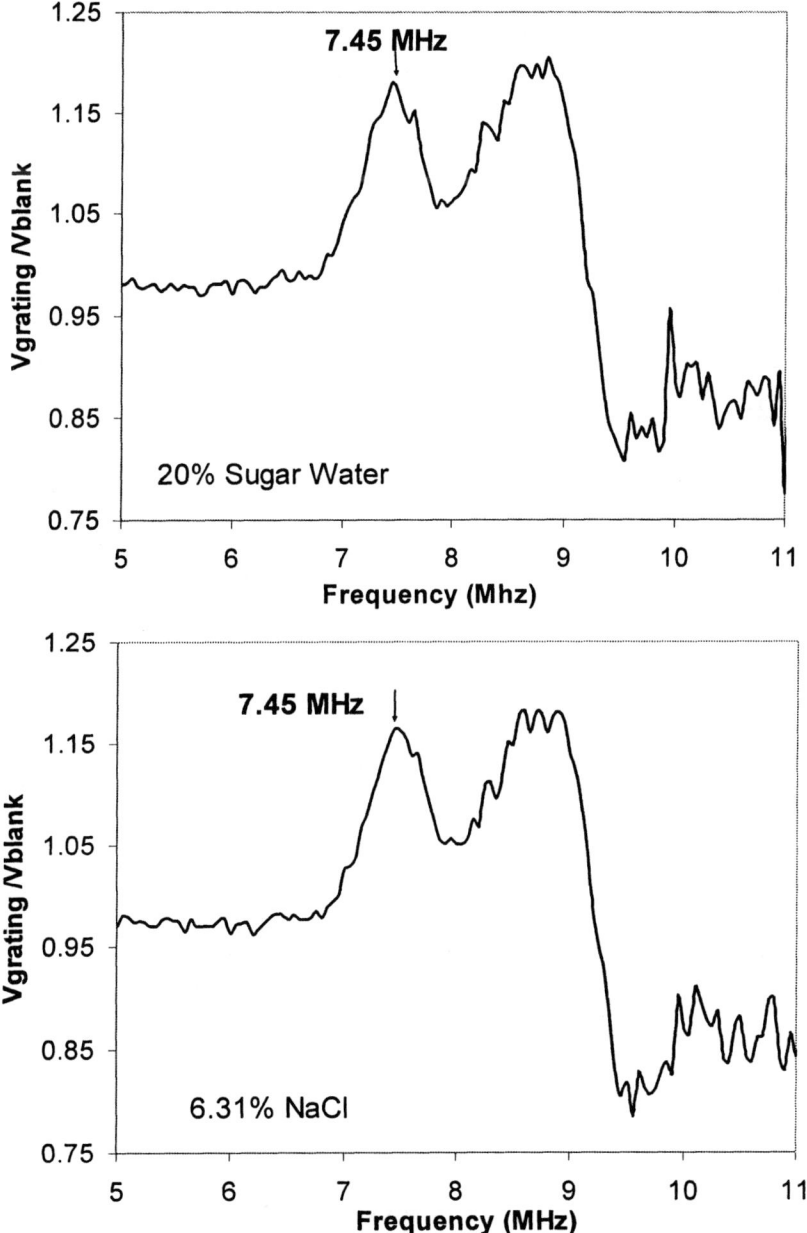

Figure 10. Scan-over-frequency data for 20% sugar water and 6.31% NaCl. The velocity of sound in 20% sugar water is 1554 m/s, and in 6.31% NaCl is 1554 m/s.

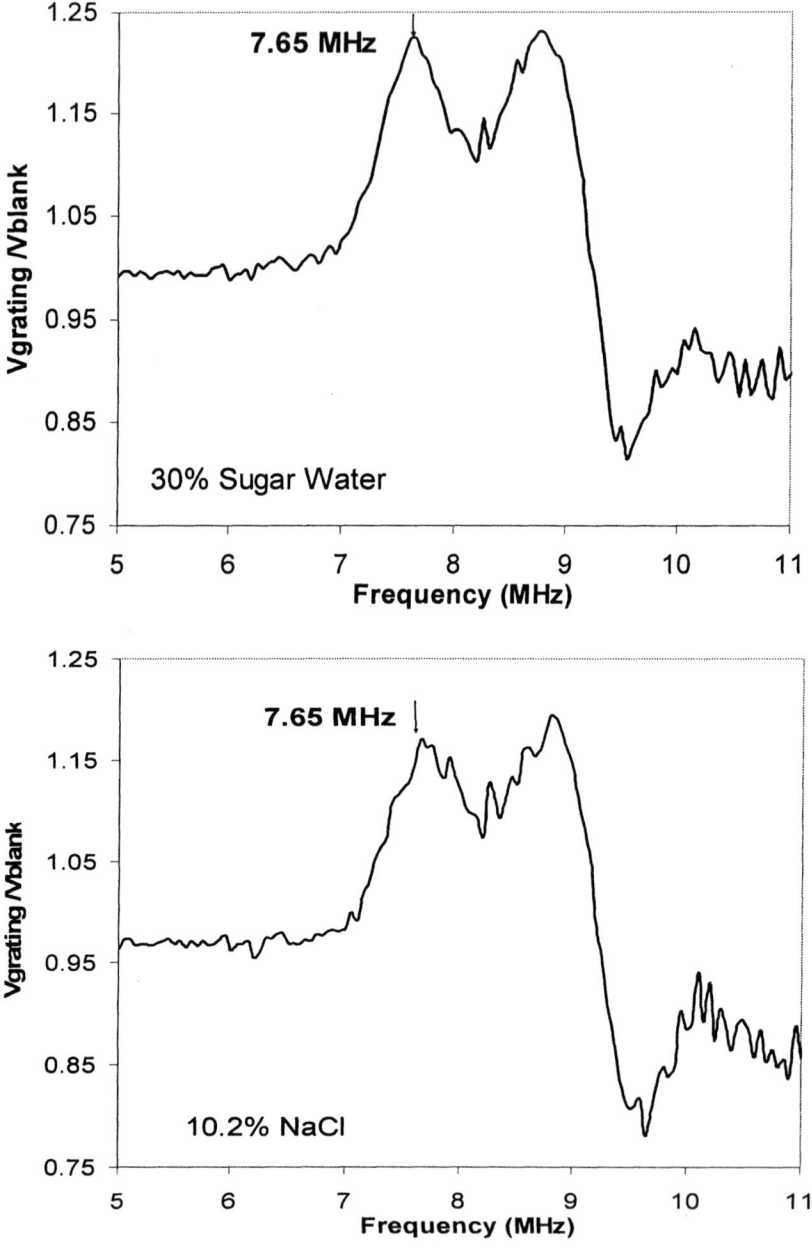

Figure 11. Scan-over-frequency data for 30% sugar water and 10.2% NaCl. The velocity of sound in 30% sugar water is 1603 m/s, and in 10.2% NaCl is 1604 m/s.

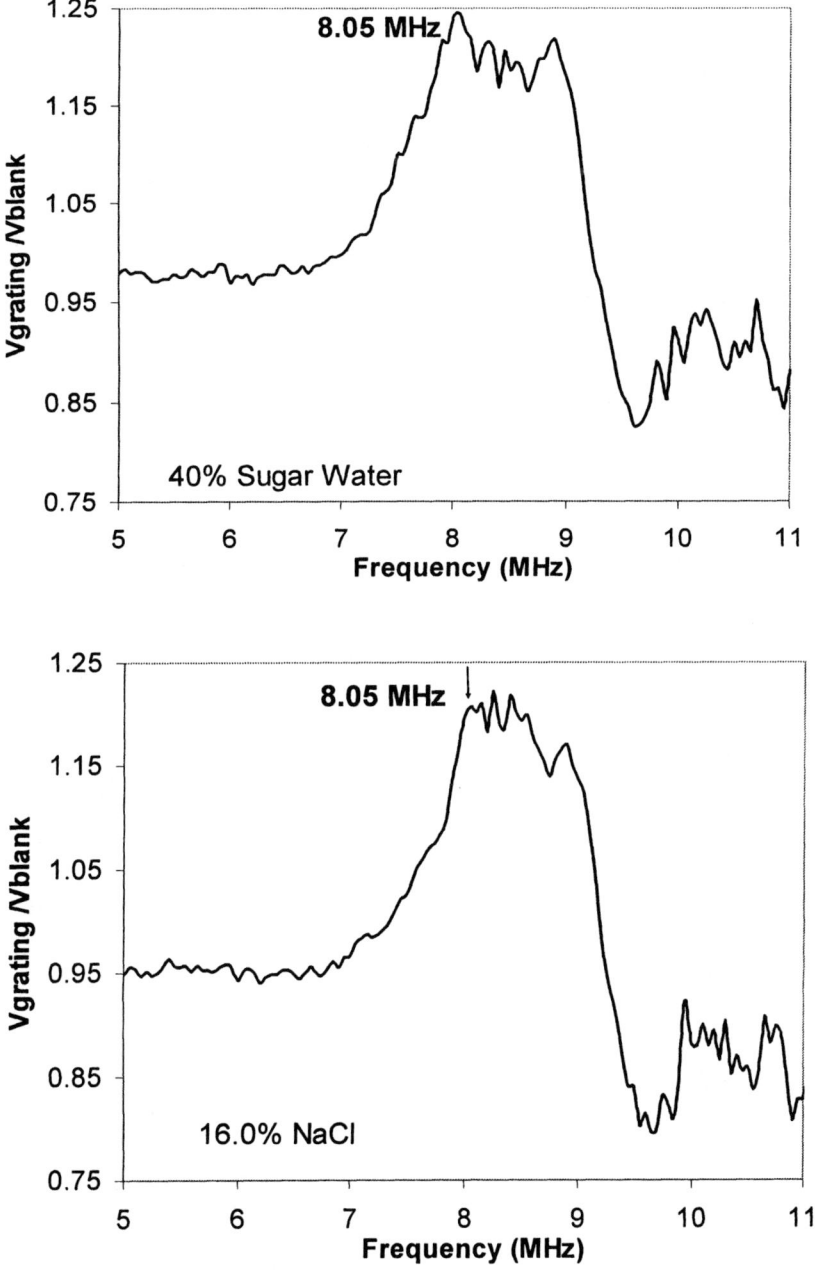

Figure 12. Scan-over-frequency data for 40% sugar water and 16.0% NaCl. The velocity of sound in 40% sugar water is 1663 m/s, and in 16.0% NaCl is 1666 m/s.

that are *not* corrected for transducer response. In these data one can see the amplitude decreasing as the frequency increases. Near the critical frequency, a small peak appears in the data, which is identified most easily by using the second derivative to note a change in curvature.

Two sets of data were obtained for SW. In each set, two trials were obtained for each solution, although the SW concentrations were not the same in both data sets. The critical frequency for each data set is the average value for the two trials. Figure 13 shows the experimental critical frequencies for the SW solutions. The asterisks show the data set #1 analyzed using the first derivative, and the diamonds using the second derivative. Data set #2 is analyzed using only the second derivative, and is shown by the square symbol. Figure 13 shows that the first and second derivative analysis methods are in good agreement. The theoretical values of the critical frequency are calculated using eq 4 and shown by the solid line.

Figure 14 shows the analysis for the NaCl solutions, using the first and second derivative, and comparison with the theory using eq 4.

Novel Method for Measuring the Velocity of Sound

Equation 4 can be inverted to obtain the velocity of sound from a measurement of the critical frequency. However, the measurement of peak height above background, as defined in Figure 9, provides another method of determining the velocity of sound (or, with slightly larger error, the acoustic impedance). Figure 15 shows the peak height above background plotted vs the velocity of sound. For a given velocity of sound, both the SW and NaCl solutions yield values of the peak height above background that are very close in value. The dependence upon density alone is ruled out because a plot of peak height above background vs density shows poor agreement.

Possible Effect of Viscosity

Comparison of the graphs of the critical frequency versus the velocity of sound for SW solutions and NaCl solutions shows that the data and theory seem to fit somewhat better for the NaCl solutions than those for the SW solutions. The data for the SW solutions show that the experimental critical frequency is smaller at larger concentrations than the theory. However, 50% SW has a viscosity of 15 cP at 20 °C, while 56% SW, 32 cP, but there is not a corresponding deviation shown between the measured and theoretical value for these two points. It is possible that the experimental error is the cause of the deviation. However, the first and second derivative analyses are in good agreement. At this point, the SW data offer a *suggestion* of the role of the

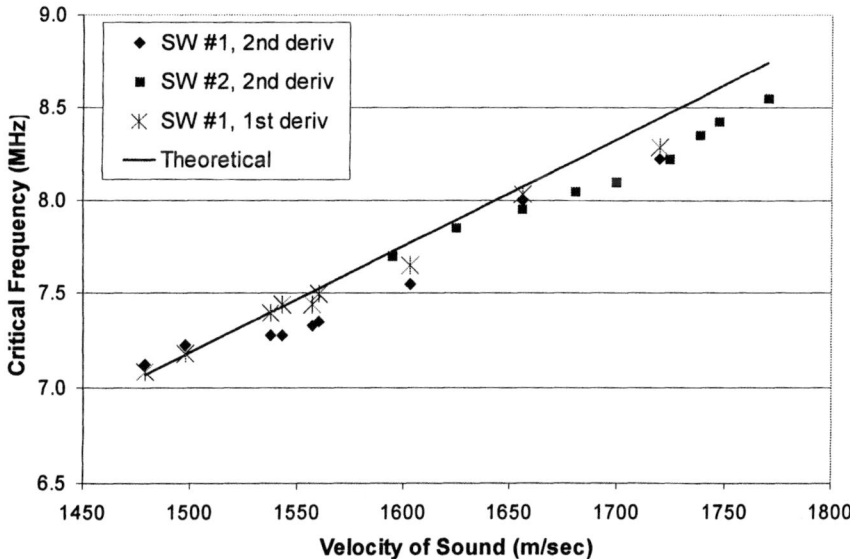

Figure 13. Comparison of experimental critical frequencies for sugar water solutions with theoretical values.

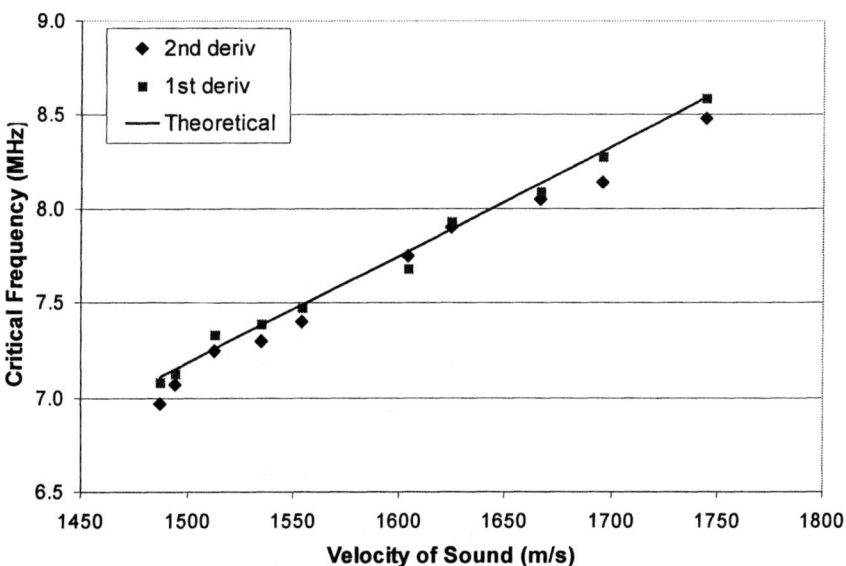

Figure 14. Comparison of experimental frequencies for NaCl solutions with theoretical values.

viscosity. Extending eq 4 to include effects of viscosity is being considered by collaborators at the University of Washington, and that work is in progress. Therefore, UDGS offers the possibility of measuring the velocity of sound and viscosity. The peak height above background yields the velocity of sound, and the deviation between the experimental and theoretical values of the critical frequency can be used to see the effect of the viscosity.

Part II. Shear Wave Reflection Techniques and the Measurement of Viscosity

Apparatus

Figure 16 shows a fused quartz wedge, with the base immersed in the liquid, for the measurement of viscosity. The horizontal shear wave transducer has a frequency of 7.5 MHz. In 1949 Mason (*14*) showed that, when ultrasound strikes the base at an angle θ with respect to a normal to the base, the effective acoustic impedance is given by

$$Z_{eff} = (\cos \theta)(\text{density})(\text{shear velocity}) \qquad (6)$$

Thus, using a large angle, such as 70°, makes the measurement of viscosity more efficient by causing a greater change in the reflection coefficient for various liquids. The plan for the research was to use the self-calibrating method developed for the density measurements (*5*) for the measurement of viscosity.

A shear wave transducer produces ultrasound in which the vibrations are perpendicular to the direction of motion, in contrast to the longitudinal transducer in which the vibrations are along the direction of motion. If one imagines an arrow pointing in the directions of vibrations, then, in Figure 16, the arrow would be perpendicular to the plane of the paper, producing the so-called horizontal shear waves.

The electronics for this measurement are quite different from those used for UDGS. In this case, a square wave pulse is sent to the transducer having a voltage of 140 V and a width of 67 nanoseconds. The output signal from the receiver was input to a 12-bit digitizer board in the computer. The data acquisition was set up to analyze the signals from the digitizer (12-bit, 100 Megasamples/sec) *automatically* using Matlab software and digitizer software.

Measurements

Figure 16 shows that the ultrasound strikes the quartz-liquid interface, where some of it is reflected to the opposite side; 100% is reflected by the quartz-air interface. It then strikes the quartz-liquid interface again, and is reflected to the transducer, where a pulse is recorded. Because shear waves do not travel easily in a liquid, a very small amount of ultrasound travels into the liquid. The reflection coefficient at the quartz-liquid interface is defined as the ratio of the amount of ultrasound reflected to the amount of incident ultrasound. As we shall see in more detail shortly, the reflection coefficient is dependent upon the viscosity. The reflection coefficient decreases as the viscosity increases because more energy is transmitted into the liquid. That is, a "stiffer" liquid can more easily support a shear wave.

The first set of data, after the electronics and computer code were all deemed to be working correctly, was obtained for water, 10% SW, 30% SW, and 50% SW. These data, shown in Figures 17 through 19, were obtained in a 3-h interval and are used to illustrate the type of data acquired. The analysis of these data is shown in Figure 20. As the reader shall see in more detail shortly, the data are analyzed by using water as a reference. The point here is that the water data and the data for the SW solutions were obtained in a relatively short period of time—3 h. The experimental data in Figure 20 show a very good fit with a straight line. An obvious question is, how do the data for water change, if at all, on another day? This is an important question for the development of a sensor. To answer this question, data were obtained over a 5-day period for water and the SW solutions and are discussed in the section entitled "Sensor Calibration and Reliability."

Figure 17 shows the signal obtained by the transducer, and Figure 18 shows the expansion of one of the echoes and the fast Fourier transform (FFT) of that signal. The analysis is carried out by obtaining the peak value of the voltage signal and the maximum value of the FFT amplitude. The data for four liquids are shown in Figure 19, where the FFT amplitude is plotted versus the echo number. These data were obtained during a 3-h time interval at room temperature (22 °C). The important point is that the data for the four samples are separated, although the separation for water and 10% SW is not obvious on this scale.

Analysis

The data are analyzed by dividing the amplitude for the liquid by the amplitude for water for each echo. The "amplitude" can refer to the FFT amplitude or the peak voltage amplitude. Thus, water is being used as the calibration liquid. The only difference between the liquid and water data is due

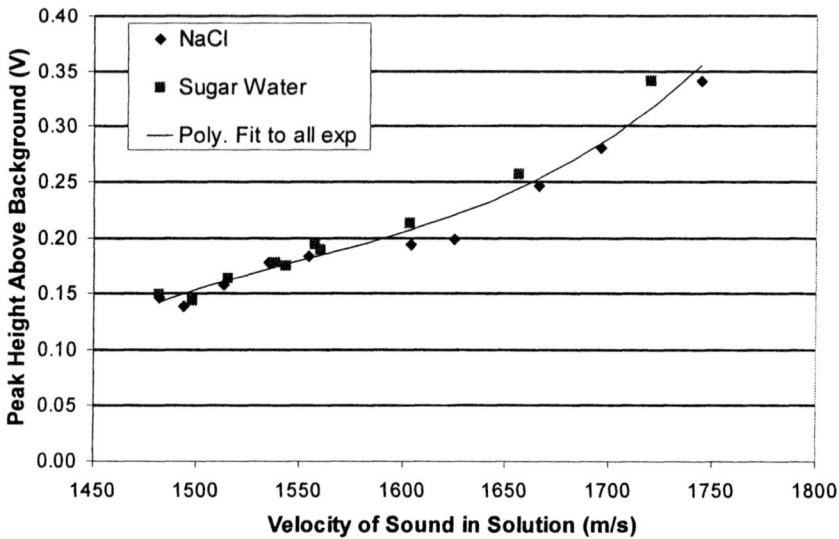

Figure 15. Peak height above background versus velocity of sound for sugar and NaCl solutions.

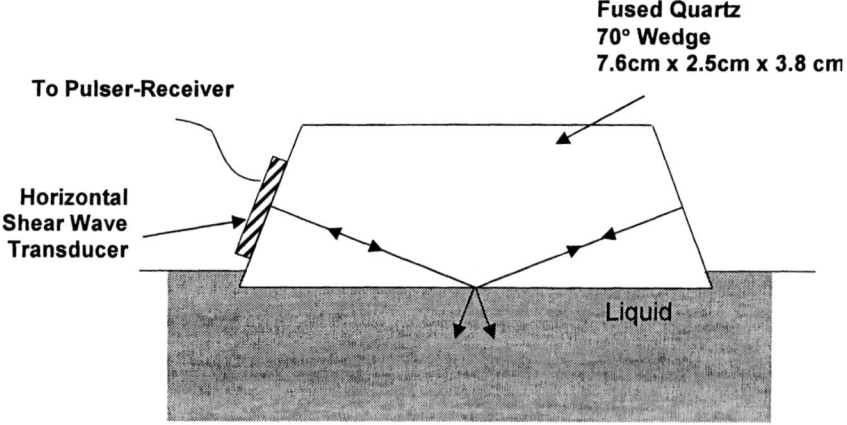

Figure 16. Diagram of apparatus for observing multiple reflections of a horizontal shear wave.

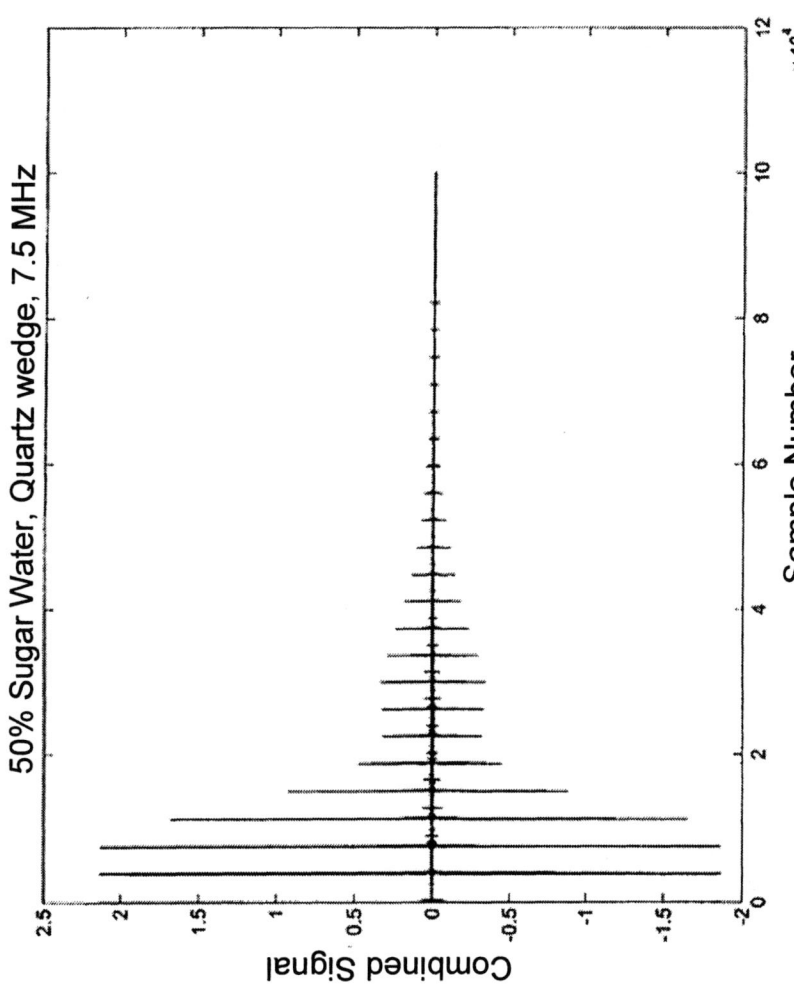

Figure 17. Signal obtained by receive transducer showing echoes produced by multiple reflections.

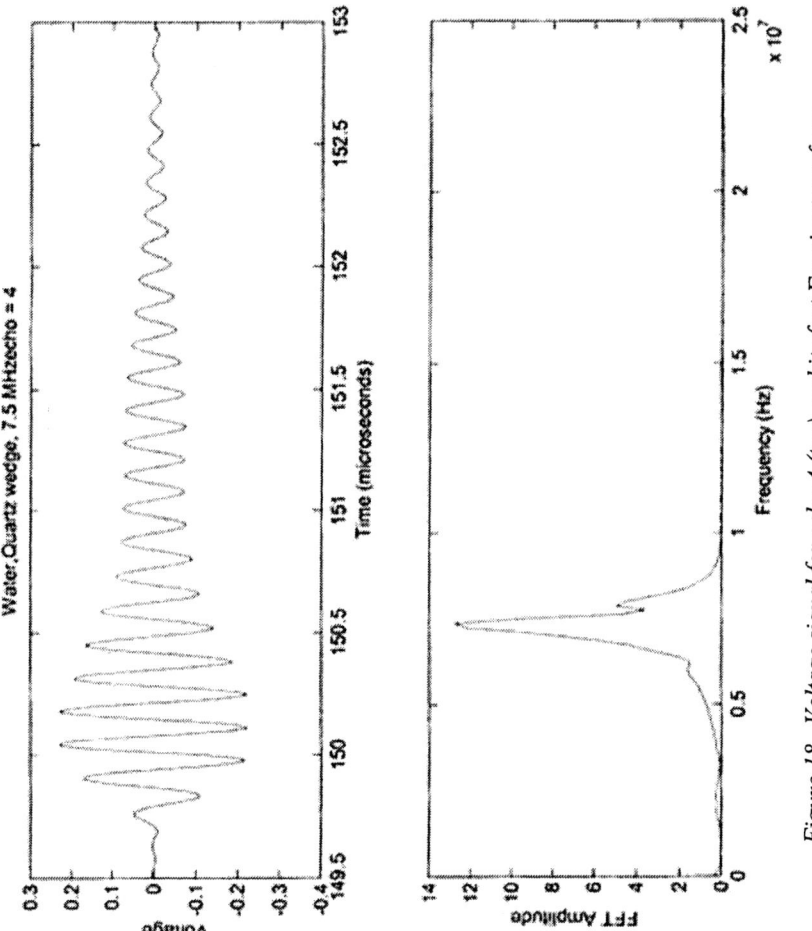

Figure 18. Voltage signal for echo 4 (top) and its fast Fourier transform.

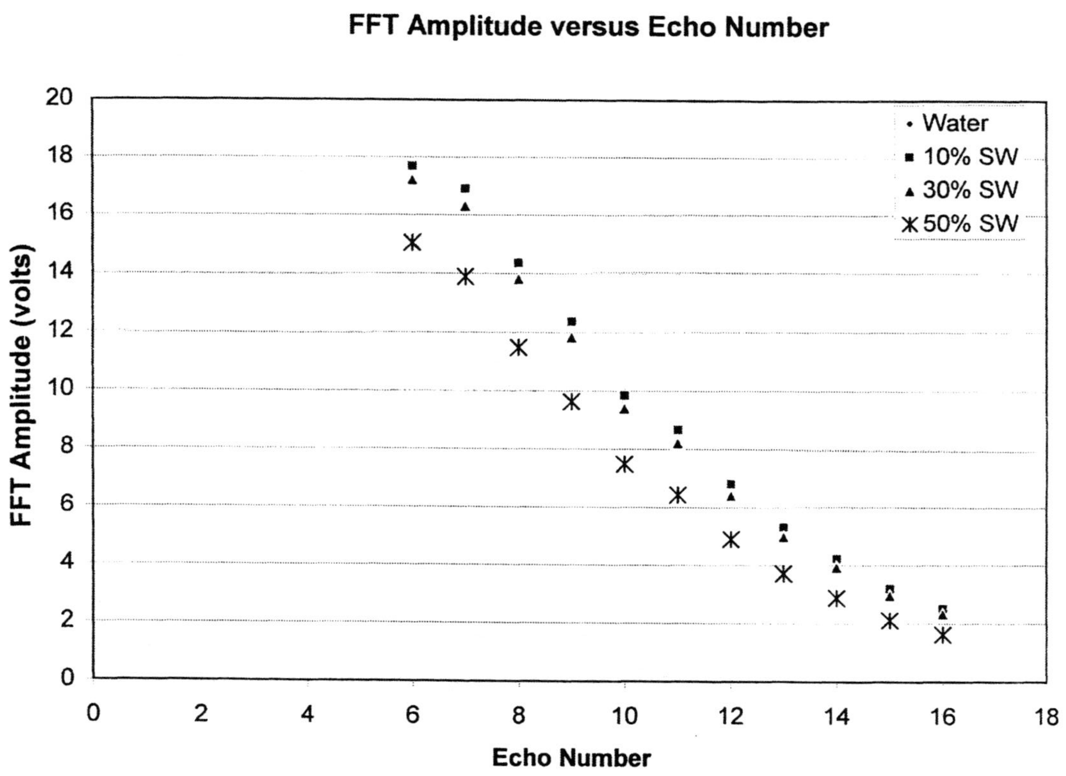

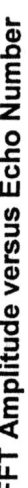

Figure 19. Data obtained for water and three sugar water solutions during a 3-h time interval.

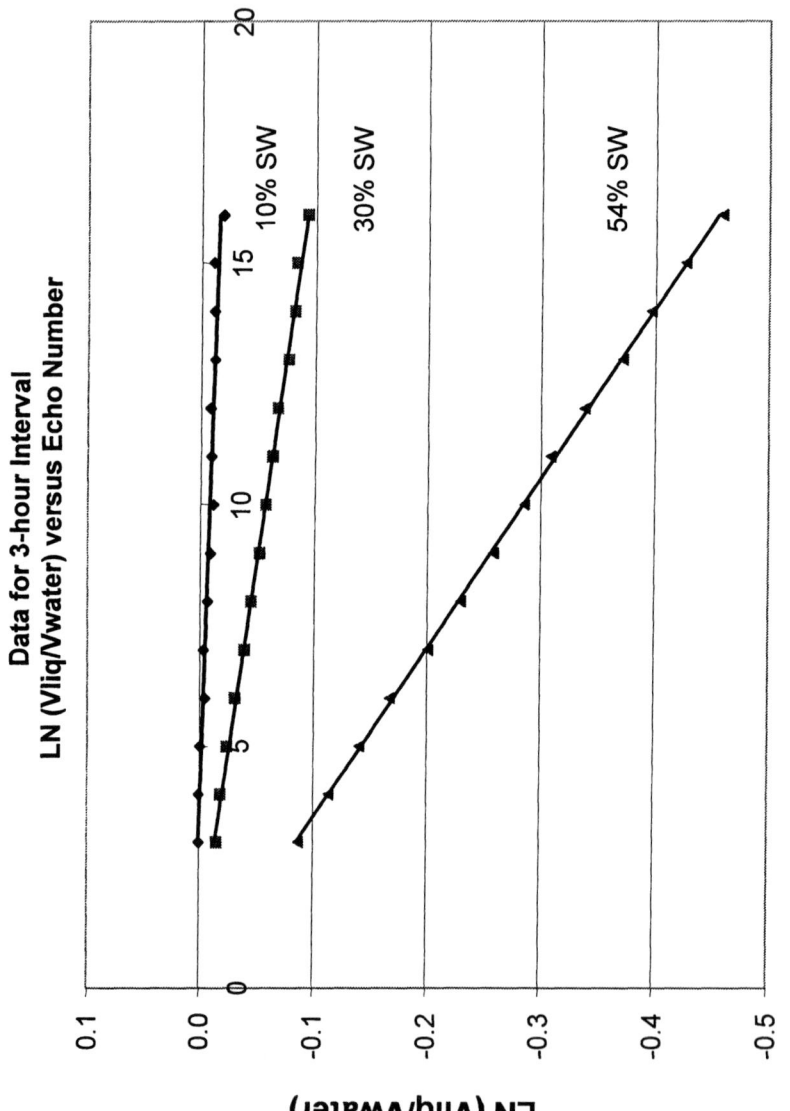

Figure 20. Data plotted on a logarithmic scale versus echo number.

to the reflection at the interface. In the division, all other factors cancel out. The natural logarithm of the ratio is plotted versus the echo number, as shown in Figure 20, and the slope of the line is obtained for each liquid.

Slope and Reflection Coefficient

To obtain the relationship between the slope and the reflection coefficient, we use the fact that the voltage of the Nth echo is given by:

$$\text{Voltage } \alpha \text{ (reflection coefficient for liquid)}^N \qquad (7)$$

Water is used for calibration of the slopes. Using eq 7 for water and another liquid, we obtain the following for the echo N_1:

$$\frac{V_{liqN1}}{V_{wtrN1}} = \frac{R_{liq}^{N1}}{R_{wtr}^{N1}} \qquad (8)$$

A similar equation can be written for echo N_2. Taking the logarithm of both equations and subtracting yields the desired relationship between slope and reflection coefficient:

$$R_{liq}/R_{wtr} = e^{slope} \qquad (9)$$

The term "slope" refers to the slope on the logarithmic plot shown in Figure 20. To determine the reflection coefficient for the liquid using the "slope," the reflection coefficient for water must be determined.

Determination of the Product of Viscosity and Density

The relationship between viscosity and the reflection coefficient is given in reference 3 as:

$$(\rho \eta)^{0.5} = \rho_s c_{TS} \cos\theta \left(\frac{2}{\omega}\right)^{0.5} \left(\frac{1-R}{1+R}\right), \qquad (10)$$

where ρ and η are the density and viscosity of the liquid, ρ_s is the density of the solid, c_{TS} is the shear wave velocity in the solid, θ is the angle of incidence with respect to the normal, R is the reflection coefficient, and ω is the angular frequency expressed in radians per second.

Solving eq 10 for the reflection coefficient R:

$$R = \frac{1-K}{1+K}, \qquad (11)$$

where K is given by

$$K = \frac{\left(\frac{\rho \eta \omega}{2}\right)^{0.5}}{\rho_s c_{TS} \cos \theta} \qquad (12)$$

Since the viscosity of water is 1 cP, eqs 11 and 12 can be used to determine the reflection coefficient for water. Then, using the slope on logarithmic plot, eq 9 can be used to determine the reflection coefficient for the liquid. The next step is to determine the product of the density and viscosity using eq 10. For fused quartz, the shear velocity is 3760 m/s, and the density is 2200 kg/m^3.

For the determination of the viscosity, the density of the liquid or slurry must be determined also.

Self-Calibrating Feature

The self-calibrating feature means that the measurement is not affected by changes in the pulser voltage. Suppose that the pulser voltage drops by, say, 5%. Then each echo will also drop by 5%. On a plot of LN (Vliq/Vwater) vs the echo number, the straight line will occur at a different place, *but the slope will be unchanged.* This concept has been tested experimentally for density (*5*). A U.S. patent was awarded for the self-calibrating feature for density and viscosity measurements (*15*).

Data Obtained during 3-H Interval

Table I shows the results of the data illustrated in Figure 20. The agreement between the data and theory is excellent, and the viscosity values are also in very good agreement with handbook values.

Table I. Results for Data Obtained during 3-H Interval

Liquid	Slope	Rc_{liquid}	Sensor Product of Density & Viscosity (cP-g/cm³)	Density by Weight (g/cm³)	Viscosity at 22 °C (centipoise)	Handbook Viscosity at 20 °C (centipoise)
10% SW	-1.28E-03	0.99594	1.401	1.038	1.35±0.04	1.33
30% SW	-6.15E-03	0.99352	3.587	1.121	3.20±0.04	3.18
54% SW	-2.83E-02	0.98257	26.313	1.253	21.0±0.2	24.6

Sensor Calibration and Reliability

The data presented here were obtained at a room temperature of about 22° C. The sensor would be calibrated by obtaining data for water, similar to that shown in Figure 19, for several temperature values over the desired temperature range. The data for each temperature would be stored in a file. For the reliable operation of the sensor, the water reference file must not change appreciably at a given temperature day after day. To test the reliability over time, five sets of data were obtained for water over a five-day period, usually one each day. For each echo, average values of the peak voltage and FFT amplitude were obtained, as well as the standard deviation of the five values. The standard deviation was converted to a percent error, which is plotted versus the echo number in Figure 21 for the peak voltage values. The five sets of data for water were used to obtain an average water reference file for the FFT amplitude and the peak voltage values. The data for 11 SW solutions were analyzed using both the FFT amplitude water reference file and also the peak voltage water reference file. The results showed that the most accurate values of the viscosity were obtained using the peak voltage water reference file. Table II shows the results using the peak voltage water reference file and using

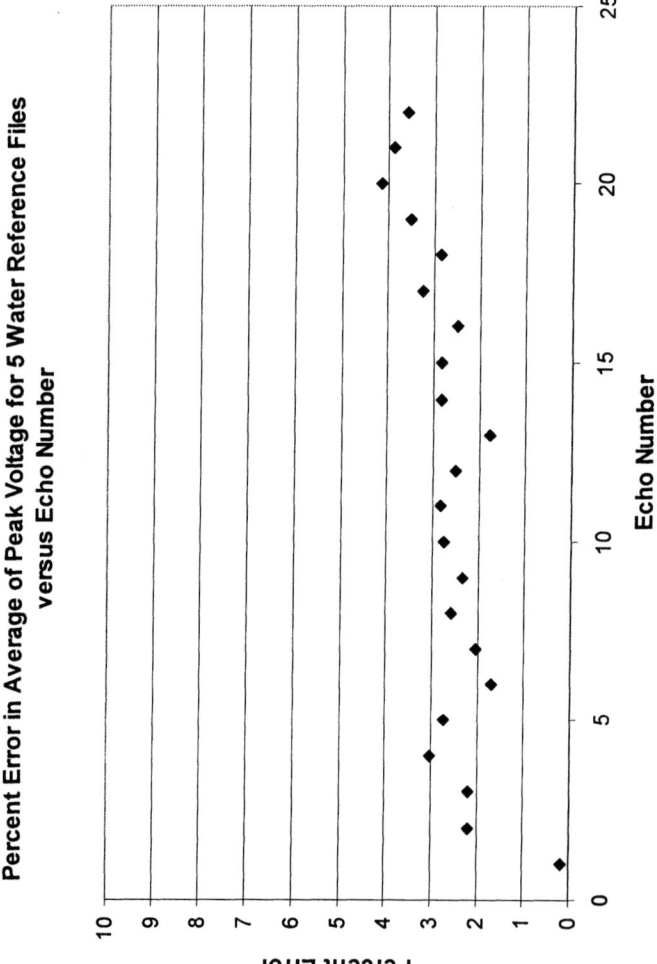

Figure 21. Percent of error in average of five sets of data.

Table II. Viscosity Measurements Obtained for Sugar Water Solutions Using Average Water Reference File

Liquid	No. of Trials	Sensor Viscosity* (cP)	Std. Deviation (cP)	Handbook Viscosity At 20 °C (cP)
5% SW	3	1.35	0.19	1.14
10% SW	2	1.64	0.80	1.33
15% SW	3	1.72	0.33	1.59
20% SW	3	2.11	0.72	1.94
25% SW	2	2.49	0.46	2.45
30% SW	1	2.78		3.18
35% SW	2	3.32	0.03	4.33
40% SW	2	6.53	0.92	6.15
45% SW	2	7.41	0.31	9.43
50% SW	2	12.5	0.26	15.4
55% SW	2	24.0	3.7	28.4
60% SW	2	45.1	0.93	42.7

Note: The density of the liquid, obtained from weighing a known volume, was substituted into the density-viscosity product measured by the sensor to determine the viscosity.

echoes 2 through 20 to obtain the slope on the logarithmic plot. Figure 22 compares the experimental measurements with the Handbook values. The agreement is quite good.

On-line Instrument to Measure Density and Viscosity

For a pipeline sensor, the density can be measured through the stainless steel pipeline wall, as has been discussed earlier (5). The fused quartz wedge

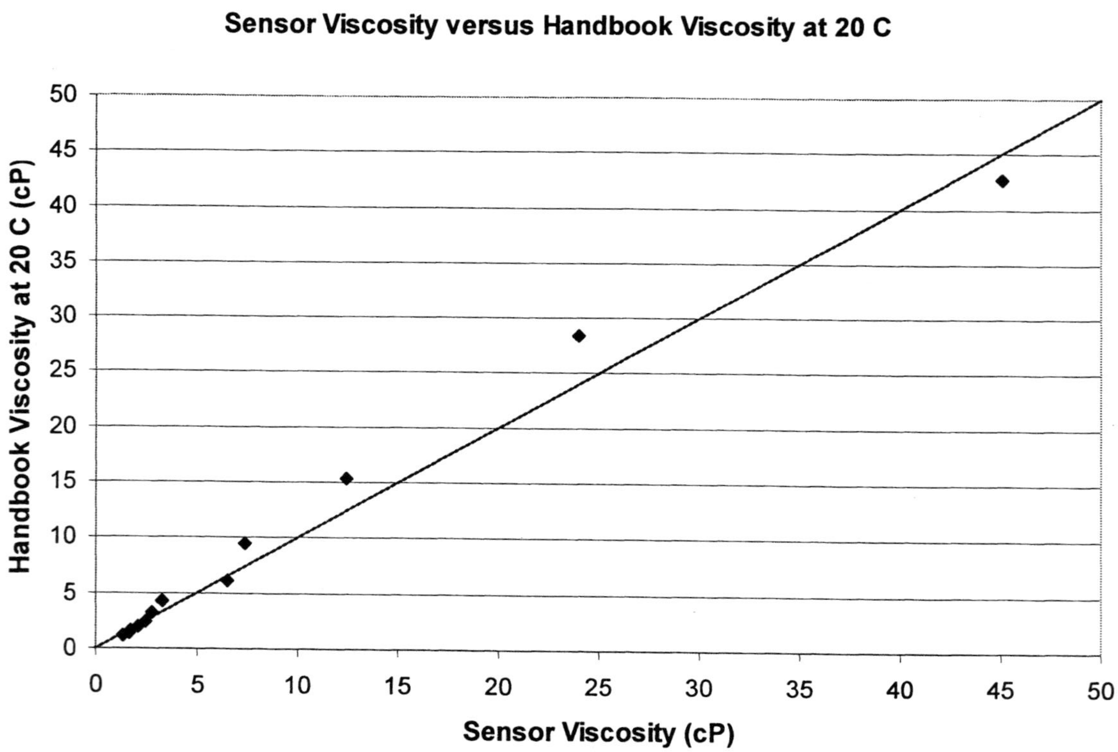

Figure 22. Comparison of viscosity obtained by sensor with handbook values.

can be configured to form part of the pipeline wall. The design can be similar to the current density sensor on the pipeline between Tanks SY101 and SY102 on the Hanford Site, where the sensor forms part of the pipeline wall.

Path Forward

The next step in the UDGS experiments is to carry out experiments with polystyrene spheres to observe the expected decrease in amplitude for various particle sizes. Experiments will be carried out for the stainless steel grating shown in Figure 8, as well as for polystyrene gratings because the evanescent wave is expected to be larger. We shall also investigate designing a quartz wedge that is smaller, so that the variation in the water reference files is reduced.

Acknowledgments

We gratefully acknowledge the significant contributions of Justus Adamson in the research presented here. During the summer of 2003, he was an Office of Science Student Undergraduate Laboratory Intern (SULI) at Pacific Northwest National Laboratory and is a graduate of Brigham Young University.

References

1. Greenwood, M. S.; Brodsky, A.; Burgess, L.; Bond, L. J. *Rev. Prog. QNDE* **2002**, *22B*, 1637–1643.
2. Greenwood, M. S.; Brodsky, A.; Burgess, L.; Bond, L. J.; Hamad, M. *Ultrasonics* **2004**, *42*, 531–536.
3. Greenwood, M. S.; Bamberger, J. A. *Ultrasonics* **2002**, *39*, 623–630.
4. Greenwood, M. S.; Bamberger, J. A. *Ultrasonics* **2002**, *40*, 413–417.
5. Greenwood, M. S.; Bamberger, J. A. *J. Fluids Engr.* **2004**, *126*, 189–192.
6. Brodsky, A. M.; Burgess, L. W.; Smith, S. A. *Appl. Spectrosc.* **1998**, *52*, 332A.
7. Anderson, B. B.; Brodsky, A. M.; Burgess, L. W. *Phys. Rev.* **1996**, *54*, 912.
8. Smith, S. A.; Brodsky, A. M.; Vahey, P. G.; Burgess, L. W. *Anal. Chem.* **2000**, *72*, 4428.
9. Greenwood, M. S.; Brodsky, A.; Burgess, L.; Hamad, M.; Bond, L. J. *Rev. Prog. QNDE* **2004**, *24B*, 1729–1736.
10. Greenwood, M. S.; Ahmed, S.; Bond, L. J. to be published in *Ultrasonics.*
11. Greenwood, M. S., U.S. Patent 6,877,375, 2004.

12. Greenwood, M. S.; Adamson, J. D.; Bond, L. J. *Rev. Prog. QNDE* **2004**, *24B*, 1690–1697.
13. Greenwood, M. S.; Adamson, J. D.; Bond, L. J. to be published in *Ultrasonics*.
14. Mason, W. P.; Baker, W. O.; McSkimin, H. J.; Heiss, J. H. *Phys. Rev.* **1949**, *75*, 936.
15. Greenwood, M. S. U.S. Patent 6,763,698, 2004.

Chapter 7

Novel Chemical Detection Strategies for Trichloroethylene and Perchloroethylene

Andrew C. R. Pipino[1], Johan P. M. Hoefnagels[2], John T. Woodward[1], Curtis W. Meuse[1], and Vitalii Silin[1]

[1]Chemical Science and Technology Laboratory, National Institute of Standards and Technology, Gaithersburg, MD 20899
[2]Department of Applied Physics, Eindhoven University of Technology, P.O. Box 513, 5600 MB, Eindhoven, The Netherlands

Novel applications of cavity ring-down spectroscopy (CRDS) to chemical detection are described. Using a linear optical resonator with an intra-cavity double-Brewster-window flow cell, CRDS is employed to probe the optical response to adsorption of the surface-plasmon resonance (SPR) of an ultra-thin (0.2 nm), nanostructured Au film. Detection limits for trichloroethylene (TCE), perchloroethylene (PCE), and NO_2 are found to be 7×10^{-8} mol/L, 2×10^{-8} mol/L, and 4×10^{-9} mol/L, respectively. As the ultra-thin nanostructured film is well described by a distribution of nanospheres with a mean diameter of 4.5 nm, Mie theory is employed to account for some aspects of the optical response. In a second implementation of CRDS, evanescent wave CRDS (EW-CRDS) is used to detect TCE, cis-dichloroethylene (cis-DCE), and trans-DCE by probing the first C-H stretching overtones in the near-IR with a monolithic folded resonator (MFR), providing spectroscopic selectivity and a reversible response. In a comparison of EW-CRDS to previous sensing technologies, the sensitivity obtained using an unclad MFR for TCE detection is found to be comparable to that obtained with a long-effective-path-length optical waveguide using a TCE-enriching polysiloxane coating. By applying an analyte-enriching, protective coating to an MFR, EW-CRDS may provide a sensitive, selective, and robust technology for long-term environmental monitoring.

© 2006 American Chemical Society

Introduction

Cavity ring-down spectroscopy (CRDS) (*1, 2*) is an emerging optical absorption technique that evolved from advances in optical fabrication and coating technology. CRDS has been applied predominantly to gas phase diagnostics, although efforts to extend the technique to surfaces (*3–8*), films (*9–12*), and liquids (*13, 14*) are increasing. All variants of CRDS use the photon decay time in a low-loss optical resonator as the absorption-sensitive observable, permitting detection of small optical losses from weak transitions or low concentrations. In general, the photon decay or "ring-down" time is given by

$$\tau(\omega) = \frac{t_r}{\Gamma_i + \Gamma_a},$$

where t_r is the round-trip transit time for light in the cavity, and the denominator contains intrinsic and sample absorption losses, respectively. By measuring the change in the ring-down time, in the presence and absence of analyte, as a function of laser frequency, the absolute absorption spectrum of the sample is obtained. The minimum detectable absorption can be expressed as the product of the intrinsic loss, which depends on the resonator design, and the fractional uncertainty in the ring-down time, which depends on the detection and digitization electronics. To reduce the intrinsic resonator loss, ultra-high-reflectivity multi-layer-dielectric mirrors are used with reflectivity as high as R = 99.9998% (*15*). Scattering losses at intra-cavity interfaces are reduced to ~ 1 x 10^{-6} through chemo-mechanical polishing techniques (*16*), which routinely permit surfaces with root-mean-square roughness of < 0.05 nm to be achieved. Further, after decades of refinement for telecommunications purposes, the transmission of fused silica has reached a minimum loss of < 1 dB/km (< 2.3 x 10^{-6} loss/cm) at 1550 nm, which permits ultra-low-loss solid-state resonators to be realized. These advances in optical technology combined with high-speed A/D converters and fast, low-noise detectors enable the use of ultra-low-loss optical resonators for high sensitivity spectroscopy, with CRDS being a relatively prevalent implementation.

In the following, we briefly summarize recent results on several novel applications of CRDS. First, trichloroethylene (TCE), perchloroethylene (PCE), and NO$_2$, which are prevalent environmental contaminants, are detected through a combination of CRDS with surface plasmon resonance (SPR). The localized SPR of a distribution of Au nanoparticles is probed by CRDS to detect trace levels of these analytes. The experimental configuration is briefly described, detection limits are evaluated, and some insight into the SPR response is provided by Mie scattering calculations. Evanescent wave cavity ring-down spectroscopy (EW-CRDS), which is a variant of CRDS employing intra-cavity total-internal reflection, is also employed to detect TCE, cis-dicholoroethylene

(DCE), and trans-DCE by direct absorption of the C-H overtone in the near IR, around 1645 nm. The experimental configuration incorporating a monolithic folded resonator (MFR) is briefly described, and a comparison to previous measurements employing a long-effective-path-length planar waveguide with a TCE-enriching polysiloxane coating is given. The results suggest that a sensitive, selective, and robust sensing technology based on EW-CRDS can be realized by employing a polysiloxane coating with a MFR as the sensing element.

Surface-Plasmon-Resonance-Enhanced Cavity Ring-Down Detection

Experimental

The CRDS measurements employed a linear resonator with an intra-cavity flow cell as shown in Figure 1, which is described in more detail elsewhere (*17*). The linear resonator is formed from a pair of concave mirrors with a radius of curvature of 1 m, a separation of 37 cm, and a maximum reflectivity of R = 99.997% at the center wavelength of 543 nm. The intra-cavity flow cell consists of a pair of fused silica optical flats, which are gasket sealed to the stainless-steel cell at the Brewster angle (55.4°). The laser source was a 10 Hz, frequency-tripled, injection-seeded-Nd:YAG-laser-pumped, optical parametric oscillator (OPO) with a 0.075 cm^{-1} line width followed by a two-stage optical parametric amplifier (OPA). The signal beam was employed at 555 nm. Ring-down transients were detected with a PMT and digitized with a 12-bit, 50 MHz PC-based data acquisition board.

The nanostructured films were produced by sputter deposition of 0.2 nm of 99.96% purity Au in a high vacuum chamber (base pressure 1 x 10^{-5} Pa) using a dc magnetron-sputtering source. After deposition the coated substrates were annealed at 700 °C for 2 min and used immediately after cooling to room temperature. Blank substrates were also subjected to the same evacuation and heating procedure to check for contamination or degradation, but showed the same base loss as typical clean substrates, within normal substrate-to-substrate variation. Atomic force microscopy (AFM) of the Au films revealed an approximate Gaussian distribution of nearly spherical particles with a mean diameter of 4.5 nm and a standard deviation of 1.1 nm. The density of particles on the surface was found to be 382 particles/μm^2 with a standard deviation of 94 particles/μm^2. A more detailed description of the films is available elsewhere (*17*).

Low concentrations (< 10^{-7} mol/L) of TCE, PCE, and NO_2 were generated by a flow system that incorporated a mass flow controller and a diffusion vial or permeation tube held at (30.3 ± 0.1) °C. Calibration of the diffusive sources was accomplished gravimetrically. The carrier gas stream was provided by boil-off from a 250 L liquid N_2 Dewar having a residual moisture content of 10^{-10} mol/L.

Au Nanoparticle SPR Response

As shown in Figure 2, the SPR of the Au nanoparticles responds sensitively to TCE, PCE, and NO_2 for freshly deposited films for concentrations of 2200 nmol/L, 660 nmol/L, and 83 nmol/L, respectively. As the response contains both reversible and irreversible components, the sensitivity is reduced with multiple exposures, as explored in more detail elsewhere (*17*). The absolute loss resulting from analyte adsorption is shown, where the base loss (film base loss plus the intrinsic loss) has been subtracted by averaging 100 points prior to dosing. Each data point is an average of 25 laser shots. From Figure 2, based on a 1-min response time and a minimum detectable ring-down time change of 0.1%, the initial step increase yields detection limits of 7×10^{-8} mol/L, 2×10^{-8} mol/L, and 4.3×10^{-9} mol/L for TCE, PCE, and NO_2, respectively. While TCE and PCE have negligible absorption at the SPR probe wavelength, the relatively high refractive indices for these chlorinated hydrocarbons induce large SPR responses. By contrast, the considerably lighter NO_2 molecule has a moderately strong absorption at the probe wavelength as shown in the inset figure, which contributes 31×10^{-6} to the optical loss at the probe wavelength of 555 nm. However, the gas phase contribution for NO_2 is small relative to the initial SPR response to adsorption. Indeed the sensitivity to NO_2 is higher than TCE or PCE, which may be an example of surface-enhanced absorption. Yet NO_2 is also known to form a bi-dentate chelate bond with Au(111) through the two oxygen atoms (*18*), which could contribute to the SPR response for these small particles which have relatively high surface-to-volume ratio.

The remarkably high sensitivity of the SPR-CRDS measurements to adsorption can be attributed in part to the sensitivity of CRDS for detection of small absorption losses but also to the intrinsic sensitivity of the Au nanoparticles to adsorption. In particular, the SPR sensitivity for a fixed mass of Au increases with decreasing particle size, which has been observed by others recently (*19, 20*). In Figure 3, Mie scattering calculations are employed to demonstrate this point. For a fixed ad-layer thickness of 0.3 nm, the fractional change in the extinction cross section is shown as a function of wavelength for three particle sizes, having radii of (A) 2.25 nm, (B) 5 nm, and (C) 10 nm. Note that the change in cross section arising from adsorption is largest for the smallest radius and peaks at 555 nm. Furthermore, the relative change in cross-section for smallest particle size is substantial (43%). A more detailed discussion is provided elsewhere (*17*).

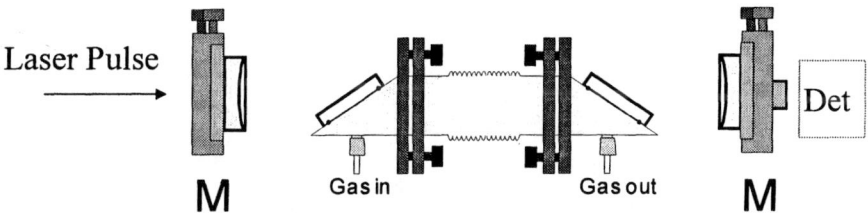

Figure 1. The optical configuration for CRDS measurements on Au nanoparticles is shown. High-reflectivity mirrors (M) form the high finesse linear resonator. An intra-cavity flow cell, which incorporates two gasket-sealed optical flats oriented at Brewster's angle, provides a low-loss means for simultaneously dosing and optically probing the nanoparticle distribution deposited on a Brewster window.

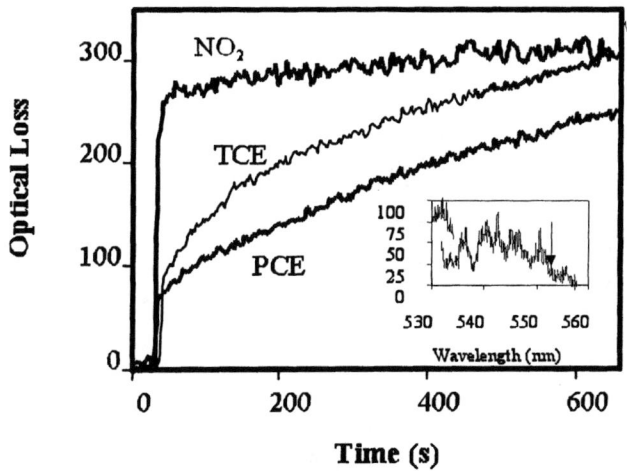

Figure 2. The SPR response (as optical loss $\times 10^6$) of a Au-nanoparticle distribution to TCE (2200 nmol/L), PCE (660 nmol/L), and NO_2 (83 nmol/L) is probed by CRDS for freshly deposited Au films. The inset shows a gas-phase NO_2 spectrum. Gas-phase absorption contributes 31×10^{-6} to the observed response loss (indicated by the vertical arrow in the inset figure).

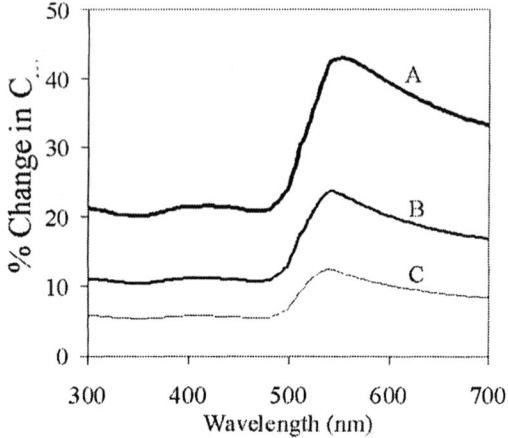

Figure 3. The sensitivity of the SPR response to adsorption for different particle sizes is examined in the monodisperse case. The fractional change in the extinction cross section arising from a 0.3 nm coating with a refractive index corresponding to PCE is shown as a function of wavelength for three sphere radii: (A) 2.25 nm, (B) 5 nm, and (C) 10 nm.

Evanescent Wave Cavity Ring-Down Spectroscopy of TCE, cis-DCE, and Trans-DCE

Experimental

Several optical resonator designs have been employed for EW-CRDS, where the key feature is the integration of one or more total-internal-reflection (TIR) interfaces into a low-loss resonator design such that the associated evanescent wave can be employed in an attenuated-total-reflectance-type measurement. Resonator designs for EW-CRDS have included extensions of the conventional linear resonator with intra-cavity elements (3, 7), a miniature TIR-ring resonator (4, 5), and the MFR (6). The TIR-ring resonator, which is a broadband device, utilizes photon tunneling for input coupling, while the MFR accepts a free-space propagating wave at the expense of a coating-limited bandwidth like a conventional CRDS resonator. Yet for detection of absorption bands with relatively narrow widths, the MFR has sufficient bandwidth, especially in the NIR, where overtone line widths are typically less than 1/10 the coating bandwidth. Further, the MFR is rugged and stable, providing the potential for a rugged sensor. As depicted in Figure 4, the MFR employed for

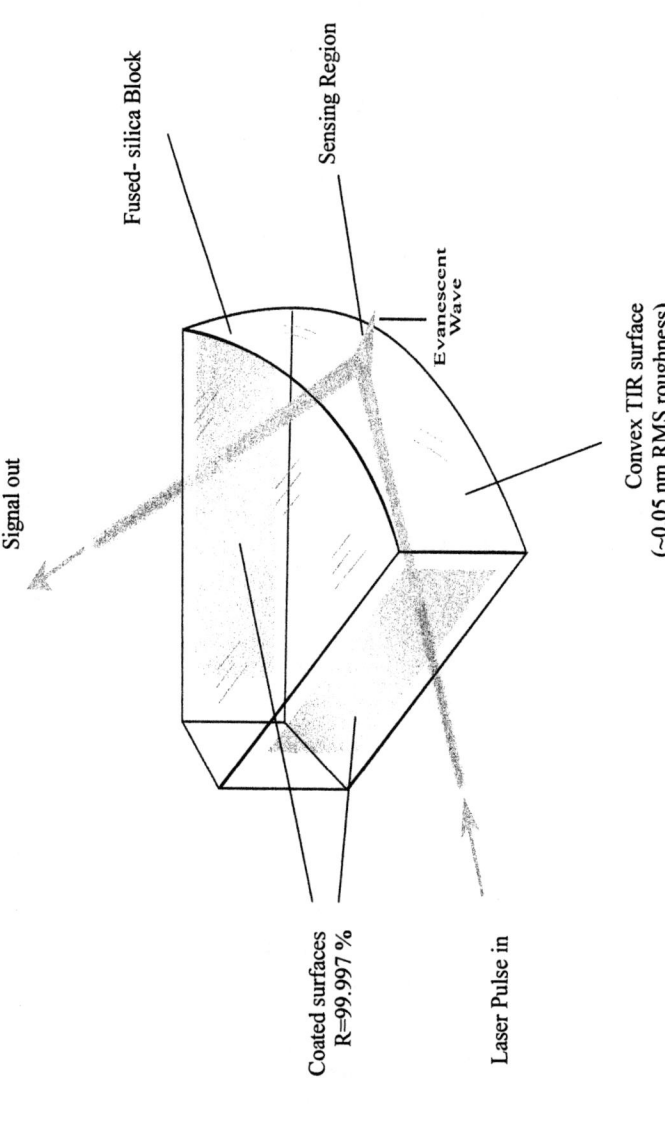

Figure 4. The monolithic, folded resonator (MFR) for EW-CRDS is depicted. Two high-reflectivity-coated planar surfaces and a convex TIR surface form the stable resonator, which is fabricated from a single block of high-purity fused silica. Light is injected into one planar surface, and the residual transmission is detected through the second planar surface. An evanescent wave emanates from the apex of the TIR surface to form the sensing region, which has dimensions defined by the resonator-mode spot size.

this study was fabricated from fused silica with a per-pass length of 3 cm and an incident angle at the TIR surface of 45°. The reflectivity of the two planar, coated surfaces was R = 99.997% at the center wavelength of 1650 nm. The convex TIR and coated surfaces were polished to ~ 0.05 nm root-mean-square surface roughness. The idler beam of the OPO/OPA was employed as the NIR source. The minimum loss per pass of the resonator was 1.1×10^{-4}, which provided an adequately long ring-down time of 1.3 µs. The resonator was mounted on a platform that provided control of all degrees of freedom relative to the incident laser beam. The weakly astigmatic Gaussian output beam of the resonator was collected with a NIR anti-reflection-coated doublet lens and imaged onto a high-speed, InGaAs detector. The detector was mounted on a sub-platform that provided control of all degrees of freedom relative to the resonator. A more detailed description of the MFR and the experimental configuration can be found elsewhere (*21*).

Evanescent Wave Spectra

The absolute s-polarized evanescent wave absorption spectra for the first C-H stretching overtones of TCE, cis-DCE, and trans-DCE as measured by EW-CRDS with an MFR are shown in Figures 5a-c, respectively. The TIR surface is dosed in each case under equilibrium conditions at the room temperature (22.0 °C ± 0.5 °C) vapor pressure of the haloethylenes (9.794×10^3 Pa, 2.705×10^4 Pa, and 4.443×10^4 Pa, for TCE, c-DCE, and t-DCE, respectively at 25 °C) (*22*) by employing a glass cell with a liquid reservoir that is gasket-sealed to the convex TIR surface. The inset of Figure 5a shows the raw data for TCE and the baseline intrinsic loss, prior to subtraction. The baseline loss was established by fitting a third-order polynomial to the average of multiple-background scans. The resulting spectra show broad underlying features with narrow peaks superimposed. Indeed a gas phase contribution to the spectra can be anticipated based on the peak absorption calculated using the absolute absorption cross-sections (*21*) and the polarization-dependent "effective thickness" for the evanescent wave (*23*). Superimposed on each evanescent wave spectrum in Figure 5 is the estimated gas-phase contribution (heavy solid lines). As detailed elsewhere (*21*), by subtracting the gas-phase contribution, absolute absorption spectra can be obtained for the adsorbed species. Using the s- and p-polarized adsorbate spectra, the average molecular orientation on the surface can be determined. Further, employing conventional CRDS to obtain the absolute absorption cross-sections for the gas-phase species, the absolute surface coverage of each species can be determined with knowledge of the surface orientation by invoking conservation of the integrated band intensity between the gas- and adsorbed phases (*21*).

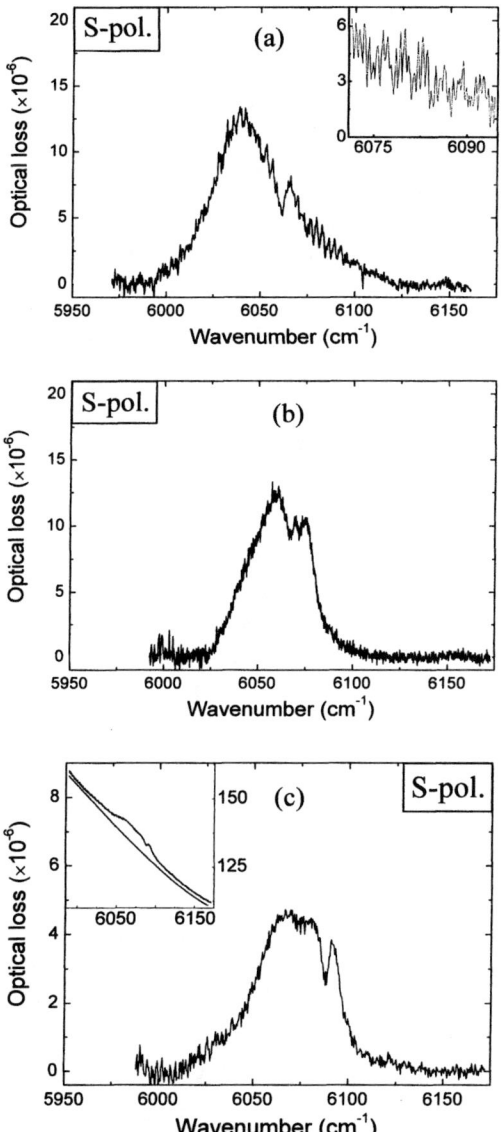

Figure 5. Absolute evanescent wave absorption spectra of (a) TCE, (b) cis-DCE, and (c) trans-DCE are shown as measured by EW-CRDS with a monolithic folded resonator. The inset in (a) shows the raw data before subtraction. Gas-phase spectra, which were measured with conventional CRDS, are also superimposed in gray. The inset in (c) shows an alignment of rotational structure between evanescent wave and conventional CRDS measurements.

Sensitivity Comparison: EW-CRDS and Planar Waveguide Technology

Ache and co-workers (24, 25) investigated the detection of TCE with the first C-H overtone using a long-effective-path-length planar waveguide. A polysiloxane coating was also employed with this device to reversibly enrich the local concentration of TCE in the evanescent wave. The chemical composition of the polysiloxane was also varied to optimize sensitivity with respect to the structure-dependent partition coefficient and refractive index (26). Furthermore, it was shown that competing absorption from C-H groups of the polymer could be essentially eliminated through efficient deuteration (26). The gas-phase detection limit for TCE obtained using the optimized planar waveguide with a polysiloxane coating was found to be 0.1 mmol/L in the gas phase at atmospheric pressure. By comparison, EW-CRDS also provides an equivalent detection limit determined by the minimum detectable absorption of 1×10^{-7} based on a 0.1% relative decay time precision and a 1×10^{-4} /pass intrinsic loss. Yet this detection limit is obtained by EW-CRDS without a TCE-enriching polymer. The addition of such a sensing layer to an EW-CRDS-based TCE detection system should lower detection limits significantly, while a deuterated polysiloxane film should permit a low intrinsic loss to be maintained (26). The polysiloxane layer, which can be thick relative to the evanescent wave decay length for an appropriate resonator design, could also serve to protect the sensing surface and eliminate interference from particulates. Furthermore, an additional improvement in sensitivity can be realized by reducing the resonator intrinsic loss relative to that found in the present work by the use of a smaller resonator that is fabricated from lower bulk-OH-content fused silica. It is worth noting that the MFR design improves in sensitivity through a reduction in size, whereas planar waveguides require an increase in length to increase sensitivity.

Conclusions

New strategies have been demonstrated for detection of TCE and PCE. The SPR-enhanced CRDS results obtained using Au nanoparticles show high sensitivity, but the response has a significant irreversible component. However, the EW-CRDS results show a significant advance since the detection limit achieved in this initial effort is comparable to that achieved with an optimized waveguide and analyte-enriching coating. By employing a low-loss, analyte-enriching coating with the rugged MFR design for EW-CRDS, a miniature, robust, sensitive, and selective chemical sensor for environmental monitoring appears feasible.

Acknowledgments

The authors gratefully acknowledge support from the Environmental Management Science Program (EMSP) of the U.S. Department of Energy under Contract DE-AI07-97ER62518 and Project #73844. J. P. M. Hoefnagels gratefully acknowledges the support of the Nederlands Foundation for Fundamental Research on Matter (FOM) and the Center of Plasma Physics and Radiation Technology.

Disclaimer

Identification of specific commercial products is provided in order to specify procedures completely. In no case does such identification imply recommendation or endorsement by the National Institute of Standards and Technology, nor does it imply that such products have necessarily been identified as the best available for the purpose.

References

1. Berden, G.; Peeters, R.; Meijer, G. *International Reviews in Physical Chemistry* **2000**, *19*(4), 565-607.
2. Wheeler, M. D.; Newman, S. M.; Orr-Ewing, A. J.; Ashfold, M. N. R. *Journal of the Chemical Society-Faraday Transactions* **1998**, *94*(3), 337–351.
3. Pipino, A. C. R.; Hudgens, J. W.; Huie, R. E. *Chemical Physics Letters* **1997**, *280*(1-2), 104–112.
4. Pipino, A. C. R.; Hudgens, J. W.; Huie, R. E. *Review of Scientific Instruments* **1997**, *68*(8), 2978–2989.
5. Pipino, A. C. R. *Physical Review Letters* **1999**, *83*(15), 3093–3096.
6. Pipino, A. C. R. *Applied Optics* **2000**, *39*(9), 1449–1453.
7. Shaw, A. M.; Hannon, T. E.; Li, F. P.; Zare, R. N. *Journal of Physical Chemistry B* **2003**, *107*(29), 7070–7075.
8. Pipino, A. C. R.; Meuse, C. W.; Hoefnagels, J. P. M.; Silin, V.; Woodward, J. T. National Institute of Standards and Technology (NIST) Internal Report 6957; Gaithersburg, MD, 2003.
9. Engeln, R.; von Helden, G.; van Roij, A. J. A.; Meijer, G. *Journal of Chemical Physics* **1999**, *110*(5), 2732–2733.
10. Kleine, D.; Lauterbach, J.; Kleinermanns, K.; Hering, P. *Applied Physics B-Lasers and Optics* **2001**, *72*(2), 249–252.
11. Logunov, S. L. *Applied Optics* **2001**, *40*(9), 1570–1573.
12. Marcus, G. A.; Schwettman, H. A. *Applied Optics* **2002**, *41*(24), 5167–5171.

13. Hallock, A. J.; Berman, E. S. F.; Zare, R. N. *Analytical Chemistry* **2002**, *74*(7), 1741–1743.
14. Xu, S. C.; Sha, G. H.; Xie, J. C. *Review of Scientific Instruments* **2002**, *73*(2), 255–258.
15. Rempe, G.; Thompson, R. J.; Kimble, H. J.; Lalezari, R. *Optics Letters* **1992**, *17*(5), 363–365.
16. Brown, N. J. *Annual Review of Materials Science* **1986**, *16*, 371–388.
17. Pipino, A. C. R.; Woodward, J. T.; Meuse, C. W.; Silin, V. "Surface-Plasmon-Resonance-Enhanced Cavity Ring-Down Detection," *J. Chem. Phys 120 (3), 1585 (2004)*.
18. Wickham, D. T.; Banse, B. A.; Koel, B. E. *Catalysis Letters* **1990**, *6*(2), 163–172.
19. Kalyuzhny, G.; Vaskevich, A.; Schneeweiss, M. A.; Rubinstein, I. *Chemistry-A European Journal* **2002**, *8*(17), 3850–3857.
20. Xu, H. X.; Kall, M. *Sensors and Actuators B-Chemical* **2002**, *87*(2), 244–249.
21. Pipino, A. C. R.; Hoefnagels, J. P. M.; Watanabe, N. "Absolute surface coverage measurement using a vibrational overtone," *J. Chem. Phys 120 (6), 2879 (2004)*.
22. Daubert, T. E.; Danner, R. P. *Physical and Thermodynamic Properties of Pure Chemicals: Data Compilation;* Amer. Inst. Chem. Eng.; Hemisphere: New York, 1989; Vol. 4.
23. Harrick, N. J. *Internal Reflection Spectroscopy;* Interscience Publishers: New York, 1967.
24. Burck, J.; Zimmermann, B.; Mayer, J.; Ache, H. J. *Fresenius Journal of Analytical Chemistry* **1996**, *354*(3), 284–290.
25. Mayer, J.; Burck, J.; Ache, H. J. *Fresenius Journal of Analytical Chemistry* **1996**, *354*(7–8), 841–847.
26. Zimmermann, B.; Burck, J.; Ache, H. J. *Sensors and Actuators B-Chemical* **1997**, *41*(1–3), 45–54.

Separations Chemistry and Technology

Chapter 8

Separation of Fission Products Based on Room-Temperature Ionic Liquids

Huimin Luo[1], Sheng Dai[2], Peter V. Bonnesen[2], and A. C. Buchanan, III[2]

[1]Nuclear Science and Technology Division and [2]Chemical Sciences Division, Oak Ridge National Laboratory, Oak Ridge, TN 37831

Imidazolium-based ionic liquids containing cis-dicyclohexano-18-crown-6 (DCH18C6) were tested for the extraction of Sr^{2+} and Cs^+. For Sr^{2+} extraction, the influence of the addition of a sacrificial cation exchanger was also investigated. A decrease in strontium distribution coefficients in ionic liquids was observed for the system using Na^+ as a sacrificial ion. This observation is rationalized by the competition of Na^+ for crown ethers. A slight increase in strontium distribution coefficients was observed for an ionic liquid system utilizing oleic acid as the cation exchanger. This moderate increase of the strontium distribution coefficient can be attributed to the proton-mediated ion exchanging process and the potential synergistic effect of the organic acid. A series of imidazolium-based ionic liquids containing 2-alkoxy-2-oxoethyl substituents on one or both of the imidazole nitrogens was synthesized and characterized by NMR spectra. The solvent extraction results using calix[4]arene-bis(*tert*-octylbenzo-crown-6) [BOBCalixC6] in these ionic liquids are reported.

Introduction

The applications of ionic liquids (ILs) as replacement solvents for various catalytic reactions and separation processes have been extensively explored ($1-13$). Figure 1 gives the structure templates of the two most common classes of ambient-temperature ionic liquids. The cation is usually a heterocyclic cation, such as a dialkyl imidazolium ion or an N-alkylpyridinium ion. The relatively large size of these organic cations compared to simple inorganic cations accounts for the low melting points observed for these organic cations when paired with a variety of anions, such as BF_4^-, PF_6^-, $CF_3SO_3^-$, or other complex anions. These ion pairs or salts are usually liquids from around -100 °C and are thermally stable to around 200 °C, depending on the specific structures of the anions and cations. Unlike conventional solvents currently in use, these ionic liquids are nonflammable, chemically tunable, and have no detectable vapor pressure (1).

We ($13-15$) and others ($7-11$) have been interested in the development of IL-based solvent extraction methods for the separation of fission products. In contrast to the high-temperature inorganic ionic liquids (molten salts), room-temperature ionic liquids can be made hydrophobic while retaining ionicity. This dual property forms the basis for using room-temperature ionic liquids as unique separation media for the solvent extraction of ionic species. Large distribution coefficients (D_M) for the extraction of metal ions have been observed with ionic liquids containing complexing ligands ($13-15$). For example, whereas conventional solvent extraction of Sr^{2+} using dicyclohexano-18-crown-6 can deliver practical D_M values of less than one, our experiments with ionic liquids as extraction solvents delivered values of D_M on the order of 10^4 (14). The enhanced distribution coefficients can be attributed to the unique solvation properties of ionic liquids for ionic species, with ion-exchange processes playing an important role as revealed in details by Dietz and Dzielawa (11), and Visser et al. ($7-9$). However, the large D_M values observed for fission products using IL-based extraction processes make it very difficult to strip the metal ions, and to recycle the crown ethers. Another drawback associated with IL-based extraction processes is the loss of ILs through ion-exchange reactions ($7-11$). These deficiencies associated with IL-based extraction processes for metal ions prompted us to explore IL systems with sacrificial cation exchangers that are also environmentally benign. Here, we report our results concerning the development of IL-based extraction systems containing Na^+ and H^+ as sacrificial cations in the extraction of strontium and cesium ions. The extractant used in this investigation for Sr^{2+} is cis-dicyclohexano-18-crown-6 (DCH18C6), while the extractant used for Cs^+ is calix[4]arene-bis(*tert*-octylbenzo-crown-6) [BOBCalixC6]. Moyer and coworkers ($17-18$) have previously conducted extensive investigations into the use of BOBCalixC6 for extraction of Cs^+ in conventional solvents.

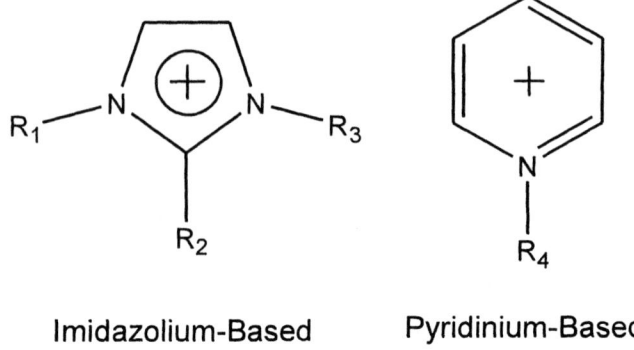

Imidazolium-Based   Pyridinium-Based

Figure 1. Structures of two most common ionic liquid cations.

Experimental Section

Materials and Methods

All chemicals and solvents were reagent grade and used without further purification unless noted otherwise. DCH18C6 was purchased from Aldrich and is a mixture of cis-syn-cis and cis-anti-cis isomers. BOBCalixC6 was obtained from IBC Advanced Technologies (America Fork, UT) and used as received (97% stated purity). 1-Alkyl-3-methylimidazolium bis[(trifluoromethyl)sulfonyl]-amide [$C_n$mim][NTf_2] ionic liquids, in which the alkyl groups were ethyl, butyl, hexyl, and octyl, were synthesized by modified literature procedures (*19*) and the details were reported elsewhere (*20*). Aqueous solutions were prepared using deionized water with a specific resistance of 18.0 megohm-cm or greater. ^{1}H-, and ^{13}C-NMR spectra were obtained in $CDCl_3$ with a Bruker MSL-400 NMR spectrometer, operating at 400.13 MHz for proton, and 100.61 MHz for carbon. Proton and carbon chemical shifts are reported relative to tetramethylsilane (TMS). The UV spectra were measured using a Varian UV-VIS-NIR spectrometer (Model 5000). Concentrations of Cs^+ and Sr^{2+} were determined using a Dionex LC20 ion chromatograph equipped with an IonPac CS-12 analytical column.

Extraction Experiments

The extraction experiments were performed in duplicate for each room temperature ionic liquid (RTIL) by contacting 1 mL of RTIL containing various concentrations of the extractant with 10 mL of cation-containing aqueous solution (1.5 mM) for 60 min in a vibrating mixer. After centrifugation, the upper aqueous phase was separated and the concentration of cations was determined by ion chromatography. Metal ion distribution coefficients were calculated as: $D_M = ([M_{initial}]-[M_{final}])/[M_{final}]$ and uncertainty in D_M is ± 5%.

General Procedure for the Synthesis of 1-(2-Alkoxy-2-oxoethyl)-3-methyl imidazolium bromide

Equal molar amounts of 1-methylimidazole and alkyl bromoacetate were mixed in acetonitrile. The mixture was stirred at room temperature for 24 h. The product was obtained in nearly quantitative yield as a viscous liquid, which was of sufficient purity to be used in the preparation of the corresponding NTf_2 salt.

Synthesis of 1-(2-Ethoxy-2-oxoethyl)-3-methylimidazolium bromide (task-specific ionic liquid [TSIL] 1)

From 1-methylimidazole (4.11 g, 50 mmol) and ethyl bromoacetate (8.35 g, 50 mmol), 10.8 g of 1-(2-ethoxy-2-oxoethyl)-3-methylimidazolium bromide was obtained (yield 87%). ^{1}H-NMR: δ, 10.08 (s, 1H), 7.83 (s, 1H), 7.71 (s, 1H), 5.51 (s, 2H), 4.25 (q, 2H), 4.12 (s, 3H), and 1.29 (t, 3H). ^{13}C-NMR: δ, 165.75 (CO), 137.70 (CH), 123.67 (CH), 122.91 (CH), 62.63 (CH_2), 50.03 (CH_2), 38.40 (CH_3), and 13.91 (CH_3).

Synthesis of 1-(2-*tert*-Butoxy-2-oxoethyl)-3-methylimidazolium bromide (TSIL 2).

From 1-methylimidazole (5.16 g, 62.9 mmol) and *tert*-butyl bromoacetate (12.26 g, 62.9 mmol), 14.8 g of 1-(2-*tert*-butoxy-2-oxoethyl)-3-methylimidazolium bromide was obtained (yield 85%). ^{1}H-NMR: δ, 9.37 (s, 1H), 7.64 (s, 1H), 7.54 (s, 1H), 5.19 (s, 2H), 3.95 (s, 3H), and 1.52 (s, 9H). ^{13}C-NMR: δ, 166.36 (CO), 138.45 (CH), 124.42 (CH), 123.96 (CH), 84.22 (C), 51.19 (CH_2), 36.95 (CH_3), and 27.99 (CH_3).

General Procedure for Synthesis of 1,3-Bis(2-alkoxy-2-oxoethyl)-imidazolium bromide

The literature method (21) for preparing 1,3-dialkylimidazolium bromide was used. Imidazole (2.0 g, 29.4 mmol) in tetrahydrofuran (THF) (20 mL) was added dropwise to 95% sodium hydride powder (0.75 g, 29.4 mmol) in

THF (20 mL) at 0 °C. The ice bath was removed, and the mixture was stirred for 2 h at room temperature. Following dropwise addition of the alkyl bromoacetate (58.8 mmol) at room temperature, the mixture was heated to reflux for 18 h. The precipitate was filtered and thoroughly rinsed with THF, followed by dichloromethane. The filtrate was evaporated in vacuum, and the residue was rinsed with diethyl ether and dried under vacuum to give the desired product as a waxy solid.

Synthesis of 1,3-Bis(2-ethoxy-2-oxothyl)-imidazolium bromide (TSIL 3)

From imidazole (2 g, 29.4 mmol) and ethyl bromoacetate (9.82 g, 58.8 mmol), 7.8 g of 1,3-di(ethyl acetate) imidazolium bromide was obtained (yield 82%). ^{1}H-NMR: δ, 9.98 (s, 1H), 7.72 (s, 2H), 5.41 (s, 4H), 4.25 (q, 4H), and 1.29 (t, 6H). ^{13}C-NMR: δ, 165.84 (CO), 138.63 (CH), 123.30 (CH), 62.95 (CH_2), 50.46 (CH_2), and 14.07 (CH_3).

Synthesis of 1,3-Bis(2-*tert*-butoxy-2-oxoethyl)-imidazolium bromide (TSIL 4)

From imidazole (2 g, 29.4 mmol) and *tert*-butyl bromoacetate (11.47 g, 58.8 mmol), 9.5 g of 1,3-bis(2-*tert*-butoxy-2-oxoethyl)-imidazolium bromide was obtained (yield 85%). ^{1}H-NMR: δ, 9.97 (s, 1H), 7.75 (s, 2H), 5.30 (s, 4H), and 1.50 (s, 18H). ^{13}C-NMR: δ, 164.43 (CO), 138.23 (CH), 123.03 (CH), 84.26 (C), 50.53 (CH_2), and 27.66 (CH_3).

General Procedure for Synthesis of 1-(2-Alkoxy-2-oxoethyl)-3-methyl imidazolium or 1,3-Bis(2-alkoxy-2-oxoethyl)-imidazolium bis[(trifluoromethyl)sulfonyl]-imide

The corresponding bromide was dissolved in 50 mL of deionized (D.I.) water and heated to 70 °C, and an equal molar amount of N-lithiotrifluoromethanesulfonimide (LiTf$_2$N) was dissolved in 100 mL of D.I. water and heated to 70 °C. The two solutions were combined and stirred to mix well. The resulting cloudy solution was cooled and then extracted with chloroform three times. The combined organic phases were washed four times successively with D.I. water to ensure that all the Li$^+$ was removed. Evaporation of the solvent produced the desired compound as a light-yellow liquid, except one compound (TSIL 8) was obtained as a solid.

Synthesis of 1-(2-Ethoxy-2-oxoethyl)-3-methylimidazolium bis[(trifluoromethyl) sulfonyl]-imide (TSIL 5)

From the bromide, TSIL 1 (10.8 g, 43.3 mmol) and LiTf$_2$N (12.44 g, 43.3 mmol), 17.7 g of TSIL 5 was obtained (yield 91%). ^{1}H-NMR: δ, 8.48 (s, 1H), 7.40 (s, 2H), 4.95 (s, 2H), 4.22 (q, 2H), 3.85 (s, 3H), and 1.27 (t, 3H). ^{13}C-NMR: δ, 166.97 (CO), 137.93 (CH), 124.37 (CH), 119.75 (CF$_3$, q, coupling constant of C-F is 321 Hz), 118.20 (CH), 63.36 (CH$_2$), 50.74 (CH$_2$), 37.08 (CH$_3$), and 14.30 (CH$_3$).

Synthesis of 1-(2-*tert*-Butoxy-2-oxoethyl)-3-methylimidazolium bis[(trifluoromethyl) sulfonyl]-imide (TSIL 6)

From the bromide, TSIL 2 (6.28 g, 22.7 mmol) and LiTf$_2$N (6.5 g, 22.7 mmol), 6.3 g of TSIL 6 was obtained (yield 58%). ^{1}H-NMR: δ, 8.75 (s, 1H), 7.38 (s, 1H), 7.30 (s, 1H), 4.89 (s, 2H), 3.38 (s, 3H), and 1.48 (s, 9H). ^{13}C-NMR: δ, 164.55 (CO), 137.34 (CH), 123.85 (CH), 123.12 (CH), 119.69 (CF$_3$, q, coupling constant of C-F is 321 Hz), 84.88 (C), 50.36 (CH$_2$), 36.31 (CH$_3$), and 27.67 (CH$_3$).

Synthesis of 1,3-Bis(2-ethoxy-2-oxoethyl)-imidazolium bis[(trifluoromethyl) sulfonyl]-imide (TSIL 7)

From the bromide, TSIL 3 (3.6 g, 11.1 mmol) and LiTf$_2$N (3.21 g, 11.1 mmol), 4.24 g of TSIL 7 was obtained (yield 73%). ^{1}H-NMR: δ, 8.98 (s, 1H), 7.40 (s, 2H), 5.04 (s, 4H), 4.30 (q, 4H), and 1.30 (t, 6H). ^{13}C-NMR: δ, 165.35 (CO), 138.47 (CH), 123.31 (CH), 119.70 (CF$_3$, q, coupling constant of C-F is 321 Hz), 63.24 (CH$_2$), 50.16 (CH$_2$), and 13.87 (CH$_3$).

Synthesis of 1,3-Bis(2-*tert*-butoxy-2-oxoethyl)-imidazolium bis[(trifluoromethyl) sulfonyl]-amide (TSIL 8)

From the bromide, TSIL 4 (3.8 g, 10.1 mmol) and LiTf$_2$N (2.91 g, 10.1 mmol), 4.3 g of TSIL 8 was obtained as a yellow solid (yield 73%). ^{1}H-NMR: δ, 8.89 (s, 1H), 7.38 (s, 2H), 4.90 (s, 4H), and 1.48 (s, 18H). ^{13}C-NMR: δ, 164.29 (CO), 137.21 (CH), 123.31 (CH), 120.45 (CF$_3$, q, coupling constant of C-F is 321 Hz), 84.96 (C), 50.50 (CH$_2$), and 27.68 (CH$_3$).

Results and Discussion

Synthesis

The TSILs containing the ester groups were synthesized to investigate the potential effect of ester groups on solvent extractions for Cs^+. The concept of TSILs was originally proposed by Davis and coworkers (7).The reaction used for synthesizing these TSILs is illustrated in Scheme 1. Eight ionic liquids (five are new) were synthesized in good yield. Compared to the reaction of 1-alkyl bromide with 1-methylimidazole, the reaction of alkyl bromoacetate with 1-methylimidazole is faster. This may be due to the electron withdrawing property of the ester group, making the nucleophilic reaction easier to proceed. The syntheses of TSILs **1, 3,** and **5** have been previously reported (22, 23). The procedure for TSILs **1** and **5** used in this paper is similar to the literature procedure except a different solvent (acetonitrile instead of THF) was used. There were no NMR data for TSILs **1** and **5** reported in reference 22; therefore, no comparison of the NMR data can be made. Our procedure for TSIL **3** involved only a one step synthesis, while the reported procedure needed two reaction steps. The NMR data for TSIL **3** are consistent with the literature data (23).

The change of the proton chemical shifts for H2 on the imidazolium ring upon changing the anion from the bromide to NTf_2 is listed in Table I. It is clear from Table I that the chemical shifts of H2 on the imidazolium ring move downfield when the bromides are converted into NTf_2 salts. This observation is consistent with data for 1-alkyl-3-methylimidazolium cations (19).

Extraction Results

Effects of Different Anions

According to the cation-exchange model proposed by Dietz and Dzielawa (11), and Visser et al. (7–9), only cations are directly involved in the extraction processes so that the distribution coefficients are expected to only weakly depend on the counter anions of extracted cations. Table II shows the D_{Sr} values as a function of different concentrations of DCH18C6 and the counter anion. From these data, it is clear that the extraction efficiency for Sr^{2+} was essentially unaffected by variation of the aqueous phase anion from nitrate to chloride at lower concentrations of DCH18C6. This finding is consistent with the results reported by Bartsch (10) and the ion-exchange model. At higher concentration of DCH18C6 the variation of the aqueous phase anion has some influence on the Sr^{2+} extraction efficiency.

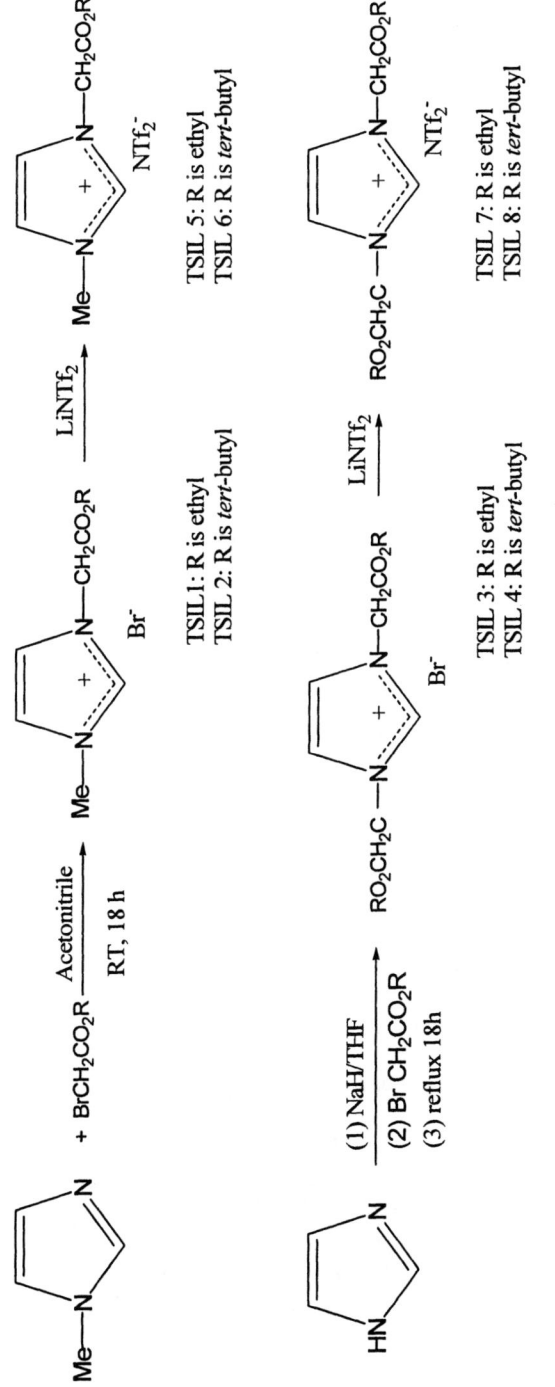

Scheme 1.

Table I. Comparison of Proton Chemical Shifts of H2 on Imidazolium Ring for 1-(2-Alkoxy-2-oxoethyl)-3-methyl imidazolium and 1,3-Bis(2-Alkoxy-2-oxoethyl)-imidazolium Cations

Imidazolium Cations	Proton Chemical Shifts (δ) of H2 on Imidazolium Ring		
	Br^-	Tf_2N^-	$\Delta\delta$
Me—N⁺—N—CH$_2$CO$_2$C$_2$H$_5$	10.08	8.48	1.60
Me—N⁺—N—CH$_2$CO$_2$C(CH$_3$)$_3$	9.37	8.75	0.62
H$_5$C$_2$O$_2$CH$_2$C—N⁺—N—CH$_2$CO$_2$C$_2$H$_5$	9.98	8.95	1.03
(CH$_3$)$_3$CO$_2$CH$_2$C—N⁺—N—CH$_2$CO$_2$C(CH$_3$)$_3$	9.97	8.89	1.08
Me—N⁺—N—CH$_2$CH$_2$CH$_2$CH$_3$	10.32	8.74	1.58

Effect of Alkyl Chain Length of Ionic Liquids and Addition of Sacrificial Cations

As seen from Table III and Figure 2, both D_{Sr} and D_{Cs} values are inversely proportional to the alkyl (R) chain length of imidazolium cations. This observation is again consistent with the previous results by Dietz (*11*) and Bartsch (*10*). This trend can be attributed to the synergistic effect of ion-exchange and solvation of charged macrocyclic complexes by ionic liquids. The alkyl chain length has a greater effect on D_{Sr} than on D_{Cs}.

Table II. Effect of Different Anions on Extraction Results (D_{Sr} values) of [C$_4$mim][NTf$_2$] Containing DCH18C6 without and with NaBPh$_4$ (0.12 M)

DCH18C6 Concentration	1.50×10^{-2} M		2.70×10^{-2} M		5.68×10^{-2} M		1.08×10^{-1} M	
	Without NaBPh$_4$	With NaBPh$_4$	Without NaBPh$_4$	With NaBPh$_4$	Without NaBPh$_4$	With NaBPh$_4$	Without NaBPh$_4$	With NaBPh$_4$
Sr(NO$_3$)$_2$	23.6	16.6	209	77.9	632	311	1084	496
SrCl$_2$	31.3	20.8	242	78.2	713	328	911	705

Table III. Effect of Carbon Chain Length of ILs and Synergistic Effect of Three Different Hydroxy Acids on Extraction Results (D_M) of [$C_n$mim][NTf_2] Containing DCH18C6 (0.02M)

Aqueous Phase	Hydroxy Acid Additive to ILs (0.1M)	C_n in [$C_n$mim][NTf_2]			
		Ethyl, C_2	Butyl, C_4	Hexyl, C_6	Octyl, C_8
CsCl	None	33.1	16.8	12.1	2.69
SrCl$_2$	None	465	74.1	15.1	2.06
	Oleic acid	466	137	16.6	2.22
	HDFN	466	64.3	15.9	1.82
	3,5-dtbp	419	58.7	12.45	1.71

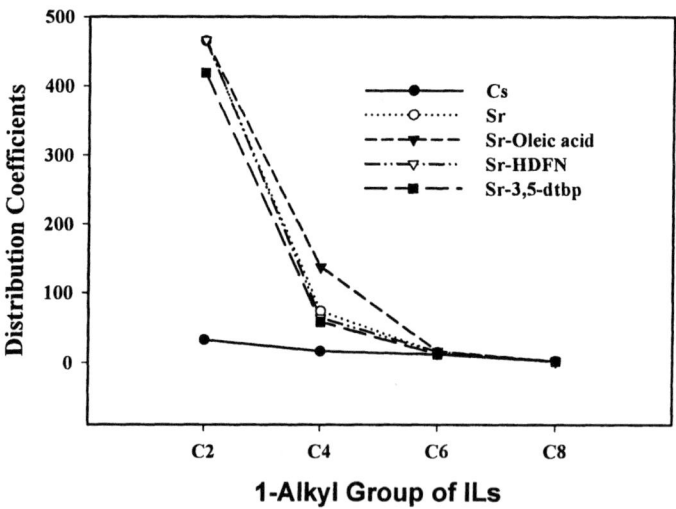

Figure 2. Effect of carbon chain length of ionic liquids and synergistic effect of hydroxy acids on extraction results of [$C_n$mim][NTf_2] containing DCH18C6.

Based on the ion-exchange extraction model, the extra loss of cations from ionic liquids into aqueous phases during solvent extraction of metal cations is through ion exchange with the aqueous metal cations. This dependence of D_{Sr} on the chain length of the alkyl group in imidazolium cations is attributed to the hydrophobicity of the corresponding organic cations. For example, the D_{Sr} value of the [C$_8$mim][NTf$_2$] based extraction system is significantly less than that of the [C$_2$mim][NTf$_2$] based extraction system. Accordingly, the addition of a sacrificial cationic species (Na$^+$ or H$^+$) should reduce the loss of ionic liquids during the solvent extraction. If the added sacrificial cations in ionic liquids have no affinities toward DCH18C6 and are more hydrophilic than organic imidazolium cations, we should expect an enhancement of D_{Sr} through the enhanced ion-exchange processes by the sacrificial cationic species.

Sodium tetraphenylborate (NaBPh$_4$) should be an excellent candidate as a sacrificial cationic exchanger. It is known that the BPh$_4^-$ anion forms insoluble compounds with big cations, such as Cs$^+$ and imidazolium cations in aqueous solutions (24). Therefore, the release of the BPh$_4^-$ anion from the imidazolium-based ionic liquids to aqueous solutions should be negligible. Our experiments showed that the solubility of NaBPh$_4$ in RTILs is reasonably high.

Table II shows the extraction results obtained using [C$_4$mim][NTf$_2$] containing various concentrations of DCH18C6 without and with NaBPh$_4$ (0.12 M). As seen from Table II, the corresponding D_{Sr} values with the addition of NaBPh$_4$ decrease significantly compared with no addition of NaBPh$_4$. This large decrease of D_{Sr} with the addition of NaBPh$_4$ indicates that Na$^+$ competes strongly with Sr^{2+} for the coordination with DCH18C6. The measurement of the concentration of C$_4$mim$^+$ in aqueous phases reveals significant substitution of C$_4$mim$^+$ by Na$^+$ in the ion-exchange process during extraction. Accordingly, the addition of NaBPh$_4$ decreases the loss of RTILs by as much as 20%, which is proved by the UV spectral measurements of C$_4$mim$^+$ in the aqueous phases (shown in Figure 3).

Another environmentally benign cation for the sacrificial ion-exchanging process is the proton. In fact, organic acids have been extensively explored for the synergistic effect on conventional solvent extractions via proton transfer (25). The synergistic effects of oleic acid and two organic hydroxy acids (structural formulas shown in Figure 4) on the IL-based solvent extractions were examined. As seen from Table III and Figure 2, the synergistic effects are minimal with the exception of oleic acid in [C$_4$mim][NTf$_2$]. In the latter case, D_{Sr} nearly doubles. No enhancement of D_{Sr} by the hydroxy acids can be attributed to the high pK_a values ($pK_a > 9$) of both acids (25). The pH value of aqueous solutions is about 6–7. Therefore, only low concentrations of proton ions are available from the hydroxy acids for ion exchange. The further investigation of the synergistic effect under different conditions (pH and temperature) is underway in our laboratory.

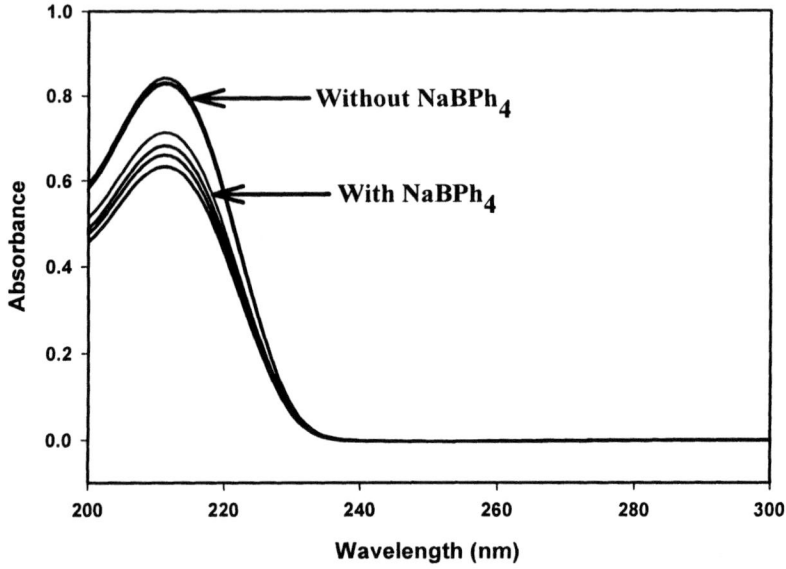

Figure 3. UV spectra of aqueous solutions equilibrated with [C₄mim][NTf₂] containing various concentrations of DCH18C6 with and without NaBPh₄.

CH₃(CH₂)₇CH=CH(CH₂)₇COOH   HF₂C(CF₂)₇CH₂OH

Oleic acid   1H, 1H, 9H-hexadecafluorononanol (HDFN)

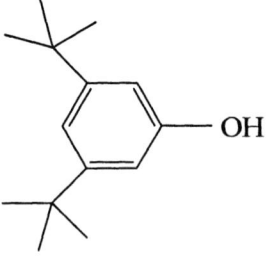

3,5-Di-*tert*-butylphenol (3,5-dtbp)

Figure 4. Structural formulas of oleic acid and two hydroxy acids.

Extraction Results of BOBCalixC6 in TSILs

The solvent extraction experiments could only be conducted with three new ionic liquids because TSIL 8 is a solid at room temperature. The results of these experiments and comparison with [C_4mim][NTf_2] are shown in Table IV. Clearly, the D_{Cs} values of TSIL 5 and new TSILs 6 and 7 without use of BOBCalixC6 are greater than that obtained with [C_4mim][NTf_2]. This observation indicates that the new TSILs have either more hydrophilic cations as compared to C_4mim$^+$ or coordinating functional groups facilitating extraction processes. However, the extraction experiments with BOBCalixC6 show that the D_{Cs} values of three TSILs with the use of BOBCalixC6 are smaller than that obtained with [C_4mim][NTf_2], but are still an order of magnitude better than without BOBCalixC6. Therefore, the synergistic effect of carbonyl-containing solvents is not a dominant factor. The observed decrease of the D_{Cs} values can be attributed to the increased hydrophobicity of the organic cations in three TSILs, which impels the ion-exchange process.

Table IV. Extraction Results of BOBCalixC6 in Task Specific Ionic Liquids

Ionic Liquids	D_{Cs} (BOBCalixC6 7.7 mM)	D_{Cs} (No BOBCalixC6)
TSIL 5	5.25	0.41
TSIL 6	5.85	0.68
TSIL 7	5.26	0.18
TSIL 8 (solid)	NM	NM
[C_4mim][NTf_2]	13.8	0.024
Chloroform	0.045	UD

Note: NM: not measured; UD: under detection limit.

Conclusions

The syntheses and characterization of five new ionic liquids are reported. The solvent extraction properties of these ionic liquids have been investigated. Sacrificial cation exchangers (Na$^+$ or H$^+$) were used to reduce the loss of the organic cations from ILs during the extraction of Sr^{2+} from aqueous solutions. In the case of Na$^+$, the competition for binding DCH18C6 by Na$^+$ results in significant reductions in the D_{Sr} values. The use of oleic acid was found to enhance the extraction efficiency for strontium by [C_4mim][NTf_2] containing DCH18C6.

Acknowledgments

This research was supported by the Environmental Management Science Program of the Office of Science and Environmental Management, U.S. Department of Energy, under Contract DE-AC05-0096OR22725 with Oak Ridge National Laboratory, managed by UT-Battelle, LLC.

References

1. Welton, T. *Chem. Rev.* **1999**, *99*, 2071.
2. Wasserscheid, P.; Kim, W. *Angew. Chem., Int. Ed.* **2000**, *39*, 3772.
3. Sheldon, R. *Chem. Commun.* **2001**, 2399.
4. Gordon, C. M. *Appl. Catal., A* **2001**, *222*, 101.
5. Dupont, J.; de Souza, R. F.; Suarez, P. A. Z. *Chem. Rev.* **2002**, *102*, 3667.
6. Blanchard, L. A.; Hancu, D.; Beckman, E. J.; Brennecke, J. F. *Nature* **1999**, *399*, 28.
7. Visser, A. E.; Swatloski, R. P.; Reichert, W. M.; Mayton, R.; Sheff, S.; Wierzbicki, A.; Davis, J. H.; Rogers, R. D. *Chem. Commun.* **2001**, 135.
8. Visser, A. E.; Swatloski, R. P.; Reichert, W. M.; Griffin, S. T.; Rogers, R. D. *Ind. Eng. Chem. Res.* **2000**, *39*, 3596.
9. Huddleston, J. G.; Willauer, H. D.; Swatloski, R. P.; Visser, A. E.; Rogers, R. D. *Chem. Comm.* **1998**, 1765.
10. Chun, S.; Dzyuba, S. V.; Bartsch, R. A. *Anal. Chem.* **2001**, *73*, 3737.
11. Dietz, M. L.; Dzielawa, J. A. *Chem. Commun.* **2001**, 2124.
12. Carda-Broch, S.; Berthod, A.; Armstrong, D.W. *Anal. Bioanal. Chem.* **2003**, *375*, 191.
13. Makote, R.; Luo, H.; Dai, S. In *Ultra Clean Solvents*; Abraham, M. A., Moens, L., Eds.; ACS Symposium Series 819; American Chemical Society: Washington, DC, **2002**; p 26.
14. Dai, S.; Ju, Y. H.; Barnes, C. E. *J. Chem. Soc. Dalton Trans.* **1999**, 1201.
15. Dai, S.; Shin, Y.; Toth, L. M.; Barnes, C. E. *Inorg. Chem.* **1997**, *36*, 4900.
16. Dozol, J.-F.; Casas, J.; Sastre, A. M. *Sep. Sci. Technol.* **1995**, *30*, 435.
17. Haverlock, T. J.; Bonnesen, P. V.; Sachleben, R. A.; Moyer, B. A. *Radiochim. Acta* **1997**, *76*, 103, and references cited.
18. Bonnesen, P. V.; Delmau, L. D.; Moyer, B. A.; Leonard, R. A. *Solvent Extr. Ion. Exch.* **2000**, *18*, 1079.
19. Bonhote, P.; Dias, A. P.; Papageorgiou, N.; Kalyanasundaram, K.; Gratzel, M. *Inorg. Chem.* **1996**, *35*, 1168.
20. Luo, H.; Dai, S.; Bonnesen, P. V.; Buchanan, A. C. III; Holbrey, J. D.; Bridges, N. J.; Rogers, R. D. *Anal. Chem.* **2004**, *76*, 3078.
21. Sergei, V. D.; Bartsch, R. A. *Chem. Commun.* **2001**, 1466.
22. Gathergood, N.; Scammells, P. J. *Austr. J. Chem.* **2002**, *55*, 557.
23. Herrmann, W. A.; Gooben, L. J.; Spiegler, M. *J. Organometallic Chem.* **1997**, *547*, 357.
24. Dupont, J.; Suarez, P. A. Z.; De Souza, R. F.; Burrow, R. A.; Kintzinger, J. P. *Chem.-Eur. J.* **2000**, *6*, 2377.
25. Levitskaia, T. G.; Bonnesen, P. V.; Chambliss, C. K.; Moyer, B. A. *Anal. Chem.* **2003**, *75*, 405.

Chapter 9

Supercritical Fluid Extraction of Radionuclides: A Green Technology for Nuclear Waste Management

Chien M. Wai

Department of Chemistry, University of Idaho, Moscow, ID 83844

Supercritical fluid carbon dioxide is capable of dissolving uranium dioxide directly with a CO_2-soluble n-tributylphosphate-nitric acid complex. The extracted uranyl compound $UO_2(NO_3)_2(TBP)_2$ has a solubility of about 0.4 mol/L in supercritical CO_2 at 40 °C and 200 atm. Cesium and strontium can also be extracted by supercritical CO_2 using a crown ether and a fluorinated counteranion. Using dicyclohexano-18-crown-6 as the ligand and pentadecafluoro-n-octanoic acid as the counteranion, Sr^{2+} can be selectively extracted over Ca^{2+} and Mg^{2+} by supercritical CO_2 at 60 °C and 100 atm. Supercritical CO_2 which can penetrate into small pores of solid materials and extract radionuclides with minimal liquid waste generation appears to provide an attractive green technology for nuclear waste management.

© 2006 American Chemical Society

Introduction

Utilizing supercritical fluids as solvents for extraction, separation, synthesis, and cleaning has been a very active research area in the past two decades (*1–3*). This technology development is mainly due to the changing environmental regulations and increasing costs for disposal of conventional liquid solvents. Carbon dioxide is widely used in supercritical fluid extraction (SFE) applications because of its moderate critical constants ($T_c = 31.1$ °C, $P_c = 72.8$ atm, $\varphi_c = 0.471$ g/mL), inertness, low cost, and availability in pure form. However, because CO_2 is a linear triatomic molecule (with no dipole moment), it is actually a poor solvent for dissolving polar compounds and ionic species. A method of dissolving metal ions in supercritical CO_2 was developed in the early 1990s using an in situ chelation technique. In this method, a CO_2-soluble chelating agent is used to convert metal ions into soluble metal chelates in the supercritical fluid phase. Quantitative measurements of metal chelate solubilities in supercritical CO_2 were first made by Laintz et al. in 1991 using a high-pressure view cell and UV/Vis spectroscopy (*4*). In this pioneering study, the authors noted that fluorine substitution in the chelating agent could greatly enhance (by 2 to 3 orders of magnitude) the solubility of metal chelates in supercritical CO_2. A demonstration of copper ion extraction from solid and liquid materials using supercritical CO_2 containing a fluorinated chelating agent bis(trifluoroethyl)dithiocarbamate was reported in 1992 (*5*). Since then, a variety of chelating agents including dithiocarbamates, ß-diketones, organophosphorus reagents, and macrocyclic ligands have been tested for metal extraction in supercritical fluid CO_2 (*6*). Highly CO_2-soluble metal complexes involving organophosphorus reagents have been found and reported in the literature (*7*). Recently, direct dissolution of metal oxides such as uranium dioxide and lanthanide oxides in supercritical CO_2 using a tri-n-butylphosphate-nitric acid complex as the extractant has also been demonstrated (*8, 9*). These studies have greatly expanded potential uses of the supercritical fluid extraction technology for metal related applications.

The in situ chelation-SFE technique appears attractive for nuclear waste management because it can greatly reduce the secondary waste generation compared with the conventional processes using organic solvents and aqueous solutions. Other advantages of using the SFE technology for nuclear waste management include fast extraction rate, capability of penetration into small pores of solid materials, and rapid separation of solutes by depressurization. The tunable solvation strength of supercritical fluid CO_2 also allows potential separation of metal complexes based on their difference in solubility and partition between the fluid phase and the matrix. This article summarizes recent developments regarding SFE of radionuclides, particularly with respect to uranium extraction to illustrate the potential of the SFE technology for removing long-lived radioisotopes from contaminated wastes.

Supercritical CO_2 Extraction of Lanthanides and Actinides

Lanthanide and actinide ions in solid and liquid materials can be extracted using a chelating agent such as a ß-diketone dissolved in supercritical CO_2 (*6, 10, 11*). Fluorine containing ß-diketones are more effective than the non-fluorinated acetylacetone for SFE of the f-block elements. In several reported SFE studies for lanthanide and actinides, thenoyltrifluoroacetone (TTA) was used as a chelating agent. One reason for using TTA is that it is a solid at room temperature (m.p. 42 °C) and is easy to handle experimentally. Other commercially available fluorinated ß-diketones, often in liquid form at room temperature, have also been used for SFE of lanthanides and actinides. A strong synergistic effect was observed for the extraction of lanthanides from solid samples when a mixture of tri-n-butylphosphate (TBP) and a fluorinated ß-diketone was used in supercritical CO_2 (*10, 11*). TBP is highly soluble in supercritical CO_2 with a solubility close to 10 mole percent under normal SFE conditions. Actually at a given temperature, TBP becomes miscible with supercritical CO_2 above a certain pressure according to a recent phase diagram study of the TBP/CO_2 system (*12*). The synergistic extraction effect is probably due to an adduct formation with TBP replacing a coordinated water molecule in the lanthanide-TTA complex, thus increasing the solubility of the adduct complex.

Uranyl and thorium ions in solids and in aqueous solutions can also be extracted by supercritical CO_2 containing fluorinated ß-diketones. For example, spiked UO_2^{2+} and Th^{4+} in sand can be extracted by supercritical CO_2 containing TTA with efficiencies around 70–75% at 60 °C and 150 atm with 10 min of static and 20 min of dynamic extraction (*11*). Using a mixture of TTA and TBP, the extraction efficiencies of UO_2^{2+} and Th^{4+} are increased to > 93%. The feasibility of extracting uranyl ions from natural samples was tested using mine wastes collected from an abandoned uranium mine. The uranium concentrations in two mine waters tested were 9.6 µg/mL and 18 µg/mL, respectively. The mine waters were extracted with a 1:1 mixture of TTA and TBP in neat CO_2 at 60 °C and 150 atm for a static time of 10 min followed by 20 min of dynamic extraction. Under these experimental conditions, the percent extraction of uranium from these samples was 81% ± 4% and 78% ± 5%, respectively. The mine waters were also added to a soil sample and dried at room temperature for SFE tests. The percent extraction of uranium using TTA/TBP was in the range of 77–82% at 60 °C and 150 atm. Recently, Fox and Mincher showed that plutonium spiked in a soil can also be extracted by TTA/TBP with efficiencies around 70–80% (*13*).

In highly acidic solutions (1–6 M HNO_3), organophosphorus reagents such as TBP and TBPO (tri-butylphosphine oxide) dissolved in supercritical CO_2 can extract uranyl ions (UO_2^{2+}) and thorium ions (Th^{4+}) effectively (*14*). Uranyl nitrate alone does not show an appreciable solubility in supercritical CO_2. But, when it is coordinated with TBP, the uranyl nitrate TBP complex

$(UO_2)(NO_3)_2 \cdot 2TBP$ becomes highly soluble in supercritical CO_2 (7). The extraction efficiencies for UO_2^{2+} and Th^{4+} using TBP saturated supercritical CO_2 are comparable to those observed in solvent extraction with kerosene containing 19% v/v TBP (14).

The solubilities of $(UO_2)(NO_3)_2 \cdot 2TBP$ in supercritical CO_2 in the temperature range 40–60 °C and pressure range 100–200 atm were measured using a spectroscopic method (7). The solubility of this important uranyl complex in supercritical CO_2 is about 0.4 mol/L at 40 °C and 200 atm. This uranium concentration is similar to that used in the conventional purex (plutonium uranium extraction) process. In comparison with $UO_2(NO_3)_2 \cdot 2TBP$, the solubility of $UO_2(TTA)_2 \cdot TBP$ in supercritical CO_2 is about an order of magnitude lower (15). This information indicates that uranium and transuranics in solid materials can be extracted directly by supercritical CO_2 containing a mixture of TBP and a fluorinated β-diketone or TBP and nitric acid.

Direct Dissolution of Uranium Dioxide in Supercritical CO_2

Direct dissolution of solid UO_2 in supercritical CO_2 is difficult because uranium at +4 oxidation state does not form stable complexes with commonly known ligands. An oxidation step is needed to convert uranium from +4 to +6 oxidation state to make it extractable in supercritical CO_2. Recent reports show that TBP forms a complex with nitric acid which is soluble in supercritical CO_2 and is capable of dissolving lanthanide oxides and uranium oxides directly (8, 9). The complex is prepared by mixing TBP with a concentrated nitric acid solution. Nitric acid dissolves in the TBP phase forming a complex of the general formula $TBP(HNO_3)_x(H_2O)_y$ that is separated from the remaining aqueous phase. The TBP-nitric acid complex is soluble in supercritical CO_2, and it is capable of dissolving solid uranium dioxide directly. The UO_2 dissolution process probably involves oxidation of the tetravalent uranium in UO_2 to the hexavalent uranyl followed by formation of $UO_2(NO_3)_2 \cdot 2TBP$, which is known to have a high solubility in SF-CO_2.

Dissolution of UO_2 by a TBP-nitric acid complex depends on the stoichiometry of the complex and the density of the supercritical fluid phase. Figure 1 shows the amount of UO_2 dissolves in the supercritical fluid stream containing the $TBP(HNO_3)_{0.7}(H_2O)_{0.7}$ complex at a flow rate of 0.4 mL/min. The supercritical CO_2 solution was produced by bulbing liquid CO_2 through a cell containing the $TBP(HNO_3)_{0.7}(H_2O)_{0.7}$ complex at room temperature (around 23 °C). The dissolution of UO_2 using the $TBP(HNO_3)_{1.8}(H_2O)_{0.6}$ complex in supercritical CO_2 is more rapid because of a higher nitric acid concentration provided by the complex. The alkali metals, the alkaline earth metals, and a number of transition metals cannot be extracted by the TBP-nitric acid complex in supercritical CO_2. Sonification can enhance the dissolution rate of tightly

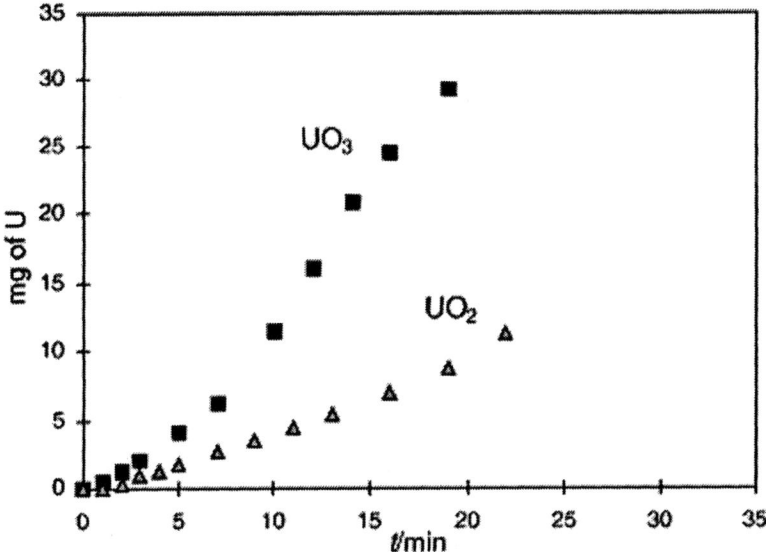

Figure 1. Dissolution of UO_2 and UO_3 in supercritical CO_2 using $TBP(HNO_3)_{0.7}(H_2O)_{0.7}$ as an extractant at 60 °C and 150 atm, flow rate = 0.4 mL/min (Reproduced with permission from reference 9. Copyright 2001 Royal Society of Chemistry.)

packed UO_2 powders significantly, probably by increasing the transport of $UO_2(NO_3)_2(TBP)_2$ from the solid surface to the supercritical fluid phase (16). Dissolution of UO_3 by a TBP-nitric acid complex (e.g. $TBP(HNO_3)_{0.7}(H_2O)_{0.7}$ or $TBP(HNO_3)_{1.8}(H_2O)_{0.6}$) in supercritical CO_2 is more efficient than the dissolution of UO_2. This can be attributed to the fact that the uranium in UO_3 is already in the +6 oxidation state, thus no oxidation is required in the formation of the uranyl complex $UO_2(NO_3)_2(TBP)_2$.

In the conventional purex process, aqueous nitric acid (3–6 M) is first used to dissolve and oxidize UO_2 in the spent fuel to uranyl ions $(UO_2)^{2+}$. The acid solution is then extracted with TBP in dodecane to remove the uranium as $UO_2(NO_3)_2 \cdot 2TBP$ into the organic phase. The SF-CO_2 dissolution of UO_2 with a TBP-nitric acid complex has advantages over the purex process since it combines dissolution and extraction into one step with minimum waste generation. This green SF process that requires no aqueous solution and organic solvent may have potential applications for processing spent nuclear fuel, decontamination of UO_2 contaminated wastes, and even for processing of rare earth ores.

Characterization of TBP-Nitric Acid Complex

The chemical nature of the TBP-HNO$_3$ complex in a SF-CO$_2$ solution is not well known. It is known that water can dissolve in TBP during mixing. The solubility of water in TBP at room temperature is about 64 grams per liter, which is close to a 1:1 mole ratio of a TBP:H$_2$O complex. The complex is most likely formed through hydrogen bonding of H$_2$O with the P=O group of TBP. When concentrated nitric acid is mixed with TBP by shaking, nitric acid also dissolves in the TBP phase forming a TBP(HNO$_3$)$_x$(H$_2$O)$_y$ complex that is separated from the remaining aqueous phase. The x and y values in the complex can be determined by acid-base titration and by Karl Fischer titration, respectively. According to a recent report, the x and y values in the complex depend on the relative amount of the acid and TBP used in the preparation. For example if 5 mL of TBP is mixed with 5 mL of concentrated nitric acid (15.5 M), the TBP phase has a composition of TBP(HNO$_3$)$_{1.8}$(H$_2$O)$_{0.6}$. If 5 mL of TBP is mixed with 1 mL of the concentrated nitric acid, the TBP phase has a composition of TBP(HNO$_3$)$_{0.7}$(H$_2$O)$_{0.7}$.

Proton NMR studies indicated that the protons of HNO$_3$ and H$_2$O in the complex showed only a single peak, suggesting a rapid exchange of the protons like in a nitric acid solution (*17*). This single proton resonance peak shifts upfield with increasing x/y ratio in the complex.

The proton resonance peak of TBP·H$_2$O complex appears at 3.85 ppm. In a TBP(HNO$_3$)$_x$(H$_2$O)$_y$ complex, the proton resonance peak shifts upfield and reaches around 11.2 ppm when the x/y ratio is 3. When the TBP(HNO$_3$)$_{0.7}$(H$_2$O)$_{0.7}$ complex is added to a low dielectric constant solvent such as chloroform (ε = 4.18 at 20 °C), fine droplets of nitric acid are formed as indicated by the NMR spectrum shown in Figure 2. The small peak at 6.49 ppm in Figure 2 is the nitric acid droplets formed in the system. The formation of these fine droplets of acid in chloroform can be attributed to an anti-solvent effect because HNO$_3$ and H$_2$O are sparsely soluble in chloroform. The same anti-solvent effect is expected to occur in supercritical CO$_2$, which has a very small dielectric constant. When the TBP(HNO$_3$)$_{0.7}$(H$_2$O)$_{0.7}$ complex was added to supercritical CO$_2$, the solution becomes cloudy, suggesting formation of small acid droplets in the fluid phase. These small droplets of nitric acid formed in the supercritical CO$_2$ phase can oxidize UO$_2$ to (UO$_2$)$^{2+}$ that is followed by the formation of UO$_2$(NO$_3$)$_2$(TBP)$_2$ which becomes soluble in supercritical CO$_2$.

Assuming the TBP-HNO$_3$ complex has a 1:1 stoichiometry, the dissolution of UO$_2$ by this complex may be expressed by the following equation:

$$UO_2(s) + 8/3 \text{ TBP-HNO}_3 \rightarrow UO_2(NO_3)_2(TBP)_2 + 3/2 \text{ NO} + 4/3 \text{ H}_2O + 2/3 \text{ TBP}$$

For complexes with other TBP:HNO$_3$ ratios, similar equations can be written.

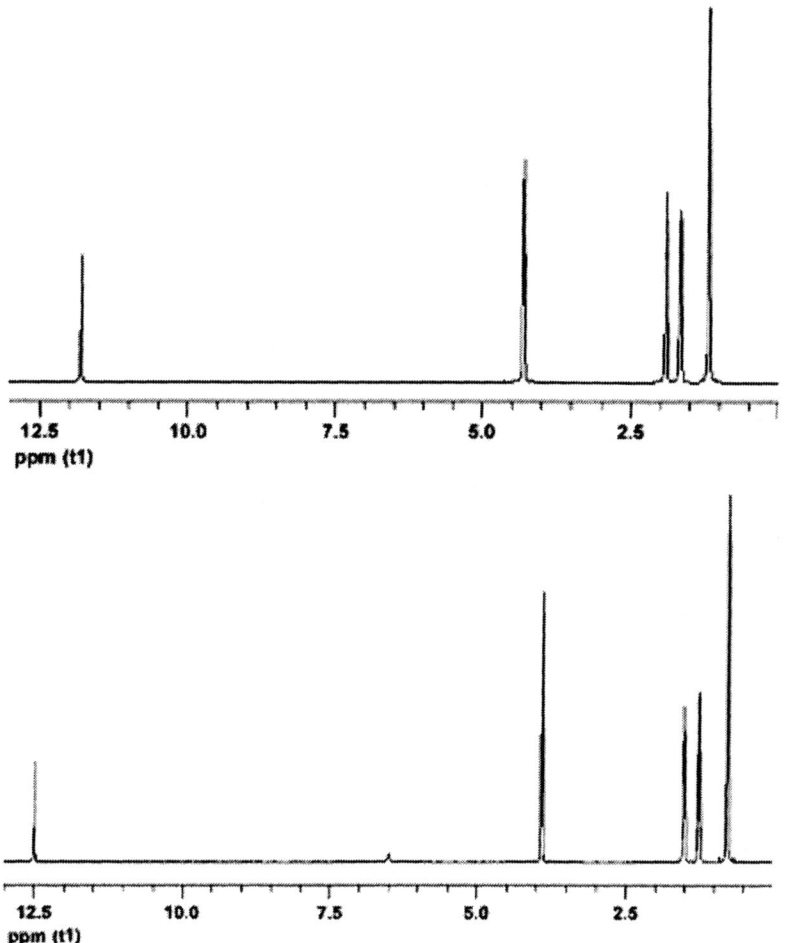

Figure 2. Proton NMR spectra of TBP(HNO$_3$)$_{0.7}$(H2O)$_{0.7}$. Top: 300 MHz NMR spectrum of the complex using an insert; bottom: 300 MHz NMR spectrum of the complex in CDCl$_3$.

Supercritical CO_2 Extraction of Cesium and Strontium

Crown ethers are selective extractants for Cs^+ and Sr^{2+} in conventional solvent extraction processes. For example, it is known that 18-membered crown ethers with cavity diameters in the range of 2.6 to 2.8 Å are the most suitable hosts for Sr^{2+} (2.2 Å cationic diameter). For Cs^+ with a 3.34 Å cationic diameter, the 21-crown-7 host with a cavity diameter in the range of 3.4-4.3 Å is suitable for selective complexation of the cation. However, these crown-metal complexes are not soluble in supercritical CO_2. One method of making the crown-metal complexes soluble in supercritical CO_2 is to use a fluorinated counteranion to extract the complexes in the supercritical fluid phase as ion-pairs. The fluorinated counteranions can neutralize the charge of the crown-metal complexes and make the resulting ion-pair soluble in supercritical CO_2.

Fluorinated counteranions such as pentadecafluroro-n-octanoic acid and perfluoro-1-octanesulfonic acid (ammonium or potassium salt) are effective for selective extraction of Cs^+ and Sr^{2+} using crown ether ligands in supercritical CO_2. The extraction efficiency depends on the amount of crown ether and counteranion used in the extraction. Table I shows some data related to the SFE of Sr^{2+} from water using dicyclohexano-18-crown-6 (DC18C6) as the ligand and pentadecafluoro-n-octanoic acid (PFOA-H) as the counteranion in supercritical CO_2 at 60 °C and 100 atm. The extraction is selective for Sr^{2+} over Ca^{2+} and Mg^{2+}. Selective extraction of Sr^{2+} in acidic solutions can also be achieved using DC18C6 and a potassium salt of PFOA with a high efficiency (*18*).

Table I. Extraction of Sr^{2+}, Ca^{2+}, and Mg^{2+} from Water by Supercritical CO_2 Containing DC18C6 and Pentadecafluoro-n-octanoic acid at 60 °C and 100 atm

Mole Ratio				% Extraction		
Sr^{2+}	: DC18C6 :	PFOA-H		Sr^{2+}	Ca^{2+}	Mg^{2+}
1	10	0		1	0	0
1	0	10		4 ± 1	1 ± 1	1 ± 1
1	5	10		36 ± 2	1 ± 1	1 ± 1
1	10	10		52 ± 2	2 ± 1	1 ± 1
1	10	50		98 ± 2	7 ± 2	2 ± 1

Note: The aqueous solution contained a mixture of Sr^{2+}, Ca^{2+}, and Mg^{2+} with a concentration of 5.6 x 10^{-5} M each; pH of water under equilibrium with SF CO_2 = 2.9; 20-min static followed by 20-min dynamic flushing at a flow rate of 2 mL/min. PFOA-H = $CF_3(CF_2)_6COOH$. (Data from reference 18.)

Dicyclohexano-21-crown-7 (DC21C7) with perfluoro-1-octanesulfonic acid as a counteranion can be used to extract Cs^+ in supercritical CO_2. However, potassium ion (K^+) can also be extracted with Cs^+ in this case (19). A more selective macrocyclic ligand is needed in order to achieve selectivity and efficiency for Cs^+ extraction in supercritical CO_2.

Potential Applications

The direct dissolution of UO_2 in supercritical CO_2 using a TBP-nitric acid complex described in this section has many potential applications. For example, this technique can be used to remove uranium from UO_2 contaminated wastes, such as those generated by nuclear fuel fabrication plants. Ash samples spiked with UO_2 can be quantitatively extracted by supercritical CO_2 containing the $TBP(HNO_3)_{1.8}(H_2O)_{0.6}$ complex at 60 °C and 200 atm using a combination of static and dynamic extraction (20). Contaminated uranium dioxide in real ash samples from a fuel fabrication plant could also be removed from the ash with efficiencies up to 85%. The possibility of using supercritical fluid solutions for reprocessing spent nuclear fuels was suggested by Smart et al. several years ago (21). Recent development in direct dissolution of UO_2 using a TBP-nitric acid complex makes this suggestion more acceptable. The possibility of dissolving UO_2 directly from spent nuclear fuels for their reprocessing in supercritical CO_2 is currently being tested by a team in Japan. The Japanese demonstration project (2002–2005) involves Mitsubishi Heavy Industries, Japan Nuclear Cycle Corp., and Nagoya University (16). The project is aimed at extracting uranium and plutonium from the mixed oxide fuel as well as the irradiated nuclear fuel using the $TBP(HNO_3)_{1.8}(H_2O)_{0.6}$ complex as an extractant in supercritical CO_2. This project will provide valuable information regarding safe handling of nuclear spent fuels in a high-pressure supercritical fluid extraction system. This information should also be useful for considering other applications of the supercritical fluid technology for treatment of a variety of nuclear waste related problems.

Acknowledgments

This work was supported by the U.S. Department of Energy Environmental Management Science Program (grant number DE-FG07-98ER14913).

References

1. McHugh, M. A.; Krukonis, V. J. *Supercritical Fluid Extraction: Principles and Practice*; Butterworth-Heinemann: Oxford, U.K., 1994.
2. Darr, J.; Poliakoff, M. *Chem. Rev.* **1999**, *99*, 495–541.

3. Phelps, C. L.; Smart, N. G.; Wai, C. M. *Chem. Edu.* **1996**, *12*, 1163–1168.
4. Laintz, K. E.; Wai, C. M.; Yonker, C. R.; Smith, R. D. *J. Supercrit. Fluids* **1991**, *4*, 194–198.
5. Laintz, K. E.; Wai, C. M.; Yonker, C. R.; Smith, R. D. *Anal. Chem.* **1992**, *64*, 2875–2878.
6. Wai, C. M.; Wang, S. *J. Chromatography A* **1997**, *785*, 369–383.
7. Carrott, M. J.; Waller, B. E.; Smart, N. G.; Wai, C. M. *Chem. Commun.* **1998**, 373–374.
8. Tomioka, O.; Meguro, Y.; Iso, S.; Youshida, Z.; Enokida, Y.; Yamamoto, I. *J. Nucl. Sci. Technol.* **2001**, *38*, 1097.
9. Samsonov, M. D.; Wai, C. M.; Lee, S. C.; Kulyako, Y.; Smart, N. G. *Chem. Commun.* **2001**, 1868–1869.
10. Lin, Y.; Wai, C. M. *Anal. Chem.* **1994**, *66*, 1971–1975.
11. Lin, Y.; Wai, C. M.; Jean, F. M.; Brauer, R. D. *Environ. Sci. Technol.* **1994**, *28*, 1190–1193.
12. Joung, S. N.; Kim, S. J.; Yoo, K. P. *J. Chem. Eng. Data* **1999**, *44*, 1034–1036.
13. Fox, R. V.; Mincher, B. J. In *Supercritical Carbon Dioxide Separations and Processes*; Gopalan, S., Wai, C. M., Jacobs, H. K., Eds.; ACS Symposium Series 860; American Chemical Society: Washington, DC, 2003; chapter 4, pp 36–49.
14. Lin, Y.; Smart, N. G.; Wai, C. M. *Environ. Sci. Technol.* **1995**, *29*, 2706–2708.
15. Wai, C. M.; Waller, B. *Ind. Eng. Chem. Res.* **2000**, *39*, 3837–3841.
16. Enokida, Y.; Sami, A. E.; Wai, C. M. *Ind. Eng. Chem. Res.* **2002**, *41*, 2282–2286.
17. Enokida, Y.; Tomioka, O.; Lee, S. C.; Rustenholtz, A.; Wai, C. M. *Ind. Eng. Chem. Res.* **2003**, *42*, 5037–5041.
18. Wai, C. M.; Kulyako, Y.; Yak, H. K.; Chen, X., Lee, S. *J. Chem. Commun.* **1999**, 2533–2535.
19. Wai, C. M.; Kulyako, Y. M.; Myasoedov, M. *Commun.* 1999, 180, 181.
20. Meguro, Y.; Iso, S.; Yoshida, Z.; Ougiyanagi, J.; Uehara, A.; Enokida, Y.; Yamamoto, I.; Tomioka, O.; Yamamoto, A.; Wada, R.; Yamaguchi, K. *Decontamination of Uranium Waste Based on Supercritical Carbon Dioxide Fluid Leaching Method*, Proceedings of Super Green Conference 2002, Kyung Hee University, Korea, Nov 3–6, 2002, pp 179–183.
21. Smart, N. G.; Wai, C. M.; Phelps, C. *Chem. Britain* **1998**, *34*, 34–36.

Chapter 10

Fundamental Chemistry of the Universal Extraction Process for the Simultaneous Separation of Major Radionuclides (Cesium, Strontium, Actinides, and Lanthanides) from Radioactive Wastes

R. Scott Herbst[1], Thomas A. Luther[1], Dean R. Peterman[1], Vasily A. Babain[2], Igor V. Smirnov[2], and Evgenii S. Stoyanov[3]

[1]Center for Advanced Energy Studies, Idaho National Laboratory, Idaho Falls, ID 83415
[2]Khlopin Radium Institute, St. Petersburg, Russia
[3]Institute of Catalysis, Novosibirsk, Russia

Scientists at the Idaho National Laboratory and Khlopin Radium Institute collaboratively developed and validated the concept of a Universal Extraction (UNEX) process for simultaneously removing the major radionuclides (Cs, Sr, actinides, and lanthanides) from acidic radioactive waste in a single solvent extraction process. The UNEX solvent incorporates three active extractants: chlorinated cobalt dicarbollide (CCD), polyethylene glycol (PEG), and diphenyl-N,N-di-n-butylcarbamoylmethylphosphine oxide (CMPO), dissolved in a suitable organic diluent to simultaneously extract target radionuclides. The process chemistry is unique, but complicated, since the extractants operate synergistically to extract the major radionuclides. Furthermore, interactions with the diluent are quite important as the diluent strongly influences the extraction properties of the solvent system. We are studying the fundamental chemical phenomena responsible for the selective extraction of the different species to understand the underlying mechanisms and facilitate enhancements in process chemistry. Our efforts to date have relied on a combination of classical chemistry techniques, IR spectroscopy, and NMR spectroscopy to identify and explain the structures formed in the organic phase, elucidate the operative chemical mechanisms, and evaluate the diluent effects on extraction properties.

Introduction

Collaborative efforts between the Idaho National Engineering and Environmental Laboratory (INL) and Khlopin Radium Institute (KRI) began in 1994 to evaluate solvent extraction technologies developed in Russia for their applicability to legacy acidic radioactive liquid wastes. The collaboration resulted in the successful demonstration of chlorinated cobalt dicarbollide (CCD) (Cs extraction) and CCD/polyethylene glycol (PEG) (Cs and Sr extraction) flowsheets on samples of actual INL tank waste (*1, 2*). Ongoing collaborative efforts focused on development and demonstration of a Universal Extraction (UNEX) process flowsheet to simultaneously and selectively extract all the major radionuclides (^{137}Cs, ^{90}Sr, and TRU elements) from INL wastes (*3–5*). In its current state of development, the UNEX process incorporates small quantities (0.01–0.1 M) of CCD (Figure 1), polyethylene glycol with an average molecular weight of 400 g/mol (PEG-400), and diphenyl-*N,N*-di-n-butylcarbamoylmethylphosphine oxide (CMPO) dissolved in a suitable diluent to simultaneously extract the major radionuclides. UNEX development efforts culminated with the identification of phenyltrifluoromethyl sulfone ($C_6H_5SO_2CF_3$, designated FS-13) as a suitable, polar diluent for the UNEX process solvent, eliminating the use of nitrobenzene and nitroaromatic-based diluents. The FS-13 diluent exhibits excellent radiolytic and chemical stability, is relatively innocuous, and provides necessary solubility of CCD, PEG, and CMPO and their respective metal complexes. Finally, the density and viscosity of the organic phase are suitable for processing applications.

The UNEX demonstrations on actual acidic waste samples are a significant milestone; it is the only demonstrated concept for the simultaneous removal of the major radionuclides from radioactive waste in a single step. Removing all major radionuclides in a single process represents a significant economic incentive by providing substantial capital and operating cost savings, compared to using multiple (two or three) unit operations to remove the different radionuclides. Recent engineering studies at the INL confirm that the UNEX process could provide substantial lifecycle cost savings compared to other treatment options for legacy INL radioactive wastes (*6, 7*).

During the development effort, results clearly established that the concentrations of CCD, PEG, and CMPO in the organic solvent dramatically impact the distribution coefficients of the radionuclides. For example, for a fixed CCD concentration, increasing the PEG concentration dramatically lowers the Cs distribution while increasing that for Sr; as the CMPO concentration is increased for fixed CCD and PEG concentrations, both the Cs and Sr distributions decrease. This project to elucidate the fundamental chemistry in the UNEX solvent will provide insight regarding the nature of these observed synergistic interactions. With this knowledge, it will be possible to establish methodologies for enhancing process efficiency and selectivity, define effective organic composition ranges, minimize waste volumes, quantify and minimize

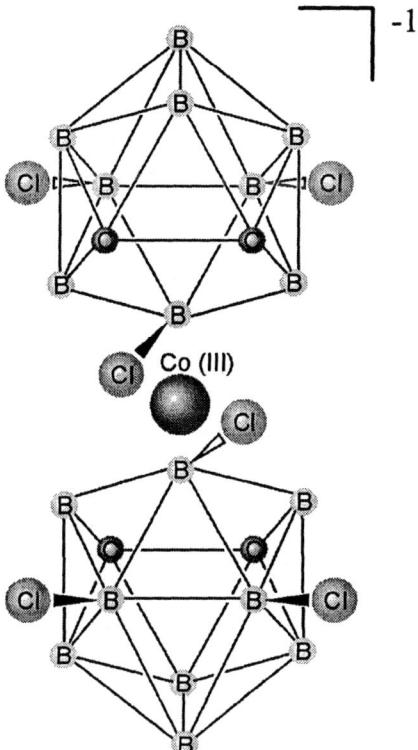

Figure 1. Hexachlorocobalt dicarbollide anion, i.e., the CCD used in the UNEX process.

solvent losses, and enhance compatibility of the process products with the final waste forms. Development of the UNEX process to its current status has progressed with minimal understanding of the complicated extraction mechanisms of the target radionuclides into the organic phase. Without detailed understanding of the fundamental physical and chemical processes occurring in the organic phase, further progress in developing the UNEX process will be difficult. This fundamental understanding is necessary to develop a highly efficient process that is suitable for use at an industrial scale over the wide range of feed compositions represented by radioactive wastes stored at the INL and elsewhere.

The methodology used in this U.S. Department of Energy (DOE) Environmental Management Science Program (EMSP) project has been to study the fundamental chemistry of the UNEX process by dividing the experimental studies into readily surmountable pieces. Consequently, the experimental work to date has focused largely on understanding the organic phase complexes and structures responsible for strontium extraction exerted by the synergistic effects of PEG-400 and CCD. At this point, a fairly complete and detailed understanding of the Sr and Cs extraction mechanism has been developed, although further experimental efforts will be directed at refining and confirming this work. Efforts in this project have very recently focused on developing an understanding of the roles of water and acid in the extraction of actinide and lanthanide elements by CMPO in polar diluents.

Results and Discussion

Cesium Extraction in CCD Systems

The selective extraction of ^{137}Cs by the CCD anion, shown in Figure 1, has been extensively studied in the United States, the Czech Republic, and Russia (8–15). Cesium extraction occurs by an uncommon liquid-liquid phase cation-exchange mechanism with complete dissociation of the solvated species in the organic phase, i.e., a neutral species is not extracted into the organic phase:

$$[M^{n+}]_{aq} + n[H^+ \cdot bH_2O]_{org} \leftrightarrow [M^{n+}]_{org} + n[H^+]_{aq} + bH_2O \qquad (1)$$

The CCD anion does not directly participate in the extraction process, but serves as the counter ion to stabilize the charge of the metal cation in the organic phase. It is of interest to note (vide infra) that the number of water molecules, b, associated with the hydrated proton in the organic phase is reported as 5.5 in the literature for CCD in nitrobenzene (16). Dipicrylamine, tetraphenylborate, polyiodide, and heteropolyacids extract cesium by the same mechanism; but, in the protonated form, only protonated chlorinated cobalt dicarbollide (HCCD) is simultaneously a strong acid and extremely hydrophobic. This combination of properties enables CCD to extract cesium from acidic media and provides low solubility of CCD in aqueous solutions. Aliphatic and aromatic nitro-compounds, such as nitrobenzene, have been the most widely studied diluents for CCD. This work has focused primarily on the use of FS-13 from a practical standpoint and 1,2-dichloroethane (DCE), from an academic standpoint, as diluents for CCD in our experiments.

For the purposes of this project, little attention has been given per se to the extraction mechanisms in systems comprised only of CCD in an appropriate diluent; this mechanism appears quite well documented and studied in the literature. However, as a by-product of preparing concentrated stock solutions of CCD in the FS-13 diluent, some noteworthy nuances of the Cs/CCD extraction mechanism have been observed experimentally. For the preparation of extractant mixtures in the appropriate diluent, a concentrated stock solution of CCD is first prepared by dissolution of the appropriate amount of CsCCD salt in the diluent. This stock solution is then converted to the protonated or acidic form of CCD, HCCD, by repeated batch contacts with equal volumes of a suitable mineral acid, typically 4 M HNO_3, 4 M H_2SO_4, or 4 M $HClO_4$. This stock solution, nominally at 0.16 M HCCD, is diluted and used in subsequent extraction experiments. It is of great practical importance that the conversion process is complete and Cs is quantitatively removed from the stock solution. Furthermore, the concentration of CCD in the organic phase is determined indirectly by potentiometric titration of the organic phase containing HCCD with standard base assuming the 1:1, H^+:CCD^- complex is formed.

The conversion of a 0.1693 M CsCCD solution in FS-13 to the HCCD form was followed by determination of organic phase acid (potentiometric titration), organic phase water (coulometric Karl Fischer titration), and ^1H NMR analysis as a function of contact number. Data from the acid and water analyses are plotted in Figure 2 for the mineral acids used in the conversion (4 M H_2SO_4 or 4 M $HClO_4$). Based on the acid concentration in the organic phase, it is apparent that multiple contacts (5 in this case) with 4 M H_2SO_4 are required to convert the Cs form of CCD to the acid form. However, the conversion is virtually complete following 1 contact with 4 M $HClO_4$. Regarding the organic phase water concentration, the data follow the same trend: the equilibrium water concentration is achieved in a single contact with 4 M $HClO_4$, multiple contacts with 4 M H_2SO_4 are required to effect the same results. It is also of interest to note that upon contacting the CCD/FS-13 solution the first time with 4 M $HClO_4$, a precipitate of insoluble $CsClO_4$ formed, providing a convenient method to remove Cs from the system. Again, this is of academic importance in that an efficient method to convert CsCCD to HCCD in solution is articulated. From a practical standpoint, the use of perchloric acid under processing conditions is likely unacceptable for a variety of reasons.

Also of interest is a comparison of the ratio of $[H_2O]/[H^+]$ in the organic phase. This ratio was determined to be 4.7 to 5.3 in the case of 4 M $HClO_4$ and 4.8 to 6.0 for 4 M H_2SO_4. This is reasonably consistent with the number of waters coordinating the hydrated proton of $b = 5.5$ (see eq 1) reported in the literature. Data on water and acid concentration in the organic phase for the system equilibrated with 4 M HNO_3 are not yet available for comparison.

Although intrinsically less precise, the proton concentration associated with the "acid" peak in the ^1H NMR taken for the above samples indicated that the number of water molecules coordinating to the acidic proton of HCCD was consistent with $b = 5$ to 6 waters of hydration. Understanding the origination of the acid peak in the NMR spectra collected during subsequent experiments with the more complicated extraction systems is also of great practical importance.

Cs and Sr Extraction in CCD/PEG-400 Systems

Mixtures incorporating PEG with CCD are of particular interest for the synergistic extraction of other components. Rais et al. (*12*) and Vanura et al.

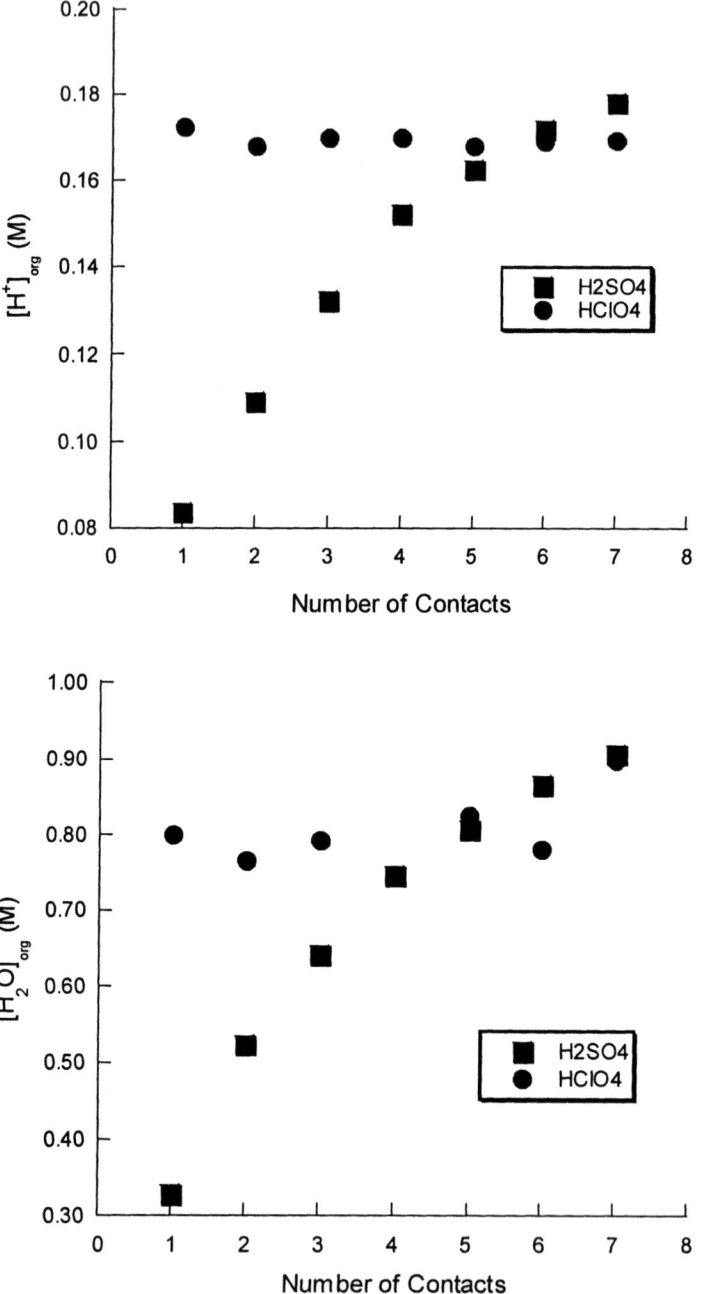

Figure 2. Concentration of acid and water in a solution of 0.1693 M CCD in FS-13 as a function of number of contacts.

(17) observed a large synergistic effect on the extraction of micro-quantities of Ca^{2+}, Sr^{2+}, Ba^{2+}, and Pb^{2+} in the presence of both PEG and CCD dissolved in nitrobenzene. It was contended that polyethylene glycol remains as a neutral molecule when associated with ionic strontium by disrupting the hydration sphere of Sr^{+2}. The result is an ionic, yet hydrophobic, species with a +2 charge. This species is transferred from the aqueous phase into the nitrobenzene phase containing the CCD anion. Prior experimental studies indicated that the charged species $PEG:Sr^{2+}$ and protonated polyethylene glycol $(PEG:H^+)$ were competing counter-ions of the CCD^- anion in the organic phase *(18)*. Furthermore, Vanura et al. *(16)* reported when nitrobenzene is used as the diluent, 1.6 H_2O molecules are co-extracted with Sr^{2+} cation. Much of our experimental efforts have focused on evaluating and confirming the nature of the organic phase complexes formed between PEG, Sr, H_2O, and CCD in either FS-13 or DCE diluents.

It is first of interest to examine the extraction behavior of Cs, Sr, Am, and Eu with CCD and PEG-400 in the FS-13 diluent. The distribution coefficients ($D \equiv [Metal]_{org} / [Metal]_{aq}$) of the radionuclides Cs, Sr, Eu, and Am are presented in Figure 3 as a function of the aqueous phase nitric acid concentration for a solution of 0.08 M CCD and 0.016 M PEG-400 in FS-13. These data were generated by equilibrium batch contacts between the organic phase and an aqueous phase of the appropriate HNO_3 concentration spiked with trace quantities of all the major radionuclides of interest (^{137}Cs, ^{85}Sr, ^{154}Eu, or ^{241}Am) at an organic-to-aqueous phase ratio of unity (O/A = 1). In all cases the organic phase was pre-equilibrated by contacting the organic three times with fresh HNO_3 of the appropriate concentration to ensure that the nitric acid was present in equilibrium quantities in the organic phase. The pre-equilibrated organic phase was then contacted with an aqueous phase containing trace amounts of all the aforementioned radionuclides and the distribution coefficients calculated from the results of gamma spectroscopy.

The resulting data are plotted in Figure 3 and indicate the extraction dependency as a function of nitric acid concentration. In all cases, the distribution coefficients are higher at the lower concentrations of nitric acid, and decrease with increasing HNO_3 concentration. At the lower acid concentrations ($[HNO_3] < 1$ M), the Cs and Sr distributions are quite high (> ~ 100), but dramatically decrease at 10 M HNO_3. This observation is consistent with the mechanism indicated in eq 1 (vide supra) where higher concentrations of H^+ drive the equilibrium to the left and decrease the concentration of metal in the organic phase. It is of interest to note that even with 10 M HNO_3, the distribution coefficients of Cs and Sr are sufficiently high to preclude the use of nitric acid to re-extract or strip these fission products from the organic phase at the process scale. The extraction behavior of Am and Eu depicted in Figure 3 indicates an interesting nuance of the CCD/PEG extraction system. At low concentrations of nitric acid, the distribution coefficients of Am and Eu are greater than or equal to unity (indicating that 50% or more is extracted to the

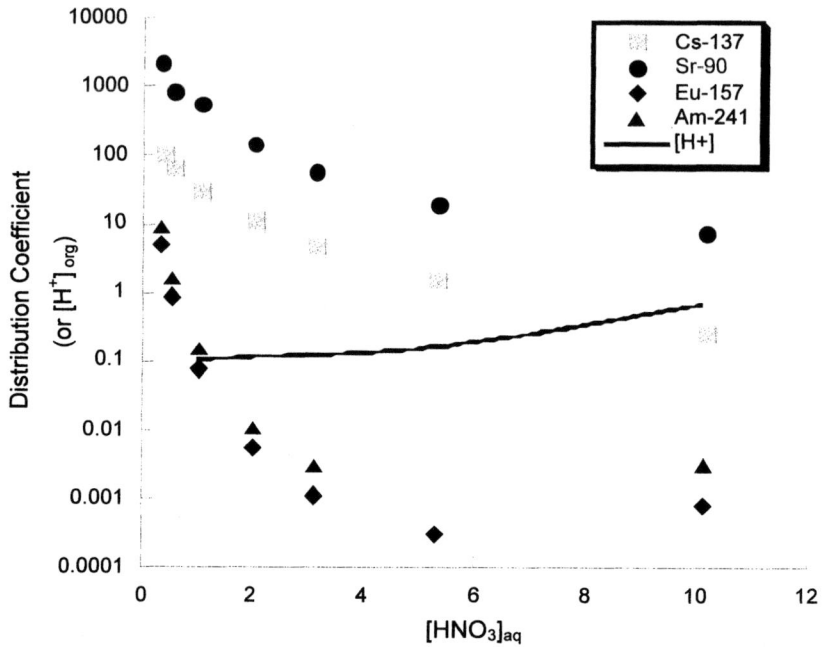

Figure 3. Extraction dependency data of several major radionuclides with varying nitric acid concentration for the system 0.08 M CCD, 0.016 M PEG-400 in phenyltrifluoromethyl sulfone (FS-13). The solid line indicates the equilibrium acid concentration in the organic phase.

organic phase at equilibrium). The nature of this extraction mechanism is unclear at this time, although the extraction of these metals in the CCD/PEG system becomes inconsequential at higher nitric acid concentrations where the distribution coefficients decrease dramatically. This indicates the necessity to include CMPO in the UNEX solvent if the actinide and lanthanide elements are to be extracted in the same process. Finally, the solid line in Figure 3 indicates the equilibrium concentration of H^+ in the organic phase as determined by potentiometric titration. These data indicate that the organic phase does not extract large amounts of acid; even with a 10 M HNO_3 aqueous phase, nitric acid is extracted to an equilibrium organic phase concentration of only ~0.9 M H^+. Note that with the trace quantities of metals used in these experiments (total metals concentration of ^{137}Cs, ^{85}Sr, ^{241}Am, and ^{154}Eu combined was less than 1.5×10^{-7} M), a very small amount of CCD and PEG are actually consumed in the organic phase by metals extraction; consequently, much of the acid in the organic phase is associated with the

CCD anion in a 1:1 molar ratio, accounting for approximately 0.08 M H^+ in the organic.

Our previous results based on the use of IR spectroscopy have indicated that the proton in the acidic form of CCD (HCCD) in DCE is hydrated in the organic phase as the $(H^+ \cdot 6H_2O)$ cation (*19*). More specifically, the IR spectroscopic data were interpreted to indicate that the species $[H_5O_2^+ \cdot 4H_2O]$ is actually formed in the organic phase. The exchange of hydrated proton $(H^+ \cdot 6H_2O)$ for strontium during the extraction process results in the transfer of Sr^{2+} from the aqueous phase to the organic phase as a hydrated species, $Sr^{2+} \cdot mH_2O$, where m = 12–15 and H_2O molecules completely fill the first coordination sphere and partially fill the second coordination sphere of Sr^{2+}. Upon addition of small quantities of polyethylene glycol (PEG, $HO(CH_2CH_2O)_n H$) to the CCD/DCE solution, the ethereal and terminal hydroxyl oxygens replace the waters of hydration in the Sr^{2+} complex, resulting in the hydrophobic (and extractable) $[Sr^{2+} \cdot PEG]$ species. The polymer chain of the PEG-400 molecule contains two terminal OH groups and the optimal number of n = 8 to 9 ethereal ($-CH_2CH_2O-$) linkages sufficient for filling the first coordination sphere of Sr^{2+} with eight atoms of oxygen. Increasing the length of the ethereal backbone, i.e., using PEG-600 (with an average of n = 13 to 14) increases the number of noncoordinating oxygens and results in a decrease in the extraction of strontium (lower equilibrium distribution coefficients). Substitution of the terminal OH groups in PEG-400 with a noncoordinating group such as -Cl, $-CH_3$, etc., also results in decreased extraction, since the terminal hydroxyl oxygens interact with Sr^{2+} cation more strongly than the ethereal oxygens. Thus, these data indicate that Sr^{2+} organizes PEG-400 into a pseudocyclic crown ether-type complex, PEG-400 contains the optimal number of coordinating oxygen atoms to form the 1:1 PEG:Sr complex, Sr^{2+} prefers PEG coordination over hydration, and PEG-400 completely fills the inner and outer coordination sphere of Sr, resulting in a very hydrophobic exterior and facilitating the extraction process. It is of interest to note that solid phase crystallography studies have reported similar 1:1 coordination by the longer chain PEG (n = 5 or 6) while 2:1 PEG:Sr^{2+} coordination was observed with the shorter chain length PEG (n = 3 or 4) (*20*).

We have since performed additional detailed investigations of systems containing HCCD, PEG-400, Sr^{2+}, or Ba^{2+} in either DCE or the more polar FS-13 diluents using predominately the techniques of IR and NMR spectroscopies. Absorption bands in the IR spectra indicating the primary regions of interest for solutions of 0.04 M HCCD and 0.02 M PEG-400 dissolved in DCE or FS-13 that have been saturated with Sr^{2+} or Ba^{2+} are listed in Table I. The absorption region around 3650 to 3500 cm^{-1} is associated with the stretching frequencies of non H-bonding OH groups of H_2O molecules and the terminal hydroxyls of PEG-400. In the anhydrous solution, the $\nu(OH)$ band for PEG is at 3555 cm^{-1} with the DCE system, while in the water-saturated solutions, this band decreases in frequency and two new, narrow bands

Table I. Comparison of the -OH Stretching Frequencies of PEG and H_2O for Sr^{2+} and Ba^{2+} Saturated Solutions of 0.04 M HCCD and 0.02 M PEG-400 in Either DCE or FS-13 for the Anhydrous "Dry" and Water-Saturated "Wet" Organic Phase

Metal	ν(OH) of PEG, "dry" extracts		ν(OH) of H_2O, "wet" extracts			
			ν_{as}		ν_s	
	DCE	FS-13	DCE	FS-13	DCE	FS-13
Sr	3555	3539	3639	3627	3564	3556
Ba	3555	3546	3639	3627	3564	3556

(ν_{as} = 3639 cm^{-1} and ν_s = 3564 cm^{-1}) of H_2O molecules bonding through the oxygen atom to the H atom in the terminal -OH group of PEG-400 are observed. The IR spectra of the wet and dry solutions are virtually identical in the frequency range of C-O-C and CO-H stretching at 1100 to 1000 cm^{-1}, further supporting the absence of H_2O molecules directly bonded with Sr^{2+} or Ba^{2+}. Consequently, additional co-extracted water molecules are held in the outer coordination sphere of the M^{2+} complex through the terminal -OH groups of PEG-400. The IR spectra of the extracted species for Sr^{2+} or Ba^{2+} in either DCE or FS-13 diluents (see Table I) indicate that in all cases the structure and composition of the Sr^{2+} and Ba^{2+} complexes are the same: the oxygen atoms from the two terminal -OH groups and six of the ethereal oxygens in the -CH_2CH_2O- groups of PEG-400 fill the first coordination sphere of the Sr^{2+} and Ba^{2+} cation. Additionally, no more than two H_2O molecules can form outer-sphere complexes by coordinating with the terminal -OH groups of PEG-400. The co-extraction of 1.6 water molecules with Sr^{2+} reported by Vanura et al. (16) is in agreement with the results of present work, which indicates that up to two water molecules can coordinate in the outer coordination sphere and co-extract with the [Sr^{2+}·PEG] and [Ba^{2+}·PEG] complexes.

The ^{13}C NMR spectra of the [Sr^{2+}·PEG] CCD_2^- and [Ba^{2+}·PEG] CCD_2^- complexes, obtained by saturating a 0.04 M HCCD + 0.02 M PEG-400 solution with strontium or barium, are in agreement with the pseudocyclic crown ether-type model. The spectra of these complexes are similar and practically coincidental for both DCE and FS-13 solvents, as indicted in Figure 4 and Table II. The chemical shifts of the signals from the C1 and C2 atoms of the terminal -CH_2CH_2OH groups and the C3 atoms of the bridging methylene groups decrease as compared with the free (uncoordinated) PEG molecule. Consequently, all of the oxygen atoms in PEG-400, including those in the terminal -OH groups, are bounded with Sr^{2+} and Ba^{2+}. The multiplicity of the C3 signal and the splitting of the C1 and C2 signals in the NMR spectra of

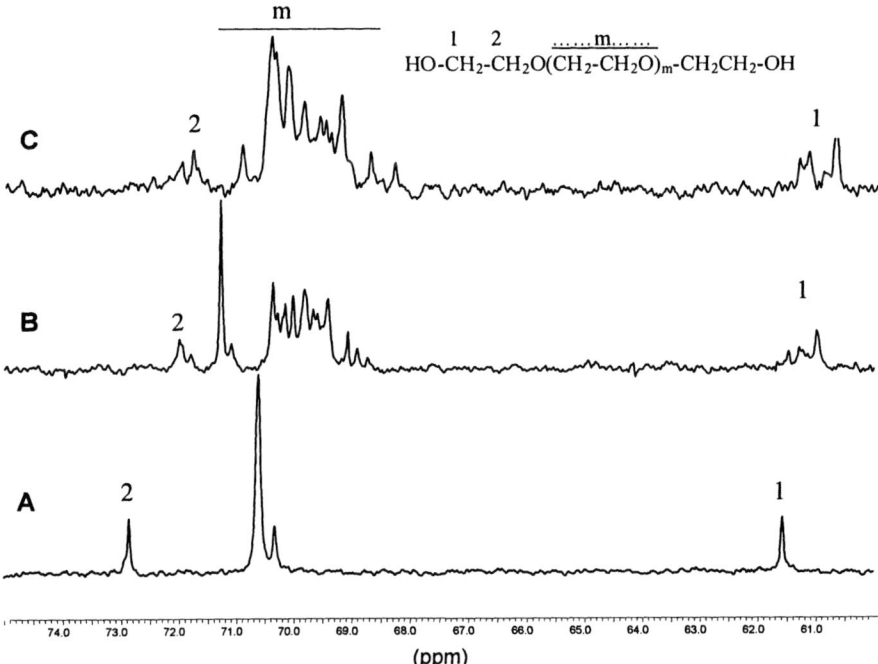

Figure 4. ^{13}C NMR spectra of a 0.02 M PEG-400 solution in DCE (A) and the $[Sr^{2+} \cdot PEG] \cdot 2CCD^-$ in the Sr^{2+} saturated solution of 0.02 M PEG + 0.04 M HCCD in DCE (B) or FS-13 (C).

$[Sr^{2+} \cdot PEG]$ and $[Ba^{2+} \cdot PEG]$ complexes are indicative of nonequivalent carbon atoms (due to coordinating oxygen atoms) owing to different conformational states of the PEG-400 molecules chelating the metal cation. The fact that the ^{13}C NMR spectra of the complexes in wet and dry solutions are virtually the same (refer to Table II) provides additional evidence that water molecules are not incorporated into the first coordination sphere of the Sr^{2+} and Ba^{2+} cations.

Perchloric Acid Complexation with CMPO in Polar Diluent Systems

In the organic phase, CMPO interacts with $HClO_4$, which is confirmed by signal displacement to lower fields in the ^{31}P NMR spectra in either DCE or FS-13. The ^{31}P NMR chemical shift and ^1H NMR chemical shift and intensity of the "acidic" proton resonance data as a function of $HClO_4$ concentration are shown in Table III. The ^{31}P NMR data exhibit a fairly linear downfield shift of

Table II. ^{13}C NMR Spectra of PEG-400 Complexes in Anhydrous (Dry) and Water-Saturated (Wet) Organic Solutions of 0.02 M PEG + 0.04 M HCCD in DCE or FS-13

Complex	Solution	Water Content	C2	C3	C1
PEG-400	DCE	dry	73.45	71.17	61.88
[H$_5$O$_2^+$·PEG]	0.04 M in FS-13	wet	71.64; 71.46	70.91–69.41*	61.62; 61.37
[H$_5$O$_2^+$·PEG]	0.04 M in DCE	wet	71.69; 71.48	70.55–69.42*	61.71
[Sr^{2+}·PEG]	0.02 M in DCE	wet	71.65*	69.99–68.84*	61.42–60.73*
[Sr^{2+}·PEG]	0.02 M in DCE	dry	71.5-71.19*	70.06–68.32*	61.09–60.91*
[Ba^{2+}·PEG]	0.02 M in DCE	wet	72.24-71.37*	70.21–68.97*	61.71; <u>61.33</u>
[Ba^{2+}·PEG]	0.02 M in DCE	dry	72.1-71.13*	70.69–69.23*	62.04; <u>61.75</u>
[Sr^{2+}·PEG]	0.02 M in FS-13	wet	71.6; <u>71.4</u>	70.09–68.85*	61.06; 60.57
[Sr^{2+}·PEG]	0.02 M in FS-13	dry	71.5; <u>71.28</u>	70.14–68.66*	61.18–60.84*

Note: Multiple signals; the strongest signals are underlined.

Table III. ^{31}P and ^{1}H NMR Chemical Shift (ppm) and Intensity Data of HClO$_4$ Contacted with CMPO in DCE or FS-13

	^{31}P δ	^{1}H δ	Intensity
11M HClO$_4$/DCE	44.6	11.6	4.0
8M HClO$_4$/DCE	38.6	10.7	2.7
6M HClO$_4$/DCE	38.9	9.1	2.6
"wet" DCE	28.3	2.4	2.1
11M HClO$_4$/FS-13	44.0	11.1	5.5
8M HClO$_4$/FS-13	38.0	10.5	3.2
6M HClO$_4$/FS-13	38.2	9.3	3.1
"wet" FS-13	28.4	3.1	2.1

the resonance, indicating a decrease in the electron density as the phosphoryl oxygen is protonated. However, the ^{1}H NMR data of the "acidic" proton resonance does not exhibit such a linear change with the increase in

HClO$_4$ concentration; instead, the change in the chemical shift of the resonance occurs rapidly with the initial rise in the acid concentration, but at higher concentrations the change in chemical shift is more gradual. The intensity of this resonance shows the opposite behavior; the intensity increases gradually with the initial rise in the acid concentration, and then it begins to increase rapidly at the higher acid concentrations.

The ^{13}C NMR data in Table IV show the chemical shift changes expected for the protonation of the phosphoryl oxygen and the carbonyl. An unexpected observation is the downfield chemical shift for the C1 resonances. This behavior was not observed previously with the HNO$_3$/CMPO system. Not only are the C1 resonances shifting downfield, the difference between the two resonances is decreasing. These data indicate that the HClO$_4$ is also protonating the nitrogen, resulting in a decrease of the diastereotopic nature of the butyl groups.

Conclusions and Future Directions

The research efforts to date have refined and broadened our understanding of the basic extraction mechanism operative in the CCD/PEG system and verified that the complexes formed in the organic phase are consistent regardless of the choice of DCE or FS-13 as the diluent. The large, hydrophobic CCD anion is dissociated (a strong acid) in the organic phase and serves as the counter-ion to stabilize the extracted cationic metal complexes. The PEG-400 molecule dehydrates the +2 metal cation by displacing waters of hydration in the inner coordination sphere of the metal through the ethereal and hydroxyl oxygens. Thus, a hydrophobic, divalent metal complex is formed with PEG-400 in a 1:1 molar ratio that is present in the organic phase. The chain length of the PEG-400 molecule is of the optimal length to form a pseudocyclic crown ether-type complex, and contains the optimal number of coordinating oxygen atoms to form the 1:1 PEG:Sr complex; Sr^{2+} prefers PEG coordination over hydration, and PEG-400 completely fills the inner and outer coordination spheres of Sr, resulting in a very hydrophobic exterior, facilitating the extraction process. Further investigations are planned to confirm this behavior.

Preparations are in progress to use Extended X-ray Absorption Fine Structure (EXAFS) to examine and confirm microscopic details such as metal coordination numbers and bond distances. Additionally, provisions are being made to obtain samples of mono-disperse PEG compounds (commercially available PEG-400 is polydisperse with the (-CH$_2$-CH$_2$-O-)$_n$ chain lengths varying from n = 6 to 12 and an average molecular weight of 400 amu) to further evaluate the stoichiometry of the PEG:Sr^{2+} complex. For lower molecular weight PEG complexes (n $\leq$ 4), two molecules of PEG would theoretically be necessary to coordinate Sr^{2+} in the proposed extraction mechanism. For chain lengths of n ~ 5 or longer, one molecule of PEG would theoretically be sufficient to complex the Sr^{2+} cation in the extraction process.

Table IV. ^{13}C NMR Chemical Shift Data (ppm) of HClO$_4$ Contacted with CMPO in DCE or FS-13

	C6 δ	C7 δ (δΔ)	C5 δ (δΔ)	C1 δ (δΔ)
11M HClO$_4$/DCE	167.5	124.1, 122.6 (1.5)	33.8, 33.1 (0.7)	51.9, 50.5 (1.4)
8M HClO$_4$/DCE	168.0	127.1, 125.7 (1.4)	34.1, 33.3 (0.8)	50.7, 49.0 (1.7)
6M HClO$_4$/DCE	166.7	128.1, 126.7 (1.4)	34.9, 34.0 (0.9)	49.9, 47.8 (2.1)
"wet" DCE	164.4	133.9, 132.6 (1.3)	38.0, 37.2 (0.8)	48.6, 45.9 (2.7)
11M HClO$_4$/FS-13	168.0	a	33.8, 33.0 (0.8)	51.8, 50.3 (1.5)
8M HClO$_4$/FS-13	168.4	a	34.0, 33.2 (0.8)	50.7, 49.1 (1.6)
6M HClO$_4$/FS-13	167.4	a	34.6, 33.8 (0.8)	50.1, 48.0 (2.1)
"wet" FS-13	164.9	a	38.3, 37.5 (0.8)	48.9, 46.0 (2.9)

[a] Chemical shift data not recorded due to interfering signals of FS-13.

We have only recently undertaken a detailed study of the extraction mechanism associated with actinide and rare earth elements in the presence of CMPO. We will continue to elucidate the mechanism and structures of the metal complexes formed in the organic phase using the NMR, IR, and wet chemistry techniques that were useful in studying the CCD/PEG systems. Ultimately, we anticipate additional confirmation and further understanding of the structures and mechanism though EXAFS experiments on extraction systems containing CMPO.

Acknowledgments

This work was carried out under financial support of the DOE EMSP project 81995 and through contract DE-AC07-99ID13727 and the U.S. Civilian Research & Development Foundation (CRDF Project RC2-2342-ST-02).

References

1. Law, J. D.; Herbst, R. S.; Todd, T. A.; Brewer, K. N.; Romanovskiy, V. N.; Esimantovskiy, V. M.; Smirnov, I. V.; Babain, V. A.; Zaitsev, B. N. *INEL-96/0192;* Idaho National Engineering Laboratory: Idaho Falls, ID, 1996.
2. Law, J. D.; Brewer, K. N.; Todd, T. A.; Wade, E. L.; Romanovskiy, V. N.; Esimantovskiy, V. M.; Smirnov, I. V.; Babain, V. A.; Zaitsev, B. N. *INEL/EXT-97-00064*, Idaho National Engineering Laboratory: Idaho Falls, ID, 1997.
3. Herbst R. S.; Law, J. D.; Todd, T. A.; Romanovskiy, V. N.; Babain, V. A.; Esimantovskiy, V. M.; Zaitsev, B. N.; Smirnov, I. V. *Sep. Sci. Technol.* **2002**, *37* (8), 1807–1831.
4. Romanovskiy, V. N.; Smirnov, I. V.; Babain, V. A.; Todd, T. A.; Law, J. D.; Herbst, R. S.; Brewer, K. N. *Solvent Extr. Ion Exch.* **2001**, *19* (1), 1–21.
5. Law, J. D.; Herbst, R. S.; Todd, T. A.; Romanovskiy, V. N.; Babain, V. A.; Esimantovskiy, V. M.; Smirnov, I. V.; Zaitsev, B. N. *Solvent Extr. Ion Exch.* **2001**, *19* (1), 23–36.
6. Banaee, J.; Barnes, C. M.; Battisti, T.; Hermann, S.; Losinski, S. J.; McBride, S. *INEEL/EXT-2000-01209*, Idaho National Engineering and Environmental Laboratory: Idaho Falls, ID, 2000.
7. Dietz, M. L.; Horwitz, E. P. In *Science and Technology for Disposal of Radioactive Tank Waste*; Schulz, W. W., Lombardo, W. W., Eds.; Plenum Publishing Co.: New York, 1998; pp 231–243.
8. Todd, T. A.; Brewer, K. N.; Herbst, R. S.; Tranter, T. J.; Romanovskii, V. N.; Lazarev, L. N.; Zaitsev, B. N.; Estimantovskii, V. M.; Smirnov, I. V. In *Value Adding Through Solvent Extraction*; Shallcross, D. C., Paimin, R., Prvcic, L. M., Eds.; University of Melbourne Press: Melbourne, 1996; pp 1313–1318.
9. Brewer, K. N.; Herbst, R. S.; Olson, A. L.; Todd, T. A.; Tranter, T. J.; Romanovskiy, V. N.; Lazarev, L. N.; Zaitsev, B. N.; Esimantovskiy, V. M.; Smirnov, I. V. *WINCO-1230;* Idaho National Engineering Laboratory, Westinghouse Nuclear Idaho Co., Inc.: Idaho Falls, ID, 1994.
10. Law, J. D.; Herbst, R. S.; Todd, T. A.; Brewer, K. N.; Romanovskiy, V. N.; Esimantovskiy, V. M.; Smirnov, I. V.; Dzekun, E. G.; Babain, V. A.; Zaitsev, B. N. *INEL-95/0500*; Idaho National Engineering Laboratory: Idaho Falls, ID, 1995.
11. Rais, J.; Selucky, P.; Kyrs, M. *J. Inorg. Nucl. Chem.* **1976**, *38*, 1376.
12. Rais, J.; Sebestova, E.; Selucky, P.; Kyrs, M. *J. Inorg. Nucl. Chem.* **1976**, *38*, 1742.
13. Makrlík, E.; Vanura, P. *J. Radioanal. Nucl. Chem., Letters* **1985**, *96*, 381.
14. Makrlík, E.; Vanura, P. *J. Radioanal. Nucl. Chem., Letters* **1985**, *96*, 451–456.
15. Esimantovskii, V. M.; Galkin, B. Y.; Dzekun, E. G.; Lazarev, L. N.; Ljubtsev, R. I.; Romanovskii, V. N.; Shishkin, D. N. In *Proceedings of the Symposium of Waste Management*, Tucson, AZ, **1992;** p 22.
16. Vanura, P.; Makrlik, E.; Rais, I.; Kyrs, M. *Collect. Czech. Chem. Commun.* **1982**, *5* (47), 1444–1464.
17. Vanura, P.; Rais, J.; Selucky, P.; Kyrs, M. *Collect. Czech. Chem. Commun.* **1979**, *44* (1), 157.
18. Selucky, P.; Vanura, P.; Rais, J.; Kyrs, M. *Radiochemical Radioanalysis Letters* **1979**, *38*, 397.
19. Smirnov, I. V.; Stoyanov, E. S.; Vorob'eva, T. P. *Czech. Journal of Physics*, **2003**, *53*, A501–A508.
20. Rodgers, R. D.; Jezl, M. L.; Bauer, C. B. *Inorg. Chem.* **1994**, *33*, 5682–5692.

Chapter 11

Dynamics of Switch-Binding by a Linear Ligand That Transforms to a Macrocycle upon Chelation to a Metal Ion: Synthesis, Kinetics, and Equilibria

Mansour M. Hassan[1], Chi Zhang[2], Jong-ill Lee[2], K. Mani Bushan[2], Anne McCasland[2], Richard S. Givens[2], and Daryle H. Busch[2]

[1]Department of Chemistry, Faculty of Higher Education, University of Aden, Aden, Yemen
[2]Department of Chemistry, University of Kansas, 1251 Wescoe Hall Drive, 2010 Malott Hall, Lawrence, KS 66045–7582

It is proposed that for chemical processes in which equilibration is too slow to achieve certain goals, switch-binding and switch-release reactions may be used to accelerate complex formation. Whereas photo triggers are well known for switch-releasing, switch binding is first described here for notably slow complex formation reactions between metal ions and macrocyclic ligands. A linear tetradentate ligand, $L^{L/C}$, was designed with functional groups that will react with each other (primary amine, carbonyl) at its extremities. Upon complexation the metal ion is predicted to cause reaction between these functional groups, producing a macrocyclic ligand that encircles the metal ion. Equilibrium studies were made with the metal ions Cu^{2+}, Hg^{2+}, Ni^{2+}, and Zn^{2+} both with the switch-binding ligand, $L^{L/C}$, and with a ligand, L^L, that is very similar except it cannot undergo the switching process. The formation constants were very similar for the two ligands, with the metal ion affinities decreasing in the order $Cu^{2+} \gg Ni^{2+} \approx Zn^{2+} > Hg^{2+}$. The kinetics of reaction were studied for formation of the nickel(II) complexes of both ligands.

Results gave a first indication that nickel very rapidly chelates to the switch-binding ligand before forming the macrocycle. The nickel(II) reacts with $L^{L/C}$ in a complicated rapid set of processes on the fractional second time scale, followed by a process in the hour time regime. In contrast, under the same conditions, only a single rapid rate process was observed for the reaction of nickel(II) with L^L. The specific rate constants for L^L are greater than those for $L^{L/C}$. Detailed analysis, strongly augmented by mass spectrometric studies, proved that the rapid process produces a nickel complex of the linear ligand and that intramolecular ring closure takes place relatively slowly and under the control of the metal ion. This proof of concept for switch binding, using the template effect, completes the model for replacing equilibration for the reversible formation of metal complexes with a switch-binding and release process.

Introduction

In principle the most powerful known ligands can capture metal ions in the most competitive of circumstances, for example, from mineralized sites, from lesser ligands, and even from extremely dilute solutions; i.e., under circumstances where ordinary ligands, such as those used in well known separations technologies, are completely ineffective. Clearly there are compelling incentives for finding ways to apply these tight-binding ligands to the management of the metallic elements under many conditions, but major hurdles must be overcome.

The applications of such tight-binding ligands have been limited by the slow rates at which their equilibria are established. It is an experimental fact that the equilibrium constants for the binding of any kind of receptee (e.g., a metal ion) to its complementary receptor (i.e., ligand) commonly vary monotonically with the rates at which the receptee is liberated from the receptee/receptor complex (1–3). Consequently, ultra tight-binding ligands, whose equilibrium constants for binding exceed ordinary values by factors of millions or billions, will release their complement at least that many millions or billions of times slower. Most often, the slowness is even more lethargic since the rate of binding is also retarded and $K_{equil} = k_{binding}/k_{release}$. In order to make best use of tight-binding receptors it is, therefore, necessary to either accommodate any specific methodology to these slow kinetic processes or to find means of accelerating the formation and dissociation rates associated with complexation.

Here we present the concept of replacing slow rates of complex formation by a chemical switching process that we will refer to as *switch-binding*. The combined chemical processes of ligand binding and release constitute metal ion/ligand equilibration, and complete replacement of that natural process with chemical switching requires both switch-binding and switch-release. Switch-release is broadly used to generate immediate sizeable infusions of biological substrates, including metal ions, in life science studies (*4–6*). Switch-release is the subject of other work in these laboratories. The concept of switch-binding and s witch-release i s in s tep w ith the g oals o f s cience t o m ove beyond what nature gives to us spontaneously.

Here switch-binding is accomplished by a change in the structure of the ligand in accompaniment to its binding to the metal. The structural change is from a rapidly reacting, more weakly binding ligand structure in the free state to a more slowly reacting, more strongly binding ligand structure in the target metal complex. *Complementarity* and *ligand constraints* determine the stabilities of metal complexes (*7–10*). Complementarity, a necessary but not sufficient condition for maximum affinity, implies a consonance between metal ion and ligand in bond type (often including charge), geometry, and size. Given equal complementarity, the constraints built into the structure of the ligand determine just how s trong its c omplexes w ill be. For the example presented here, the topological constraint built into the ligand changes from that of a linear tetradentate ligand to that for a macrocycle Scheme 1. Given equal complementarity, the stabilities of complexes increase dramatically with increasing topological constraint: (simple ligand < chelate < macrocycle < cryptand). Further, as described above, the rates of their binding to and dissociation from metal ions decrease as complex stability increases.

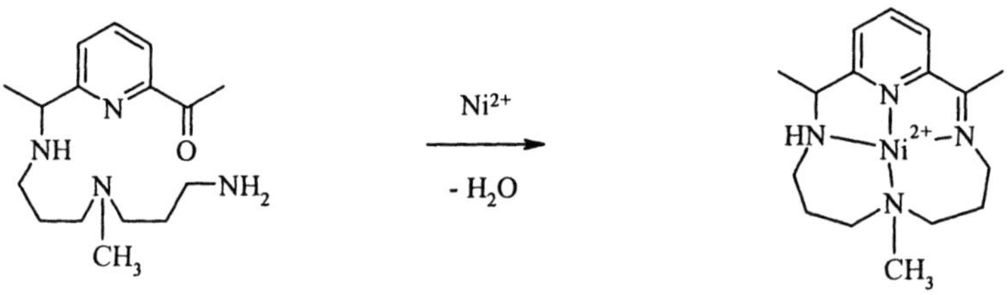

$L^{L/C}$ $\quad\quad\quad\quad\quad\quad\quad\quad [Ni(L^C)]^{2+}$

Scheme 1. Switch-Binding Process in which a Rapidly Reacting Linear Chelate becomes an Inert Macrocyclic Ligand Coordinated to a Nickel(II) Ion.

The switch-binding ligand investigated in this work, $L^{L/C}$, was designed to have complementary functional groups at its ends, primary amino and ketone groups, which react to close the macrocycle, L^C, about the metal ion. This Schiff base condensation, produces a macrocyclic ligand that is known to be nicely complementary to the nickel(II) ion. The square planar complex, $[Ni(L^C)]^{2+}$, has been prepared by a more traditional route and characterized (*11*). Because the expected product of the complete switch-binding process has been characterized, this is a good system to test the switch-binding concept. For comparison, a very similar ligand (L^L, I) that is not capable of undergoing cyclization was also synthesized, and details of the syntheses are reported. Detailed kinetic and mechanistic information has been obtained for the complexation reactions of both ligands, $L^{L/C}$ and L^L, with nickel(II). The work also includes the determination of the protonation and complex formation constants for the two ligands.

I.

Experimental

Synthesis of $L^{L/C}$ and L^L

{3-[(3-Amino-propyl)-methylamino]-propyl}-carbamic Acid Benzyl Ester (*12*)

3,3'-Diamino-N-methyldipropylamine (14.53 g, 0.10 mol) is dissolved in water (25 ml) containing bromocresol green (0.01%) as the indicator. Methanesulfonic acid (~ 19.2 g, 0.20 mol) in water (15 ml) is added slowly until the blue to yellow color transition is just achieved. The mixture is then diluted with ethanol (70 mL) and vigorously stirred at room temperature, while a solution of benzyl chloroformate (15.2 g, 0.089 mol) in dimethoxyethane (25 mL) and 25% w/v aqueous potassium acetate (~ 60 mL) is added dropwise simultaneously at rates which maintain the correct reaction pH (yellow-green indicator coloration). After the additions are complete, the mixture is stirred for an additional 4 h at room temperature. The volatiles are then removed under vacuum, and the residue is extracted with water (250 ml) and filtered to remove small quantities of the *bis*-derivative that is formed as a byproduct. The filtrate is washed with methylene chloride (3 × 100 ml), made basic with excess 40% aqueous NaOH solution, and extracted with methylene chloride (2 × 150 ml). The organic layer is washed once with saturated aqueous sodium chloride (100 ml) and dried over magnesium sulfate. The solution is filtered to remove the drying agent, and the solvent is removed by rotary evaporation to yield a yellow oil which is dried under vacuum overnight. Yield: 8.6 g (35%). ^{1}H NMR (CDCl$_3$, 400 MHz): δ 7.37(m, 5H), 5.96(s, 1H), 5.09(s, 2H), 3.27(m, 2H), 2.73(t, 2H), 2.39(m, 4H), 2.18(s, 3H), 1.63(m, 6H) ppm.

[3-({3-[1-(6-Acetyl-pyridin-2-yl)-ethylideneamino]-propyl}-methylamino)-propyl]-carbamic Acid Ester (13)

A solution of {3-[(3-amino-propyl)-methylamino]-propyl}-carbamic acid benzyl ester (2.79 g, 10 mmol) in 30 ml of dry benzene is added to a solution of 2,6-diacetylpyridine (DAP) (1.64 g, 10 mmol) in dry benzene (20 ml). The reaction mixture is heated to reflux overnight under N$_2$ with stirring. During the heating at reflux, the solution gradually becomes bright orange; with cooling and concentration of the benzene layer under vacuum, a light orange oil is obtained. Yield: 1.96 g (46%). ^{1}H NMR (CDCl$_3$, 400 MHz): δ 8.21(m, 1H), 8.03(m, 1H), 7.83(t, 1H), 7.37(m, 5H), 5.94(s, 1H), 5.08(s, 2H), 3.57(t, 2H), 3.30(d, 2H), 2.80(d, 3H), 2.40(m, 4H), 2.26(s, 3H), 1.95(m, 2H), 1.69(d, 2H), 1.60(s, 3H) ppm.

[3-({3-[1-(6-Acetyl-pyridin-2-yl)-ethylamino]-propyl}-methylamino)-propyl]-carbamic Acid Benzyl Ester (14, 15)

[3- ({3- [1- (6-Acetyl-pyridin-2-yl) -ethylamino] -propyl} -methylamino)-propyl]-carbamic acid benzyl ester (1.50 g, 3.54 mmol) is dissolved in 1,2-dichloroethane (20 ml), and then sodium triacetoxyborohydride (1.05 g, 4.95 mmol) is added slowly under N$_2$ atmosphere. Upon completing the addition of the solid, the reaction mixture is stirred for 24 h at room temperature. The reaction is quenched by adding an excess of 40% aqueous NaOH (30 ml), followed by extraction with methylene chloride (3 × 50 ml), washed with brine solution (100 ml), and dried over magnesium sulfate. The inorganic salts are filtered, and the filtrate evaporated under vacuum. The resulting yellowish-orange oil is dried on the vacuum line overnight. Yield: 1.05 g (70%). The crude product is purified by flash column chromatography (hexane/EtOAc = 1/3, then EtOAc/MeOH = 1/1) to give the purified product as a yellow oil. ^{1}H NMR (CDCl$_3$, 400 MHz): 7.85(d, 1H), 7.70(s, 1H), 7.40(d, 1H), 7.31(m, 5H), 5.99(s, 1H), 5.08(s, 2H), 3.85(m, 1H), 3.22(m, 2H), 2.70(s, 3H), 2.56(m, 1H), 2.41(m, 1H), 2.36(m, 4H), 2.15(s, 3H), 2.05(s, 1H), 1.64(m, 4H), 1.36(d, 3H) ppm. ^{13}C NMR (CDCl$_3$, 100 MHz): 200.97, 164.65, 156.88, 153.50, 137.61, 137.30, 128.87, 128.70, 128.43, 128.38, 125.05, 120.07, 66.75, 59.47, 56.73, 56.34, 46.73, 42.47, 42.17, 41.32, 40.72, 28.25, 27.51, 26.91, 26.16, 23.20 ppm. Major infrared absorbances (cm^{-1}): 3333(N-H stretch), 1398(C=O), 1530(N-H bend), 1452, 1357(-CH$_3$ bend). FAB-MS: 427.2.

1-[6-(1-{3-[(3-Aminopropyl)-methylamino]-propylamino}-ethyl)-pyridin-2-yl]-ethanone, $L^{L/C}$ (16)

A saturated solution (15 ml) of dry hydrogen bromide in acetic acid (30%) is added to cleave the N-carbenzoxy protecting group from [3-({3-[1-(6-acetylpyridin-2-yl)-ethylamino]-propyl}-methylamino)-propyl]-carbamic acid benzyl ester (0.60 g, 1.42 mmol), and stirred in a three-necked, 50-mL round bottomed flask. Evolution of carbon dioxide is evident upon acidification, indicative of effective decarboxylation. The resulting deep orange-red solution is allowed to stir for 2 h at room temperature. Dry ethyl ether (120 ml) is subsequently added directly to the flask and the yellow-white hydrobromide salt separates from the solution. Due to the extremely hygroscopic nature of the product, a syrupy material forms upon standing during vacuum filtration. The crude product is recrystallized by precipitating the solid by addition of ether to a solution of a minimal amount of dry ethanol. The purified product is dried under vacuum for 2 d to give a white-yellow solid. Yield: 0.485 g (59%). ^{1}H NMR (CDCl$_3$, 400 MHz): δ 8.08(m, 2H), 7.70(t, 1H), 4.73(m, 1H), 3.25(m, 6H), 3.09(t, 4H), 2.91(m, 4H), 2.76(s, 3H), 2.12(m, 4H), 1.95(m, 1H), 1.66(d, 3H) ppm. ^{13}C NMR (D$_2$O, 100 MHz): 203.35, 165.09, 155.26, 139.87, 126.87, 123.13, 58.27, 53.51, 53.37, 42.85, 39.95, 39.87, 36.71, 26.03, 22.30, 21.50 ppm. FAB-MS: 293.2. HRMS: 293.2333 (calc.: 293.2341).

[3-({3-[1-(pyridin-2-yl)-ethylideneamino]-propyl}-methylamino)-propyl]-carbamic Acid Benzyl Ester

A solution of {3-[(3-aminopropyl)-methylamino]-propyl}-carbamic acid benzyl ester (4.74 g, 17 mmol) in 30 ml of dry benzene is added to a solution of

2-acetylpyridine (2.08 g, 17 mmol) in dry benzene (30 ml). The reaction mixture is heated to reflux overnight under N_2 with stirring. During this time the solution gradually becomes bright orange. Heating is stopped when no more water forms. Upon cooling and concentrating the benzene layer under vacuum, a light orange oil is obtained. Yield: 5.67 g (87%). ^{1}H NMR (CDCl$_3$, 400 MHz): δ 8.59(d, 1H), 8.05(d, 1H), 7.69(t, 1H), 7.36(m, 5H), 7.30(t, 1H), 5.93(s, 1H), 5.08(s, 2H), 3.54(t, 2H), 3.30(t, 2H), 2.50(m, 4H), 2.33(s, 3H), 2.26(s, 3H), 1.95(m, 2H), 1.70(m, 2H). ^{13}C NMR (CDCl$_3$, 100 MHz): 166.70, 157.79, 156.49, 149.00, 148.25, 136.86, 136.29, 128.45, 128.02, 127.11, 124.02, 121.67, 120.82, 66.37, 55.96, 50.31, 42.07, 40.79, 28.41, 26.46, 25.83, 14.03 ppm. FAB-MS: 383.2.

[3-({3-[1-(pyridin-2-yl)-ethylamino]-propyl}-methylamino)-propyl]-carbamic Acid Benzyl Ester

[3- ({3- [1- (pyridin-2-yl) -ethylamino] -propyl} –methylamino) -propyl]-carbamic acid benzyl ester (5.67 g, 14.83 mmol) is dissolved in 1,2-dichloroethane (40 ml), and then sodium triacetoxyborohydride (4.63 g, 20.76 mmol) is added slowly under N_2 atmosphere. Upon completing the addition of the solid, the reaction mixture is stirred for 24 h at room temperature. The reaction is quenched by addition of an excess of 40% aqueous NaOH (50 ml), followed by extraction with methylene chloride (3 × 50 ml), washed with saturated aqueous sodium chloride (150 ml), and dried over magnesium sulfate. The inorganic salts are filtered, and the filtrate evaporated under vacuum. The resulting yellowish-orange oil is dried on the vacuum line overnight. The crude product is purified by flash column chromatography (hexane/EtOAc = 1/3, then EtOAc/MeOH = 1/1) to give the purified product (4.67 g, Yield: 82%). ^{1}H NMR (CDCl$_3$, 400 MHz): 8.49(d, 1H), 7.54(t, 1H), 7.28(m, 5H), 7.24(d, 1H), 7.06(t, 1H), 6.31(s, 1H), 5.03(s, 2H), 4.48(s, 1H), 3.82(m, 1H), 3.18(m, 2H), 2.59(m, 1H), 2.32(m, 5H), 2.12(s, 3H), 1.62(m, 4H), 1.34(d, 3H) ppm. ^{13}C NMR (CDCl$_3$, 100 MHz): 163.13, 156.42, 149.15, 136.73, 136.41, 128.20, 127.78, 127.68, 121.88, 121.23, 66.06, 58.91, 56.59, 55.16, 46.32, 41.60, 39.55, 26.41, 26.33, 23.75, 22.09 ppm. FAB-MS: 385.1. HRMS: 385.2602 (calc.: 385.2604).

3-({3-[1-(Pyridin-2-yl)-ethylamino]-propyl}-N-methyl)-1-aminopropane, L^L

A saturated solution (20 ml) of dry hydrogen bromide in acetic acid (30%) is added to cleave the N-carbobenzoxy protecting group from [3-({3-[1-(pyridin-2-yl)-ethylamino]-propyl}-methylamino)-propyl]-carbamic acid benzyl ester

(2.43 g, 6.31 mmol), stirred in a 100-mL round bottomed flask. Evolution of carbon dioxide is evident upon acidification, indicative of effective decarboxylation. The deep orange-red solution is allowed to stir for 2 h at room temperature. Dry ethyl ether (80 ml) is subsequently added directly to the flask, and the yellow-white hydrobromide salt separates from the solution. Due to the extremely hygroscopic nature of the product, a syrupy material forms upon standing during vacuum filtration. The crude product is recrystallized by precipitating the solid from a minimal amount of dry ethanol with the addition of ether. The purified product is dried under vacuum for 2 d to give a white-yellow solid. Yield: 2.89 g, (86%). ^{1}H NMR (CDCl$_3$, 400 MHz): δ 8.70(d, 1H), 8.17(t, 1H), 7.75(t, 1H), 7.68(d, 1H), 4.72(m, 1H), 3.26(m, 6H), 3.09(m, 4H), 2.92(s, 3H), 2.18(m, 4H), 2.07(s, 3H), 1.73(d, 3H) ppm. ^{13}C NMR (D$_2$O, 100 MHz): 153.03, 148.22, 141.69, 125.99, 124.23, 58.08, 53.51, 53.32, 43.12, 40.01, 36.78, 22.33, 21.42, 17.94 ppm. FAB-MS: 251.3 [M+H]$^+$. HRMS: 251.2244 [M+H]$^+$ (calc.: 251.2236); 331.1489 [M+HBr] (calc.: 331.1497).

Kinetic Measurements

The kinetics of the reaction of nickel(II) with the ligands L$^{L/C}$ and L^L were measured spectrophotometrically at their absorption maxima, 390 and 360 nm, respectively, using a Hi Tech MG-6000 Rapid Diode Array Stopped Flow Spectrophotometer (Model SF-41) interfaced with an IBM PC. Data acquisition and processing were carried out using Hi Tech systems kinetics software. Rate constants were calculated by the computer program's IS software from Hi Tech. The ionic strength was adjusted to 0.2 M with KNO$_3$ in all of the reactions, and the temperature was kept constant at 25 ± 0.1 °C. MES (2-[N-morpholino] ethane sulfonic acid) buffer (0.05 M) was used to maintain a constant pH. Pseudo-first order conditions ([L$^{L/C}$]$_{tot}$ ≥ 10[Ni(II)]$_{tot}$) or ([Ni(II)]$_{tot}$ ≥ 10[L^L]$_{tot}$) were maintained in all of the reactions. The reaction monitored is shown in eq 1.

$$\text{Ni(II)} + \text{H}_n\text{L}_T \longrightarrow \text{NiL}_T + n\text{H}^+ \quad (1)$$

Where H$_n$L$_T$ is the total ligand concentration of L$^{L/C}$ or L^L. The kinetics were fit to the rate expression,

$$\text{Rate} = k_f [\text{Ni(II)}] [\text{H}_n\text{L}_T]$$

where k_f is the rate constant for complex formation. For each of the systems studied, a series of kinetic runs were carried out at several pH values (5.6 – 8.5) under conditions where $[L_T] \gg [Ni(II)]$ or $[Ni(II)] \gg [L_T]$ to give the first order expression,

$$d[NiL_T]/dt = k_{obs} [Ni(II)]$$

or,

$$d[NiL_T]/dt = k_{obs} [L_T]$$

The observed first order rate constants k_{obs} were obtained from the *A/t* data (A = absorbance; t = time) which could be computer-fitted to eq 2 very well (least-squares method).

$$A = (A_o - A_\infty) \exp(-k_{obs}t) + A_\infty \qquad (2)$$

(A_o and A_∞ refer to t = 0 and t = ∞, respectively)

Potentiometric Equipment and Measurements

Reagents and Standard Solutions

Analytical grade metal nitrates were used and solutions were prepared in doubly distilled water and standardized against EDTA (AR, Aldrich) using a copper-selective electrode and calomel reference electrode. The ligand stock solution was standardized potentiometrically against a copper(II) nitrate solution using the copper/calomel electrode system. The titration was carried out in ammonia buffer at pH 10. The NaOH solution (0.201M; Aldrich) was standardized potentiometrically against KHP (potassium hydrogen phthalate, Fisher) and stored under solid mixture of anhydrous $CaCl_2$ and sodium hydroxide to minimize carbonate formation.

Potentiometric Titrations

These were performed under N_2 at 25.0 °C on a Brinkmann Metrohm 736GB Titrino equipped with an ORION Ross combination electrode (model 81-02). The electrode was standardized by three-buffer calibration using the Titrino's internal standardization method. The potentiometric equilibrium measurements were made on 50.0 ml of ligand (0.5×10^{-3} M), first in the absence of metal ions and then in the presence of each metal ion for which [L]:[M] ratios were 1:1. The ionic strength was maintained constant at

0.1 M (KNO$_3$). The pH data were collected after 0.01 ml incremental additions of standard NaOH solution (with a 20–60 s equilibration time) while the titration data (pH vs ml base) were captured in the Titrino's built-in software. Direct pH meter readings were used for calculation of the protonation and stability constants. The constants determined are mixed constants (also known as Brønsted constants) which involve the hydrogen ion activity and the concentrations of the other species. The protonation and stability constants were calculated by fitting the potentiometric data with the SUPERQUAD program (*17*). Species distribution diagrams were generated with the aid of the program Hyss (*18*).

Results

Strategies for the Synthesis of the Switch-Binding Ligand [1-[6-{3-[(3-amino-propyl)-methylamino]-propylamino}ethyl]-pyridin-2-yl]ethanone, L$^{L/C}$, and 3-({3-[pyridine-2-yl]-ethylamino}-propyl)-N-methyl-1-aminopropane, L^L, as the Hydrobromide Salts

The strategy is outlined in Schemes 2 and 3.

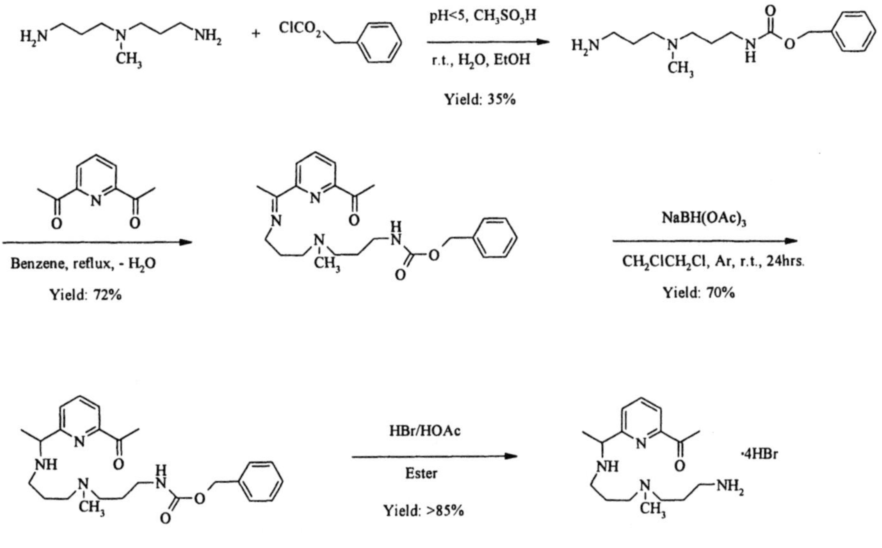

Scheme 2. Strategy for Preparation of the Switch-Binding Ligand L$^{L/C}$.

N-Methyl-3,3'-*N,N*-dipropylamine was coupled (*12*) with benzyl chloroformate in a pH controlled solution of ethanol-H$_2$O to generate {3-[(3-aminopropyl)-methylamino]-propyl}-carbamic acid benzyl ester whose unprotected primary amine was converted to the imine (*13*) of [3- ({3- [1- (6-acetylpyridin-2-yl) -ethylideneamino] -propyl} -methylamino)-propyl]-carbamic acid benzyl ester by reacting with 2,6-diacetylpyridine. Sodium triacetoxyborohydride reduced (*14, 15*) selectively the imine to an amine of [3-({3-[1-(6-acetyl-pyridin-2-yl)-ethylamino]-propyl}-methylamino)-propyl]-carbamic acid benzyl ester without reduction of acetyl group. The carbobenzyloxy group was removed by treatment with hydrobromic acid to get L$^{L/C}$·4HBr (*16*).

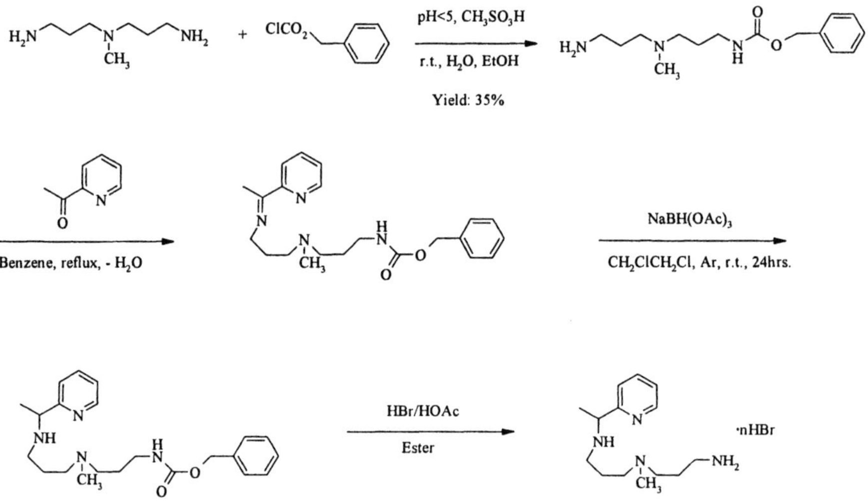

Scheme 3. Strategy for the Synthesis of the Unreactive Surrogate Ligand L^L.

N-Methyl-3,3'-*N,*N-dipropylamine was coupled (*12*) with benzyl chloroformate in pH controlled solution of ethanol-H$_2$O to generate {3-[(3-amino-propyl)-methylamino]-propyl}-carbamic acid benzyl ester whose unprotected primary amine was converted to imine (*13*) of [3-({3-[1-(pyridin-2-yl)-ethylideneamino]-propyl}-methylamino)-propyl]-carbamic acid ester by reacting with 2-acetylpyridine. Sodium triacetoxyborohydride reduced (*14, 15*) the imine to the amine of [3-({3-[1-(pyridin-2-yl)-ethylamino]-propyl}-methylamino)-propyl]-carbamic acid benzyl ester. The carbobenzyloxy group was removed by hydrobromic acid to get L^L·xHBr.

Equilibrium Studies

Protonation constants for L^L and $L^{L/C}$ for formation of their acid salts and their complex stability constants were calculated by computer fitting of the potentiometric data. The initial results of the computations were obtained in the form of overall protonation constants

$$\beta^H_i = [H_iL]/[L][H]^i$$

or overall stability constants

$$\beta_{pqr} = [M_pL_qH_r]/[M]^p[L]^q[H]^r$$

Differences between the various $\log \beta^H_i$ or the $\log \beta_{pqr}$ give the stepwise protonation constants or the stepwise formation and protonation constants of the complex reactions. The titration curve for $L^{L/C}$ was found to be reproducible and reversible despite the expectation that it would quickly form a variety of products in solutions, due to its innate ability to condense inter- or intramolecularly at a variety of pH levels. The anticipated variety of products and intermediates was expected to greatly complicate the potentiometric measurements since the identity of the ligand might constantly shift among numerous linear and cyclic structures. The protonation constants for L^L and $L^{L/C}$, together with literature data for related ligands, are summarized in Table I.

Table I. Protonation Constants for the Ligands $L^{L/C}$ and L^L at 25 °C and I = 0.1 M (KNO_3) and Data Reported for Related Ligands

	$LogK_1$	$LogK_2$	$LogK_3$	$LogK_4$
$L^{L/C}$	10.02(3)	8.53 (4)	6.61(6)	3.98(8)
L^L	10.34(2)	8.61(3)	6.65(8)	4.41(9)
Trien[1]	9.80	9.08	6.55	3.25
Py[14]aneN_4[2]	9.74	8.67	4.67	< 1

Note: Standard deviations are given in parentheses.
[1] *(19)*
[2] *(20)*

The higher values of protonation constants correspond to the protonation of the primary and secondary amine nitrogens in the acyclic ligands and to the secondary nitrogens in analogous opposing positions in the macrocycle, py[14]aneN$_4$. For both L$^{L/C}$ and L^L, the protonation constant for the secondary nitrogen is higher than that for the primary nitrogen atom, representing the typically greater basicity of secondary amines over primary amines *(21)*. The third and fourth values (logK$_3$ and logK$_4$) for L$^{L/C}$ and L^L correspond to the protonation of tertiary nitrogens bearing methyl groups and the pyridine nitrogens, respectively. The last constant for L^L was not determined because its high acidity prevented a sufficiently accurate determination by potentiometric measurements. The species distribution diagram for L$^{L/C}$ is shown in Figure 1.

Potentiometric titration curves (with and without the metal ion present) of the acid salts of L$^{L/C}$ and L^L with NaOH are shown in Figures 2a and 2b. Computer calculations revealed the presence of two complex species, [MLH]$^{3+}$ and [ML]$^{2+}$ for the switch-binding ligand, L=L$^{L/C}$. Also two complex species are formed by the surrogate ligand (L = L^L), the first of which is [ML]$^{2+}$ for all metal ions studied and the second varies as the metal ion is changed. Formation constants for these species are given in Table II. It is

Table II. Formation Constants for Metal Complexes for the Ligands L$^{L/C}$ and L^L with Several Divalent Metal Ions at 25.0 °C and I = 0.1 M KNO$_3$

log β$_{MLH}$	*Cu^{2+}*	*Hg^{2+}*	*Ni^{2+}*	*Zn^{2+}*
		L$^{L/C}$ Ligand		
log β$_{110}$	14.79(2)	7.72(6)	8.88(1)	8.80(1)[1]
log β$_{111}$	19.26(3)	15.57(2)	15.39(2)	14.72(1)[1]
		L^L Ligand		
log β$_{110}$	14.08(2)	7.91(4)	9.53(8)	7.97(1)
log β$_{111}$	20.47(4)	15.56	16.01(2)	14.62(8)

Note: Standard deviations are given in parentheses.

[1] **Iterated independently.**

noteworthy that in the titration of L$^{L/C}$ with base in presence of Cu(II) ion, a color change from blue to violet took place as the pH value exceeded 6.5. It is believed that at this pH the copper first complexes with the acyclic ligand (< pH 6) and that the resulting complex (CuL$^{L/C}$) then slowly transforms into the corresponding macrocyclic complex (CuL^C). The conversion to the macrocyclic derivative accelerates as the pH increases. Cu(II) solutions that have been titrated with base can be reversibly titrated with acid, completely removing the ligand from the metal ion, a result that is unlikely for macrocyclic complexes.

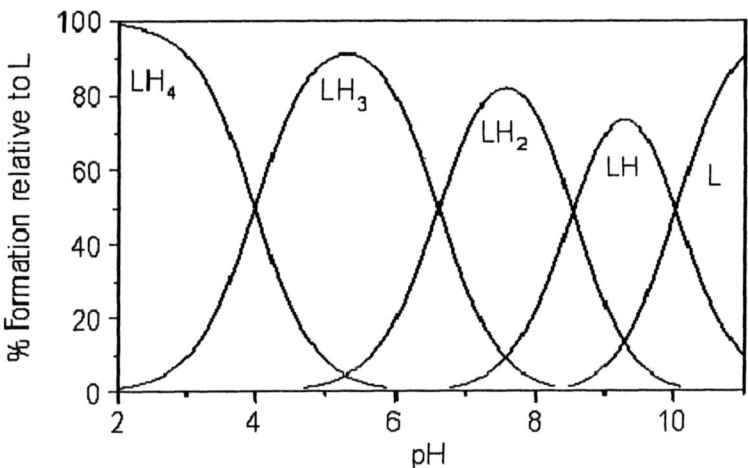

Figure 1. Speciation curves for L^{LC}, at 25 °C and I = 0.1 M KNO_3.

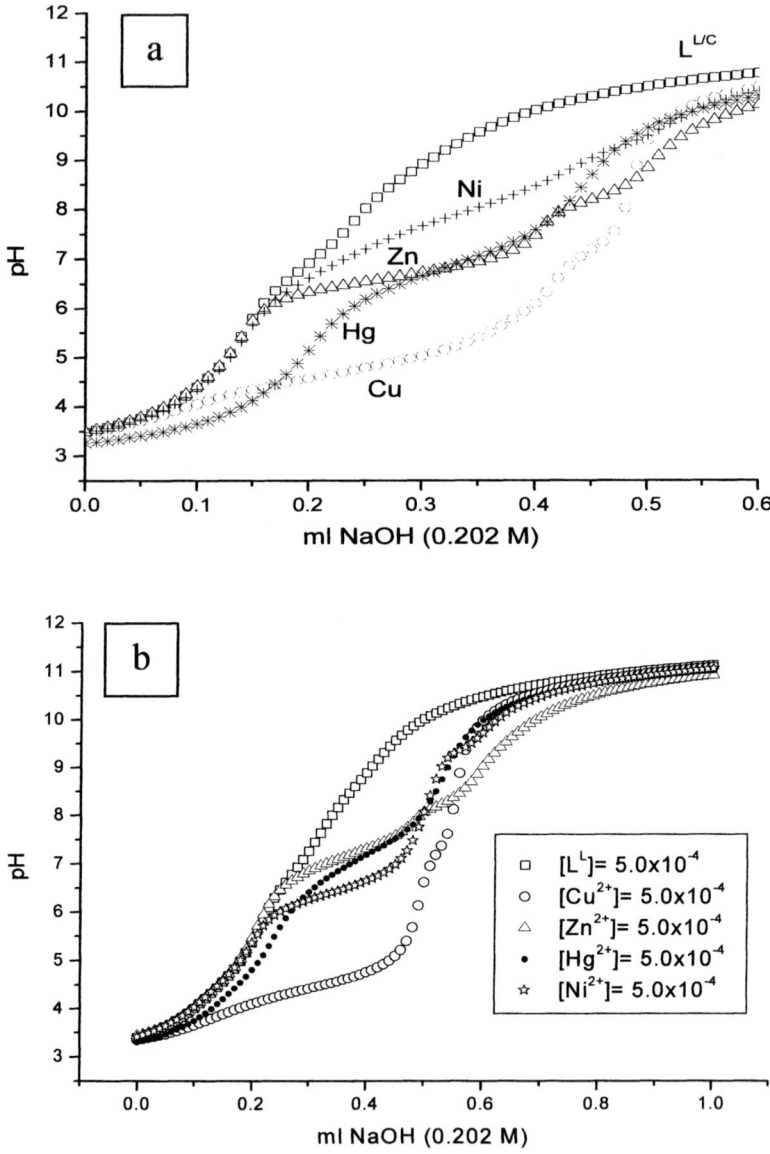

Figure 2. Potentiometric titration curves for $L^{L/C}$ (a) and L^L (b) with and without metal ions at 25 °C and I = 0.1 M (KNO_3).

The fact that the $L^{L/C}$/Cu(II) formation constant for the 1:1 complex is lower than expected is attributed to complications that follow from the uncertainty over whether equilibrium had been completely attained at each titration point (specifically above pH 6.5). The same uncertainty is associated with results for the other metal ions.

Kinetic Studies

Kinetics of Formation of $Ni(L^{L/C})^{2+}$

The rate of complexation of nickel(II) with $L^{L/C}$ was studied both by UV-vis spectrophotometer and with a stopped-flow spectrophotometer at 25 °C in the presence of 0.2 M KNO_3 from pH 5 to 7.5. All kinetic measurements were carried out under pseudo first order conditions ($[L^{L/C}] \geq [Ni(II)]$) over the pH range 5.0 to 7.5. Reaction sequences were observed in two distinct time regimes, the first in fractions of seconds and the second in hours. Both processes are accompanied by increasing absorbance (λ_{max} = 390 nm), as expected for formation of a square planar nickel(II) complex, Figure 3. The two processes can be explained by the pseudo first order rate constants k_{obs1} and k_{obs2}, eqs 3 and 4.

$$Ni(II) + L \xrightarrow[\text{fast}]{k_{obs1}} [ML]_{int}^{2+} \qquad (3)$$

$$[ML]_{int}^{2+} \xrightarrow[\text{slow}]{k_{obs2}} [ML] \qquad (4)$$

where $[ML]_{int}^{2+}$ is the intermediate complex species associated with the fast process.

Fast Process (k_{obs1})

The initial kinetic process between the Ni(II) ion and $L^{L/C}$ is complex, giving evidence for three apparent sequential steps (see Figure 4), the first of which was found to be pH and concentration dependent, whereas the other two

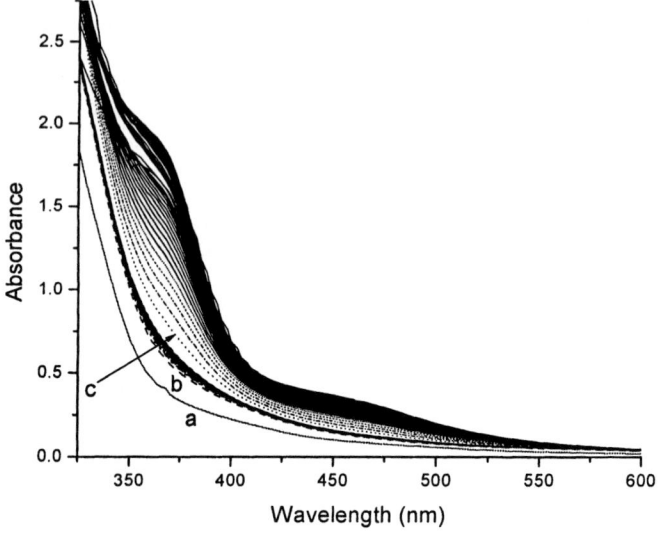

Figure 3. Time dependent spectra recorded for the complexation of $L^{L/C}$ to nickel(II) in aqueous solution at 25 °C with I= 0.2 M (KNO_3), pH = 7.15, $[L^{L/C}] = 10[Ni^{2+}] = 0.01$ M. a= spectrum of the free ligand. b = spectra of the complex (15 scans; time interval between scans = 2 min). c = spectra of the complex (40 scans; time interval between scans = 1h).

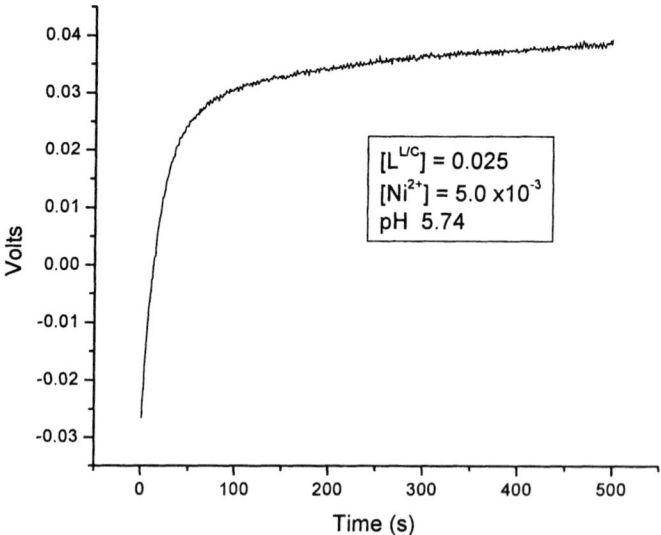

Figure 4. Spectral changes associated with the complexation of $L^{L/C}$ to Ni(II) at 25 °C (fast process, λ_{max} = 390 nm).

steps were overlapping, difficult to separate, and exhibited only small spectral changes. The first events almost certainly involve the usual reversible binding of the first few ligand sites to the metal ion, eq 5,

$$Ni^{2+} + LH_n^{n+} \underset{k_d}{\overset{k_f}{\rightleftharpoons}} NiHL^{2+} + (n-1)H^+ \quad (5)$$

(As discussion progresses, T_L will be used to represent the sum of all unprotonated and protonated forms of the ligand: L, HL^+, H_2L^{2+}, etc.)

where k_f and k_d are the rate constants for complex formation and the spontaneous dissociation of the nickel(II) complex, respectively.

At constant pH, the reaction kinetics, for the fast reaction were fitted to the differential expression

$$d[NiHL^{3+}] / dt = k_f [Ni^{2+}][LH_n] - k_d [NiHL^{3+}] \quad (6)$$

Kinetic determinations were carried out at each of several pH values under conditions where $T_L \gg T_{Ni}$ to yield the pseudo first-order rate expression

$$d[NiHL^{3+}] / dt = k_{obs1} [Ni^{2+}] \quad (7)$$

$$(k_{obs1} = k_d + k_f T_L) \quad (8)$$

for which the observed first-order rate constant, k_{obs1}, could be obtained from the A/t data (A = absorbance; t = time) computer fitted to eq 9

$$A = (A_o - A_\infty) \exp(-k_{obs1}t) + A_\infty \quad (9)$$

(A_o and A_∞ refer to t = 0 and t = ∞, respectively)

The pH dependence of k_{obs1} for the complexation of $L^{L/C}$ with Ni(II) ion is shown in Figure 5. Plots of k_{obs1} vs the total concentration of $L^{L/C}$ are linear, Figure 6, with slope k_f and intercept k_d. Linear regression analysis yielded k_d values that were not statistically different from zero.

Resolution of Specific Rate Constants

The range of pH values investigated (5.6 – 7.0) was generally limited at the upper end to pH 7 – 7.4 due to concerns regarding the possible interference caused by precipitation of $Ni(OH)_2$. At low pH values, a very large excess of

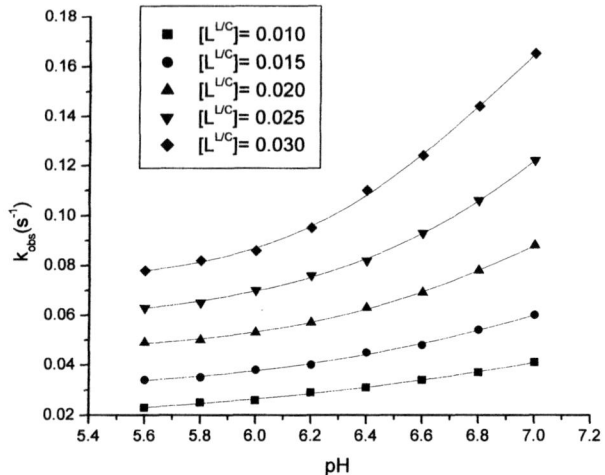

Figure 5. The pH dependence of k_{obs1} at various ligand concentrations, for the initial fast reaction of Ni(II) with $L^{L/C}$ at 25 °C.

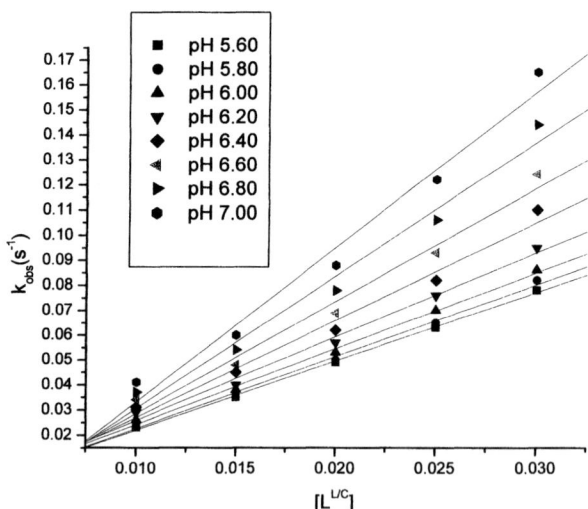

Figure 6. The dependence of k_{obs1} on T_L at various pH values, for the initial fast reaction of Ni(II) with $L^{L/C}$ at 25 °C.

the one reactant was required to force the reaction to completion. The rates of complex formation between $L^{L/C}$ and the nickel ion are proportional to $[L]_T$ and $[M]_T$. In addition they are functions of pH. Thus it is presumed that all variations with pH observed for k_f values are due to varying ratios of the existing species of $L^{L/C}$ over the pH range of the study: L, LH^+, and LH_2^{2+}. The triprotonated ligand species, LH_3^{3+}, was considered to have no significant kinetic contribution (electrostatic consideration) (see Figure 1). A possible kinetic scheme for the fast complex formation step described by eq 10 could involve the steps:

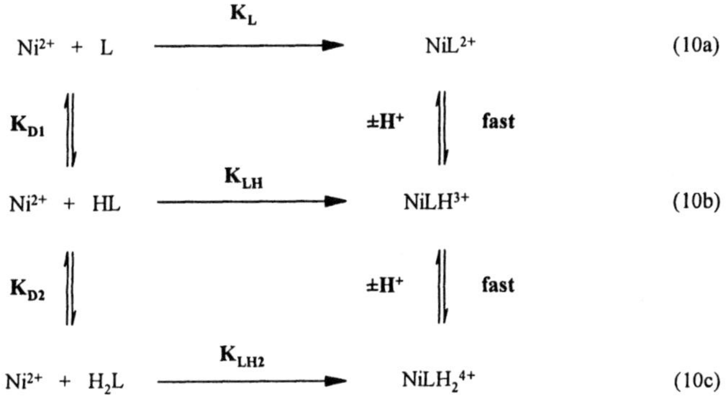

$$Ni^{2+} + L \xrightarrow{K_L} NiL^{2+} \quad (10a)$$

$$K_{D1} \updownarrow \qquad \pm H^+ \updownarrow \text{ fast}$$

$$Ni^{2+} + HL \xrightarrow{K_{LH}} NiLH^{3+} \quad (10b)$$

$$K_{D2} \updownarrow \qquad \pm H^+ \updownarrow \text{ fast}$$

$$Ni^{2+} + H_2L \xrightarrow{K_{LH2}} NiLH_2^{4+} \quad (10c)$$

Scheme 4.

where K_{D1} and K_{D2} are the first and second deprotonation constants of $L^{L/C}$ and have the values 9.55×10^{-11} M and 2.95×10^{-9} M, respectively (see Table I). By appropriate substitution it can be readily shown that

$$k_f = \frac{k_L K_{D1} K_{D2} + k_{LH} K_{D2}[H^+] + k_{LH2}[H^+]^2}{K_{D1} K_{D2} + K_{D2}[H^+] + [H^+]^2} \quad (11)$$

where k_f is the experimental second order rate constant and k_L, k_{LH}, and k_{LH2} are the resolved specific rate constants. Values of k_L, k_{LH}, and k_{LH2} were obtained by curve fitting (using the GRAFIT program (22), see Figure 7) of k_f to eq 12,

$$k_f = \frac{k_L A + k_{LH} B[H^+] + k_{LH2}[H^+]^2}{A + B[H^+] + [H^+]^2} \quad (12)$$

$(A = K_{D1}K_{D2} \text{ and } B = K_{D2})$

derived on the basis of eq 11 giving, $k_L = -832 \pm 184$ $M^{-1}s^{-1}$, $k_{LH} = 275 \pm 9$ $M^{-1}s^{-1}$ and $k_{LH2} = 2.52 \pm 0.03$ $M^{-1}s^{-1}$. This finding indicates that L of the ligand $L^{L/C}$ makes no kinetic contribution (same as LH_3^{3+}) and that the only reactive ligand species are LH^+ and LH_2^{2+}, with LH^+ being 110 times more reactive than LH_2^{2+}. As a result, 4 can now be approximated by eq 13.

$$Ni^{2+} + HL^+ \longrightarrow NiHL^{3+} ; \quad k_{LH} = 275 \pm 9 \ M^{-1}s^{-1} \quad (13)$$

Slow Process (k_{obs2})

The kinetic measurements for the slow process were made using a conventional spectrophotometer under experimental conditions similar to that used for studying the fast reaction. The pH range was 6.8 to 7.9, with $[Ni^{2+}]$ = 1.0×10^{-4} M in a 12 to 70 fold excess of $L^{L/C}$ over Ni^{2+} (I = 0.2 M, 25 °C). The kinetics are nicely pseudo first-order and the rate is insensitive to $[L^{L/C}]$. A linear increase in rate with pH was observed, Figure 8. The reaction scheme that can accommodate both the initial fast step and the subsequent slower one is

$$Ni^{2+} + LH^+ \xrightarrow{k_{obs1}} [NiLH]^{3+} \xrightarrow{k_{obs2}} [NiL]^{2+} \quad (14)$$
$$\text{fast} \qquad\qquad \text{slow}$$

At room temperature this slow reaction proceeds with a half life of 5 h around neutral pH. The reaction is attributed to the monoprotonated precursor complex, $NiLH^{3+}$. Further, because the equilibrium constant for the deprotonation step (eq 14) has been found to be 1.1×10^{-7} M (see below), a substantial amount of NiL^{2+} is available for reaction. The conversion of $[NiLH]^{3+}$ into $[NiL]^{2+}$ is accompanied by a large increase in absorbance (see Figure 3). The model that quantitatively describes the slow reaction involves a rapid proton dissociation followed by a rate-determining rearrangement step, eqs 15 and 16.

$$[NiLH]^{3+} \xrightleftharpoons{K_{MLH}} [NiL]_{int}^{2+} + H^+ \quad ; \text{fast} \quad (15)$$

$$[NiL]_{int}^{2+} \xrightarrow{k_{ML}} [NiL]^{2+} \quad\qquad ; \text{slow} \quad (16)$$

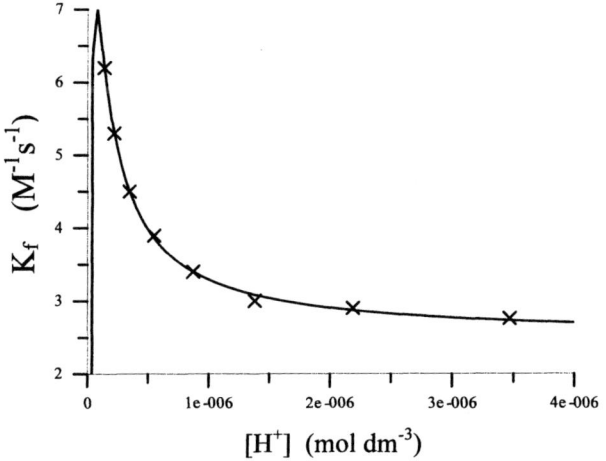

Figure 7. Plot of k_f vs $[H^+]$ (eq 12). The solid line is the calculated curve and the solid points are experimental.

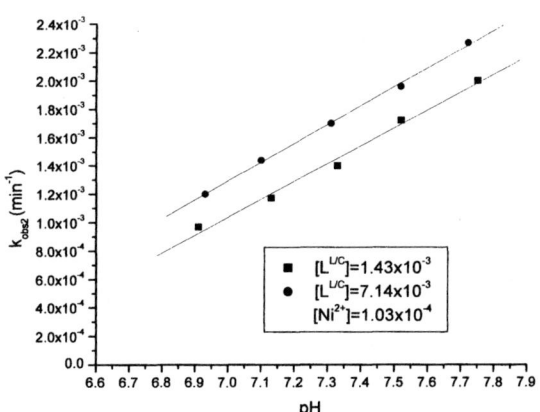

Figure 8. Plot of k_{obs2} vs pH for the slower reaction of Ni(II) with $L^{L/C}$ at 25 °C.

The rate constant k_{obs2} can be quantitatively described as a function of the deprotonation constant, K_{MLH} ($K_{MLH} = [NiL^{2+}][H^+] / [NiLH^{3+}]$) and the specific rate constant k_{ML} eq 17.

$$k_{obs2} = \frac{k_{ML} K_{MLH}}{K_{MLH} + [H^+]} \tag{17}$$

Equation 17 can be rearranged to the double reciprocal, eq 18.

$$\frac{1}{k_{obs2}} = \frac{1}{k_{ML}} + \frac{1}{k_{ML} K_{MLH}}[H^+] \tag{18}$$

The double reciprocal plot of eq 18 was then applied, using the reciprocal of k_{obs2} and the calculated H^+, Figure 9. The linear regression analysis of equation 18 yielded $k_{ML} = (4.2 \pm 0.2) \times 10^{-5}$ s^{-1} and $K_{MLH} = 1.1 \times 10^{-7}$ M.

Understanding the product of this slow reaction presents a substantial challenge. Very commonly macrocyclic ligands first form complexes in conformations that do not produce the most stable final product. Consequently, slow reactions often follow macrocyclic complex formation. The question then becomes: is the slower reaction observed in this study a rearrangement reaction or is it the ring closure step? Since the rapid and slow processes in the system under study are so well separated in time, mass spectrometric studies were conducted to demonstrate the compositions of the intermediates and products; see Scheme 1. This scheme shows that the ligand loses a mole of water when macrocyclization occurs. During short reaction times (milliseconds to minutes) the nickel-containing intermediates were shown to contain only the beginning ligand, $L^{I/C}$ (the mole of water is still in the ligand composition). On the time scale of hours, the mass spectrometry clearly establishes the appearance and eventual dominance of the macrocyclic complex. Therefore for this system, macrocyclization is slow compared to initial complex formation. The positive ion FAB-MS taken on solutions (pH ~ 7.4) following complexation experiments shows, as the only nickel complex, a peak (m/z = 331.3) corresponding to the nickel(II) complex with the fully ring closed macrocycle, L^C (Figure 10). At earlier times, the complex of the ring-open ligand is also detected (Figure 11).

Kinetics of Formation of Ni(L^L)$^{2+}$

The kinetics of the formation of Ni(L^L)$^{2+}$ were studied in the pH range 6.0 – 8.5 under conditions where (a) the total Ni(II) concentration is $\geq$ 10 times

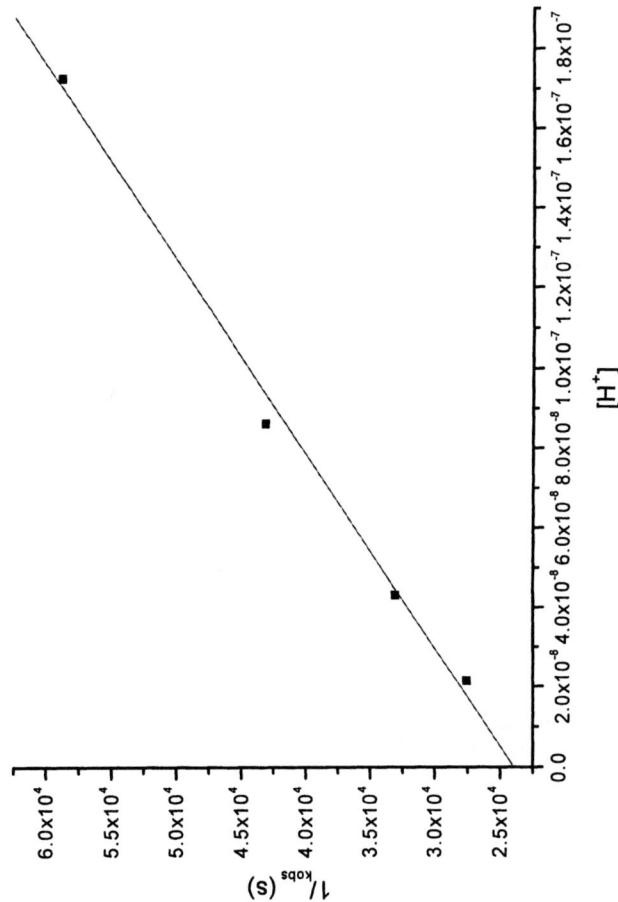

Figure 9. Double reciprocal plot of eq 11.

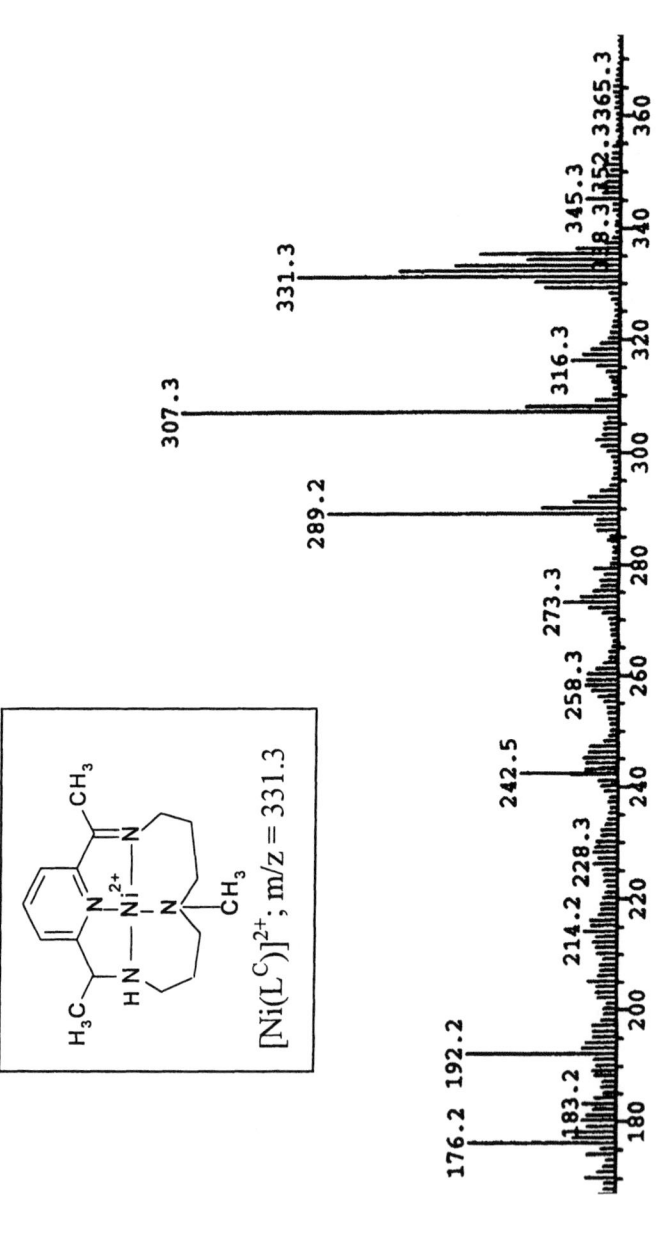

Figure 10. FAB-MS of the aqueous solution sequestered in 3-nitrobenzyl alcohol matrix after UV-vis spectrophotometric studies on the complexation of $L^{L/C}$ with nickel(II). The spectrum was taken 24 h after the combination of the metal and ligand solutions. The peak at 331.3 (m/z) confirms the in situ formation of $[Ni(L^C)]^{2+}$ (the peaks 289.2 and 307.3 are matrix lines).

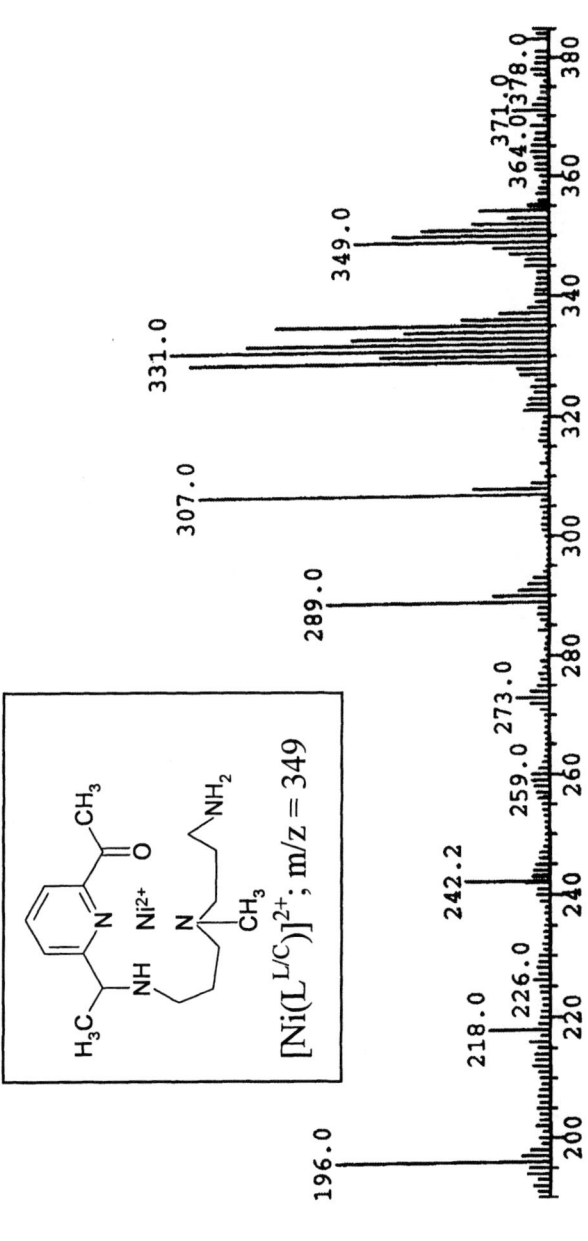

Figure 11. FAB-MS of the aqueous solution sequestered in 3-nitrobenzyl alcohol matrix after UV-vis spectrophotometric studies on the complexation of $L^{L/C}$ with nickel(II). The spectrum was taken 12 h after the combination of the metal and ligand solutions. The peak at 349 (m/z) is for the non-cyclic complex $[Ni(L^{L/C})]^{2+}$ (the peaks 289.2 and 307.3 are matrix lines).

the total ligand concentration and (b) the total ligand concentration (L^L) is ≥ 10 times excess over the total Ni(II) concentration.

Kinetics under Conditions Where $[Ni(II)] \geq 10[L^L]$

When L^L was mixed with excess Ni^{2+} in aqueous solution, a biphasic reaction was observed in all of the experiments ($\lambda = 360$ nm). The initial fast step (5–30 s) is associated with an absorbance increase and the subsequent slower reaction (50–1000 s) to an absorbance decrease involving a relatively small spectral change (Figure 12). Only the initial fast step is deemed to be significant; therefore, the apparent slower step will not be considered further in this work.

The dependence of the pseudo first-order rate constant, k_{obs}, on pH at various nickel(II) concentrations is shown in Figure 13. The plot of k_{obs} vs $[Ni^{2+}]_T$ was found to be polynomial, (Figure 14) conforming to the expression (at a given pH) given in eq 19.

$$k_{obs} = a + b[Ni^{2+}]_T + c[Ni^{2+}]_T^2 \qquad (19)$$

A possible kinetic model for the complex formation described by eq 19 is shown in Scheme 5:

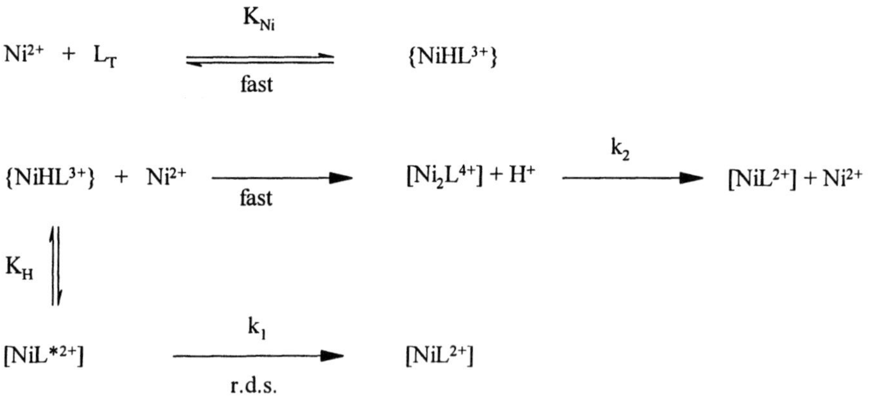

$$Ni^{2+} + L_T \underset{fast}{\overset{K_{Ni}}{\rightleftharpoons}} \{NiHL^{3+}\}$$

$$\{NiHL^{3+}\} + Ni^{2+} \xrightarrow{fast} [Ni_2L^{4+}] + H^+ \xrightarrow{k_2} [NiL^{2+}] + Ni^{2+}$$

$$K_H \updownarrow$$

$$[NiL^{*2+}] \xrightarrow[r.d.s.]{k_1} [NiL^{2+}]$$

$(L_T = L, HL^+, H_2L^{2+}, etc.)$

Scheme 5.

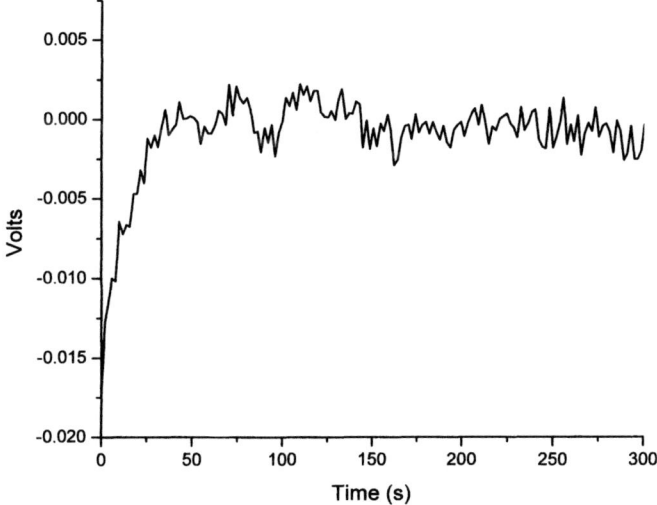

Figure 12. Optical change with time at 360 nm for the complexation of L^L (1.0×10^{-4} M) with nickel(II) (1.0×10^{-3} M) at 25 °C (the inserted plot is the corresponding spectral change over a shorter time interval).

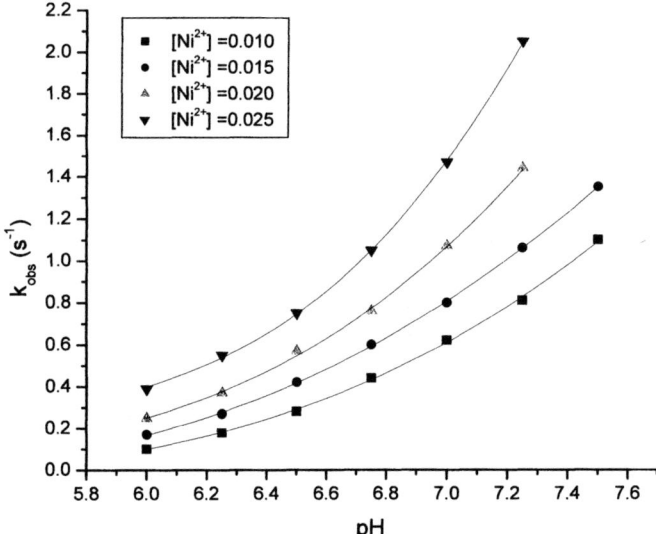

Figure 13. pH dependence of k_{obs} for the complexation of L^L with Ni(II) at 25 °C and I = 0.2 M, $[L^L]$ = 0.001M.

In the presence of excess nickel(II) ion, the binding of a second nickel ion competes with completion of the chelation process. The second path, completion of chelation, dominates at higher pH where it is accelerated by proton removal. It can be readily shown (5) that the experimentally observed rate constant, k_{obs}, is that given in eq 20.

$$k_{obs} = k_o + k_1 K_H K_{Ni}[Ni^{2+}]/[H^+] + k_2 K_{Ni}[Ni^{2+}]^2 \quad (20)$$

(assuming that, under the experimental conditions, $K[Ni^{2+}] \ll 1$)

Where k_o, $K_H K_{Ni}$, and $k_2 K_{Ni}$ = a, b, and c in eq 19, respectively (k_o corresponds to the spontaneous dissociation of [NiL^{2+}]).

Kinetics under Conditions Where $[L^L] \geq 10[Ni(II)]$

The observed reaction of Ni(II) with L^L under the conditions where $[L^L] \geq 10[Ni(II)]$ showed, to a large extent, a behavior similar to that observed for the system Ni(II)$L^{L/C}$, as described above. The dependencies of k_{obs} on pH (at fixed $[L^L]$) and on L^L concentration (at a given pH) are shown in Figures 15 and 16, respectively. In all cases, graphs of k_{obs} vs $[L^L]$ were linear conforming to the expression given in eq 21.

$$k_{obs} = k_f [L^L] + k_d \quad (21)$$

The rate constants k_d (= intercept, eq 21) were found to be ~ zero indicating that all reactions proceeded virtually to completion.

If it is assumed that **4** above also applies to this case, then for each reaction, the rate constant k_f (= slope, eq 21) can be expressed as

$$k_f = \frac{k'_L K'_{D1} K'_{D2} + k'_{HL} K'_{D2} [H^+] + k'_{H2L} [H^+]^2}{K'_{D1} K'_{D2} + K'_{D2} [H^+] + [H^+]^2} \quad (22)$$

The equilibrium constants K'_{D1} and K'_{D2} are the first and second deprotonation constants of L^L and have the values 4.47×10^{-11} M and 2.45×10^{-9} M, respectively (Table I). Computer fitting of the rate expression, eq 22, yielded $k'_L = (1.7 \pm 2.2) \times 10^3$ M^{-1} s^{-1}, $k'_{LH} = 556 \pm 30$ M^{-1} s^{-1}, and $k'_{LH2} = 29 \pm 2$ M^{-1} s^{-1}. Specific rate constants obtained in this study for nickel(II) reacting with the various species of $L^{L/C}$ and L^L in aqueous solution together with some structurally related linear and macrocyclic ligands are given in Table III.

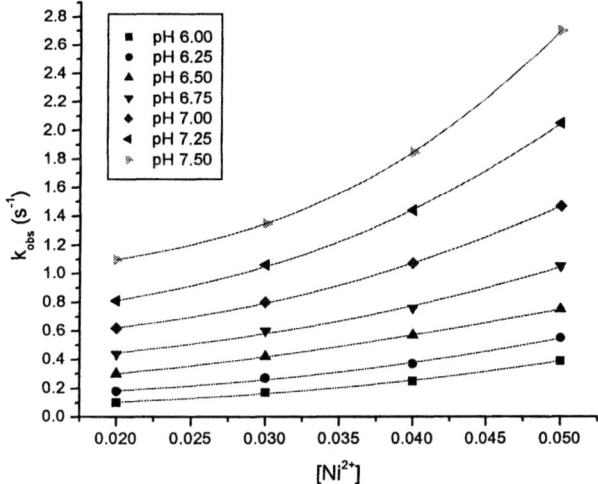

Figure 14. Plots of k_{obs} vs the total concentration of nickel(II) for the reaction with L^L at 25 °C and I = 0.2 M, $[L^L]$ = 0.001M.

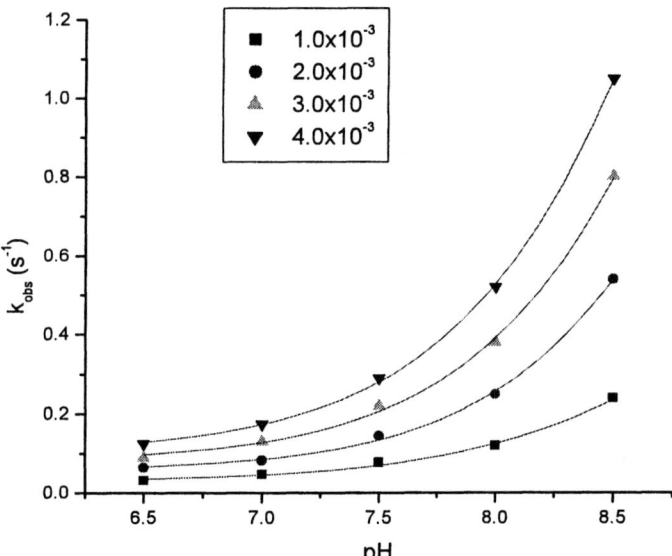

Figure 15. The pH dependence of k_{obs}, at various ligand concentrations for the reaction of Ni(II) with L^L at 25 °C.

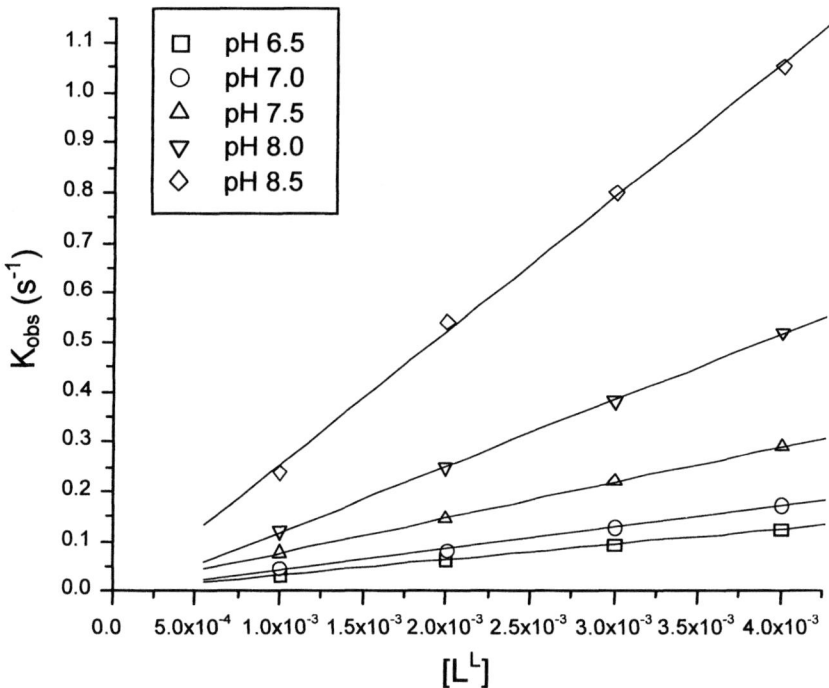

Figure 16. k_{obs} vs T_L plots at various pH values, for the reaction of Ni(II) with L^L at 25 °C.

Table III. Specific Rate Constants for Nickel(II) Reacting with Various Ligand Species of $L^{L/C}$ and L^L and Other Structurally Related Ligands in Aqueous Solution at 25 °C

Ligand	Specific Rate Constants ($M^{-1} s^{-1}$)			
	k_{LH3}	k_{LH2}	k_{LH}	k_L
$L^{L/C}$	...	2.5±0.03	275±9	...
L^L	...	29±2	556±30	(1.7±1)x10^3
Trien[1]	...	97	9300	...
Cyclam[2]	...	...	57	...
Py[13]aneN$_4$[3]	...	...	73	...

[1] (23)
[2] (24)
[3] (25)

Discussion

Equilibrium Studies

In the absence of a complementary metal ion, the novel ligand, $L^{L/C}$, is expected to quickly condense with itself forming both a variety of rings and a variety of linear oligomers. It is a property of ligands of this kind that metal ions that are complementary in size and coordination geometry can serve as templates and cause the ligand to form a single product, a macrocycle that encloses the metal ion (26). Because Schiff base formation is often rapid, $L^{L/C}$ was expected to present an ever changing molecular composition during an equilibrium titration. Remarkably, a clean, reproducible titration curve was obtained with equilibration times between base additions of 20–60 seconds. Treatment of those data gave the distinctive set of pK_a values listed in Table I. The very similar values obtained for the linear ligand L^L add credibility to the pK_a values obtained for the self-reacting ligand $L^{L/C}$. No complications were expected in the case of the linear ligand L^L. Another remarkable observation was that all four of the values determined for $L^{L/C}$ are within a few tenths of a pK unit of those for the linear tetramine, triethylenetetramine (trien).

The protonation pattern for the free ligand anticipates the early events in the binding of $L^{L/C}$ to a metal ion. The last proton to be removed from $L^{L/C}H_4^{4+}$ is the secondary amine between the pyridine moiety and the tertiary amine.

Internal hydrogen bonding may be expected to augment the inductive effects that make this amine most basic. These considerations, and the resulting relative isolation of the primary amine, point to that group as the nucleophilic center that first binds to the metal ion in the complex formation process, assuming (vide infra) that $L^{L/C}H^+$ is the dominant reacting ligand species.

Formation constants were determined for several metal ions (Cu^{2+}, Hg^{2+}, Ni^{2+}, Zn^{2+}). For pertinence to the rate studies that will be discussed later, it must be remembered that the titrations were all carried out with 20-s to 60-s equilibration times. Under the conditions of the experiments, $L^{L/C}$ forms two kinds of complexes, $[M(L^{L/C})]^{2+}$ and $[M(L^{L/C}H)]^{3+}$. The surrogate ligand L^L forms the expected complex with the neutral ligand $[M(L^L)]^{2+}$, but the monoprotonated complex is found only for copper(II).

The binding constants β_{110} are remarkably similar for the two ligands, $L^{L/C}$ and L^L, so it is apparent that both ligands chelate as linear tetradentate molecules on the time scale of the titrations. Clearly the ligand $L^{L/C}$ has not cyclized prior to binding to the metal ion in any case.

Detailed kinetic investigations have been carried out only for the reactions of the nickel(II) ion with the two ligands $L^{L/C}$ and L^L. The study of the $L^{L/C}$ system reveals a pattern that is, in general, common for reactions between metal ions and macrocyclic ligands, but in detail the pattern is unusual. As would be expected, kinetic events occur in two time regimes, fractions of seconds and hours. A single well-behaved process on the fast time scale is accompanied by what appear to be two additional, but poorly resolved processes that involve only small color changes. The first rapid process is well behaved and reveals that, under conditions of these experiments, nickel(II) undergoes complex formation with only two of the possible ligand species, $L^{L/C}H_2^{2+}$ and $L^{L/C}H^+$, and that reaction with the monoprotonated ligand is dominant. Significant concentrations of the fully deprotonated ligand species are not present at pH values sufficiently low to assure the solubility of $Ni(OH)_2$. The additional unexplained rate processes mentioned above may be associated with binding steps subsequent to the linking of the first $L^{L/C}$ donor atom to the metal ion; however, there is a second likely source. These processes might reflect parallel reactions by partially self-condensed ligands; this additional possibility is unique to self-condensable ligands. Since the surrogate ligand does not show these complications, this second explanation is favored.

At neutral pH the fast reaction is solely due to the monoprotonated ligand, $L^{L/C}H^+$, forming the intermediate complex, $[Ni(L^{L/C}H)]^{3+}$. That species converts to the final product $[Ni(L^C)]$ by what appears to be a classic pre-equilibrium proton dissociation step followed by rate determining ligand rearrangement (eq 12). Analysis of the data reveals a half life of about 5 h and a proton ionization constant of 1.1×10^{-7} M. The acidity of this last proton (from $L^{L/C}H^+$) has been increased by about 3 orders of magnitude by chelation of the ligand to the metal ion. From the equilibrium data, it is reasonable to suggest that the ligand has not undergone ring formation during the rapid reaction steps, but such a conclusion, based on those results alone, must be regarded as tentative at best. If the macrocycle had already formed, it would not be unusual to observe a very slow rearrangement reaction. Often the initial kinetic product of complex formation with macrocyclic ligands is not the thermodynamic product. A common readjustment is the slow inversion of coordinated quaternary nitrogen atoms. The strongest support, from the rate studies, for macrocyclization during the slow step is the large increase in absorbance in the region normally found for low spin square planar nickel(II) complexes.

The kinetics of binding of the surrogate ligand, L^L, supports the view that the slow reaction in the case of $L^{L/C}$ involves ring closure. The ligand L^L displays only relatively rapid kinetic processes. For experiments conducted under the same conditions as those used in the study of $L^{L/C}$, i.e., in the presence of excess ligand, the L^L system behaved in a similar fashion to the rapid reaction in the Ni/$L^{L/C}$ system. The specific rate constants for the ligand, L^L, are both larger than those for the more complicated ligand, $L^{L/C}$. For the simpler surrogate ligand, kinetic studies were also carried out in the presence of a large excess of nickel(II) ion. The excess nickel fostered formation of a 2:1 nickel:ligand complex at lower pH values.

Mass spectrometry has provided definitive proof that the macrocylization occurs during the slow kinetic time regime. The difference in mass between the nickel(II) complexes with the linear tetradentate precursor ligand, $L^{L/C}$, and the macrocyclic ligand, L^C, is ~ 18 mass units due to the loss of one water molecule. The dominant nickel(II) species at times approximating the completion of the rapid and slow processes, respectively, differ in mass by just that amount (see Figures 10 and 11). Therefore the slow reaction is ring closure, and it produces the macrocyclic complex within a precursor complex containing the switch-binding ligand. It should be recalled that there is no corresponding slow kinetic process in the case of the ligand L^C. These results establish the switch-binding concept based on metal ion templating of specially designed ligands. Further, the dynamics of this first case have been revealed in substantial detail.

Whereas no one has previously suggested the possible advantages of switch-binding for ultra tight-binding ligands, a small number of examples exist in the literature where similar topological changes in ligand structure (linear to macrocyclic) have been observed. In fact, these constitute examples of what we labeled the kinetic template effect many years ago (*27, 28*).

The first example was found as Verbruggen et al. explored numerous peptide derivatives as ligands for complexes with technetium(IV) in radiopharmaceutical development (*29–33*). Curious in situ chemistry was observed when the tetramer of L-alanine was reacted with technetium(IV), yielding an unstable complex that converted into a monooxotechnetium(V) complex of the cyclic tetra-L-alanine as shown in Scheme 6.

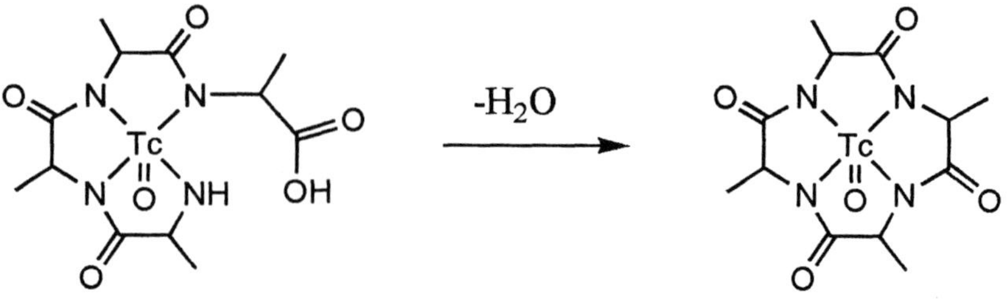

Scheme 6. Cyclization of [Tc(IV)(tetra-L-alanine)] Complex.

In this complex, the carboxyl group was forced into the vicinity of the free amino group, thereby facilitating amide formation and cyclization. Formation of the complex with the tetraamide in aqueous solution was fast, but ring closure was quite slow (4.9×10^{-3} min^{-1} at 25 °C, typically), being most rapid at pH 6. The acyclic complex was characterized by HPLC analysis; this intermediate complex was not isolated.

In their investigations on base-catalyzed imine formation, Danby and Hay successfully isolated and determined the crystal structure of the copper(II) complex of the non-cyclized β-aminoketone in Scheme 7 *(34)*.

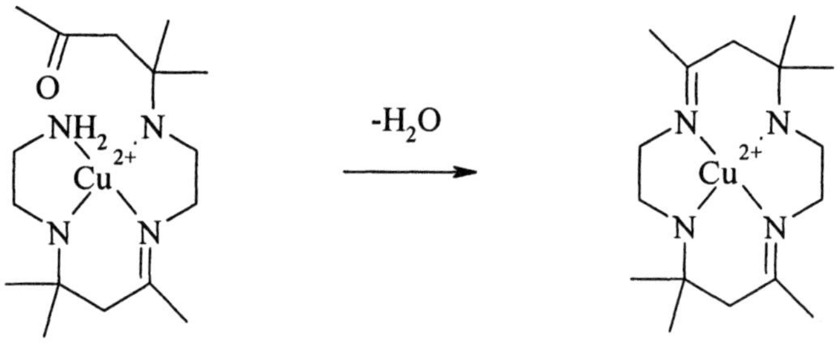

Scheme 7. Base-Catalyzed Imine Formation of trans-[14]-diene.

In basic solution, the complex undergoes ring closure to give the macrocyclic copper(II) complex of 5,7,7,12,14,14-hexamethyl-1,4,8,11-tetra-azacyclotetradeca-4,11-diene (*trans*-[14]-diene). The observed rapid ring closure occurs via an intramolecular reaction involving the hydroxocomplex [Cu(L)OH]$^+$ at a k_{obs} of 4.98×10^4 s^{-1} at 30.1 °C *(35)*.

From the results reported here and the observations recorded in the literature, it is clear that a slow macrocyclization reaction can be fostered following rapid binding of linear polydentate ligands. Thus the switch-binding principle enjoys substantial generality. This opens the way to many new uses for the chemistry of ultra tight-binding ligands. Among the perceived advantages peculiar to these novel switch-binding ligands is the ability of the rapidly binding ligand to invade strong metal ion binding sites and eventually remove metal ions as transportable species in which the metal ion is sequestered by a ligand of extreme binding affinity. These and many other possibilities remain to be investigated.

Acknowledgments

Support of this research by the U.S. Department of Energy (DOE) Environmental Management Science Program (EMSP) Grant DE-FG07-96ER14708 is deeply appreciated.

References

1. Cox, B. G.; Schneider, H.; Stroka, J. *J. Am. Chem. Soc.* **1978**, *100*, 4746.
2. Liesegang, G. W.; Eyring, E. M. In *Synthetic Multidentate Macrocyclic Compounds*; Izatt, R. M., Christensen, J. J., Eds.; Academic Press, Inc.: London, 1978; pp 245–289.
3. Eyring, E. M.; Petrucci, S. In *Cation Binding by Macrocycles*; Inoue, Y., Goekel, G. W., Eds.; Marcel Dekker, Inc.: New York, 1990; pp 179–203.
4. *Biological Applications of Photochemical Switches;* Morrison, H., Ed.; John Wiley: New York, 1993.
5. Lester, H. A.; Gurney, A. M. *Physiol. Rev.* **1987**, *67*, 583.
6. Givens, R. S.; Weber, J. J. F.; Jung, A. H.; Park, C.-H. In *Methods in Enzymology on Caged Compounds: Chemistry, Instrumentation, and Applications*; Marriott, G., Ed.; Academic Press Inc.: London, 1998, Vol. 291; pp 1–29.
7. Busch, D. H. In *Transition Metal Ions in Supramolecular Chemistry;* Fabbrizzi, L., Ed.; Kluwer Academic Publishers: Norwell, MA, 1994; pp 55–79.
8. Busch, D. H. In *Werner Centennial Volume;* ACS Symposium Series 565; American Chemical Society: Washington, DC, 1994; pp 148–164.
9. Busch, D. H. *Chem. Rev.* **1993**, *93*, 847–860.
10. Busch, D. H. *Chem. Eng. News*, June 29, 1970, p 9.
11. Barefield, E. K.; Lovecchio, F. V.; Tokel, N. E.; Ochiai, E.; Busch, D. H. *Inorg. Chem.* **1972**, *11*, 283.
12. Atwell, G. J.; Denny, W. A. *Synthesis* **1984**, 1032.
13. Moffett, R. B.; Leonard, N. J.; Miller, L. A. *Organic Synthesis*, Coll. Vol. 4, 605.

14. Abdel-Magid, A. F.; Carson, K. G.; Harris, B. D.; Maryanoff, C. A.; Shah, R. D. *J. Org. Chem.* **1996**, *61*, 3849.
15. Abdel-Magid, A. F.; Maryanoff, C. A.; Carson, K. G. *Tetrahedron Lett.* **1990**, *31*, 5595.
16. Ben-Ishai, D.; Berger, A. *J. Org. Chem.* **1952**, *17*, 1564.
17. Gans, P.; Sabatini, A.; Vaca, A. *J. Chem. Soc., Dalton Trans.* **1985**, 1195.
18. Alderighi, L.; Gans, P.; Ienco, A.; Peters, D.; Sabatani, A.; Vacca, A. *Coord. Chem. Rev.* **1999**, *184*, 311.
19. Martell, A. E.; Smith, R. M. *Critical Stability Constants*; Plenum Press: New York, 1982.
20. Costa, J.; Delgado, R. *Inorg. Chem.* **1993**, *32*, 5257.
21. McCasland, A. K. Ph.D. Dissertation, University of Kansas, Lawrence, KS, 1999.
22. Erithacus Software Limited © 1988.
23. Moss, D. B.; Lin, C-T.; Rorabacher, D. B. *J. Am. Chem. Soc.* **1973**, *8*, 5179.
24. Wu, Y.; Kaden, T. A. *Helv. Chim. Acta* **1984**, *67*, 1868.
25. Hassan, M. M.; Marafie, H. M.; El-Ezaby, M. S. *Coordination Chemistry* **2003**, *8*, 709.
26. Hubin, T. J.; Busch, D. H. *Coord. Chem. Rev.* **2000**, *200-202*, 5.
27. Thompson, M. C.; Busch, D. H. *Chem. Eng. News,* Sept 17, 1962, p 57.
28. Thompson, M. C.; Busch, D. H. *J. Am. Chem. Soc.* **1964**, *86*, 3651.
29. Bormans, G.; Peters, O. M.; Vanbilloen, H.; Blaton, N.; Verbruggen, A. *Inorg. Chem.* **1996**, *35*, 624.
30. Grummon, G.; Rajagopalan, R.; Palenik, G. J.; Koziol, A. E.; Nosco, D. L. *Inorg. Chem.* **1995**, *34*, 1764.
31. Nosco, D. L.; Beaty-Nosco, J. A. *Coord. Chem. Rev.* **1999**, *184*, 91.
32. Reichert, D. E.; Lewis, J. S.; Anderson C. J. *Coord. Chem. Rev.* **1999**, *184*, 3.
33. Thunus, L; Lejeune, R. *Coord. Chem. Rev.* **1999**, *184*, 125.
34. Hay, R. W.; Danby, A. M.; Miller, S.; Lightfoot, P. *Inorg. Chim. Acta.* **1996**, *246*, 395.
35. Danby, A. M. Ph.D. Thesis, University of St. Andrews, St. Andrews, Scotland, 1996.

Chapter 12

Organofunctional Sol-Gel Materials for Toxic Metal Separation

Hee-Jung Im[1], Terry L. Yost[1], Yihui Yang[1], J. Morris Bramlett[1], Xianghua Yu[1], Bryan C. Fagan[1], Leonardo R. Allain[1], Tianniu Chen[1], Craig E. Barnes[1], Sheng Dai[2], Lee E. Roecker[3], Michael J. Sepaniak[1], and Zi-Ling Xue[1,*]

[1]Department of Chemistry, The University of Tennessee, Knoxville, TN 37996
[2]Chemical Sciences Division, Oak Ridge National Laboratory, Oak Ridge, TN 37831
[3]Department of Chemistry, Berea College, Berea, KY 40404

Inorganic-organic silica sol-gels grafted or encapsulated with organic ligands were prepared and found to selectively and reversibly remove target metal ions such as Cu^{2+}, Cd^{2+}, and Sr^{2+}. These organofunctional sol-gel materials were hydrophilic and showed fast kinetics of metal uptake. The sol-gels were easily regenerated and used in multicycle metal removal. In our search for new ligands for metal removal, we found that the reactions of thioacetal ligands with Hg^{2+} gave $Hg(SCH_2COOH)_2$.

Novel approaches to separation and removal of toxic metals are of intense current interest. Toxic metals such as Hg^{2+} and $^{90}Sr^{2+}$ are among the principal components of hazardous wastes generated from weapon materials production (1), and their separation and removal are one of the most pressing environmental restoration problems. The removal of these metal ions from waste solutions is traditionally accomplished through ion exchange or metal-ligand complexation in organic polymer beads and solvent-extraction agents (1–3). The use of inorganic ion exchangers such as crystalline silicotitanates (CSTs) (1, 2) and ion

© 2006 American Chemical Society

exchange resins (*3*) relies on the substitution of cations (e.g., Na^+ with Cs^+) with ions from the surrounding fluid medium (*1–3*). Ligands with high affinity for target metals have been immobilized to organic polymers to remove toxic metals (*3*).

Separation of toxic metals by inorganic-organic hybrid materials has been extensively studied to improve affinity or selectivity for target ions in the cleanup of multicomponent waste media (*4–8*). These materials often contain hydrophilic inorganic backbones and organic ligands which selectively bind target metal ions. Inorganic solid materials have been functionalized through reactions between ligands and surface hydroxyl (-OH) groups. Such ligand-grafted materials mainly utilize the surface of preformed silica gels. In an effort to enhance the capacity of the bulk of SiO_2-based materials to bind Hg^{2+} ions, for example, novel mesoporous materials grafted with mercapto ligands have been developed for enhanced Hg^{2+} adsorption (*4, 5*). These mesoporous materials, having well-defined pores and established pore shapes, were prepared using micellar surfactant templates, followed by either calcinations (*4*) or solvent extraction (*5*). The tethered mercapto ligands were then loaded onto the inner and outer pore surfaces of these solids. These ligand-coated mesoporous materials, which are usually sub-micron powers, were found to have high capacity and selectivity for Hg^{2+} (*4, 5*). The enhanced capacity was largely achieved through the use of inner pores of the mesoporous materials, and the affinity of the mercapto ligand for Hg^{2+} led to the observed high selectivity.

We have developed, with support from the U.S. Department of Energy (DOE) Environmental Management Science Program (EMSP), a one-step process to give silica gels grafted or encapsulated with ligands with high affinity and capacity for target metal ions such as Hg^{2+}, Cu^{2+}, and Sr^{2+} (*9–14*). These gel monoliths are easily prepared from off-the-shelf chemicals in ca. 1 h, and could be ground into granules for metal ion uptake. They are also readily regenerated for multicycle use. The granular form is easy to handle, and especially desirable in large-scale applications. Search of ligands for Hg^{2+} removal led us to thioether carboxylic acids such as $o\text{-}C_6H_4[CH(SCH_2COOH)_2]_2$ (*15, 16*) and silica gels believed to be grafted with dithioacetal derivatives (*17*). Our studies reveal that reactions of Hg(II) salts with thioether carboxylic acids in water lead to the decomposition of these ligands with the formation of mercury(II) mercaptoacetate $Hg(SCH_2COOH)_2$ and aldehydes (*18, 19*). These studies that have been primarily carried out in our research group at the University of Tennessee are summarized here. Other collaborative studies supported by this EMSP project have been reported (*20–25*).

Grafted Sol-Gels for Selective Cu(II) and Hg(II) Separation

The grafted sol-gels used in this approach are prepared from the co-hydrolysis of $Si(OR)_4$ and $L-Si(OR)_3$ containing a functional ligand L to graft L onto the solids (eq 1) (*26–30*). $Si(OR)_4$ plays the role of a crosslinking agent to react with $L-Si(OR)_3$.

$$Si(OR)_4 + L-Si(OR)_3 + H_2O \rightarrow \underset{\underset{L}{|}}{Si-O-Si=} \qquad (1)$$

This approach to organic-inorganic hybrid materials is different from coating functional monomers on preformed silica gel or mesoporous SiO_2 surface (*4, 5*). In the latter, the surface -OH groups of preformed silica gel beads or mesoporous SiO_2 react with $L-Si(OR)_3$, thus grafting the ligands onto their surface.

Cu(II) Removal (*9, 12*)

Silica sol-gels anchored with amino [NH_2-CH_2CH_2-NH- (TMSen), -NH-CH_2CH_2-NH- (Bis-TMSen)] and mercapto (HS-) ligands (Scheme 1), which are commercially available, were studied for Cu^{2+} separation. Sol-gels grafted with amine groups have been extensively studied and used in separation (*31–36*). The TMSen-anchored gels were found to be selective for Cu^{2+} in the presence of Cd^{2+} and Zn^{2+}. In comparison, mercapto-anchored gels remove both Cu^{2+} and Cd^{2+} from a mixture of Cu^{2+}, Cd^{2+}, and Zn^{2+}. These ligand-anchored materials are easily prepared from off-the-shelf chemicals in ca. 1 h, and are hydrophilic. Their hydrophilicity is desirable as this may give potentially fast kinetics of metal upload and removal (from the gels) in aqueous solutions. The ligand-anchored sol gels loaded with Cu^{2+} ions were regenerated with acid, and the materials could then be used in multicycle metal removal.

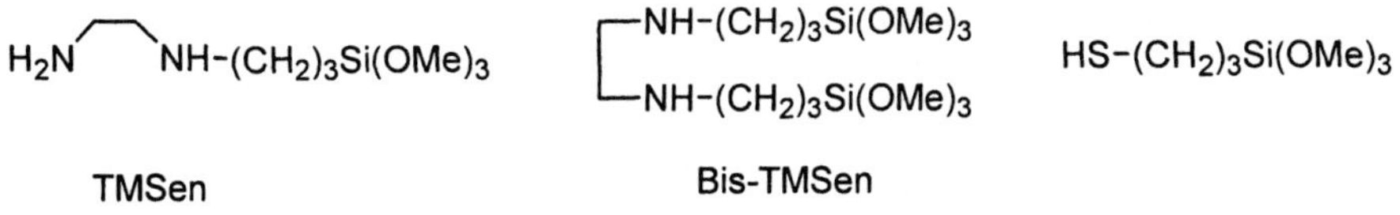

H_2N⌒NH-$(CH_2)_3Si(OMe)_3$ 　　[$-NH$-$(CH_2)_3Si(OMe)_3$ / $-NH$-$(CH_2)_3Si(OMe)_3$]　　HS-$(CH_2)_3Si(OMe)_3$

　　　TMSen　　　　　　　　　　　Bis-TMSen

Scheme 1.

Ethylenediamine and mercapto ligands were chosen for their high affinity for Cu^{2+} (37, 38). The sol-gels absorb Cu^{2+} from a solution, and addition of acid protonates the amine moieties, and releases Cu^{2+} ions. A cycle of Cu^{2+} removal is shown in Scheme 2.

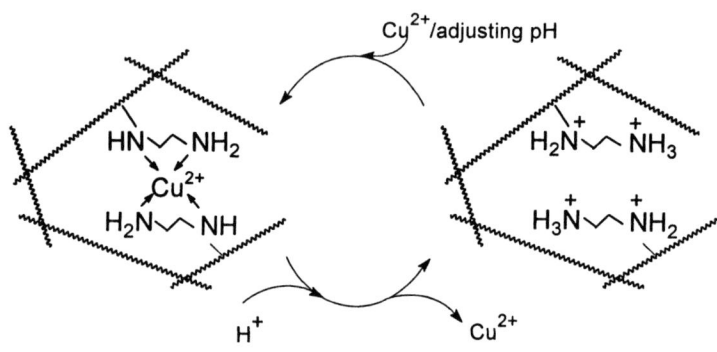

Scheme 2.

Both TMSen- and HS-anchored gels were found to remove Cu^{2+} from 7-9 ppm (ca. 0.1 mM) to beneath the AA detection limit (2 ppb) and the drinking water standard of 1 ppm (39). In (100 mM) of Cu^{2+} solution, the gel grafted with 7.4 mol % of TMSen (1) was found to be saturated with Cu^{2+} ions with the formation of a monoethylenediamine complex $Cu(TMSen)^{2+}$. In the removal of Cu^{2+}, the distribution coefficients K_d (40) of gels grafted with 16.7 mol % of TMSen- (2) and 17.0 mol % of HS- (3) ligands are 1.62×10^3 and 1.3×10^4, respectively. The distribution coefficient of the Bis-TMSen-grafted gel (4) is slightly lower. The Cu^{2+}-uptake capacity of gel 1 is comparable to other commercial or reported materials for Cu(II) removal (41–45).

In studies of competitive binding of Cu^{2+}, Cd^{2+}, and Zn^{2+} by 2, it was found to remove Cu^{2+} to below the AA detection limit in a $CuCl_2$ solution $\{[Cu^{2+}] = 6.57(0.05)$ ppm$\}$ containing Cd^{2+} and Zn^{2+} as competing soft metal ions. In comparison, a gel grafted with mercapto ligand removes both Cu^{2+} [64.3(0.2) ppm] and Cd^{2+} (119.2 ppm) from a solution containing Zn^{2+} to below the AA detection limit. Thus, the TMSen-grafted sol-gels are more selective for Cu^{2+} separation. Bis-TMSen anchored gels were found to have Cu^{2+} binding capacity similar to that of TMSen anchored gels, although the distribution coefficient K_d of the former is lower.

The kinetics of Cu^{2+} uptake by TMSen-grafted gel 1 and its removal from the gels by HCl were studied. In the first 10 min of the Cu^{2+} uptake ($[Cu^{2+}]$ = ca. 64 ppm), the gel reached ca. 36% of total Cu^{2+} absorption. In 65 min, the uptake reached 65%. The Cu^{2+} removal from the gel by HCl was essentially complete in 2 min. When more concentrated Cu^{2+} solution was used, faster uptake kinetics and higher Cu^{2+} loading capacities were observed.

Hg(II) Removal (*11*)

Mercapto (HS-) ligand has been grafted on preformed mesoporous SiO_2, and the materials have been found to have high selectivity and capacity for Hg(II) (*4–8*). These mesoporous materials, which are usually sub-micron powers with well-defined pores and established pore shapes, are prepared using micellar surfactant templates, followed by either calcinations (*4*) or solvent extraction (*5*). The mercapto ligand [HS-$(CH_2)_3$Si$(OMe)_3$] was then loaded onto the pore surfaces of the solids. The preparation of the mercapto-coated mesoporous powders usually involves at least two steps. The enhanced capacity of these mesoporous materials is largely achieved through the use of inner pores of the materials, and the affinity of the mercapto ligand for Hg^{2+} leads to the observed high selectivity.

We have compared Hg^{2+} uptake capacities of the ligand-grafted mesoporous powders prepared with surfactants, and the ligand-grafted silica granules prepared in our one-step, surfactant-free process in eq 1. The work shows that, under the conditions used here, the silica granules (500-595 μm) prepared in one step without surfactants are also mesoporous with a narrow pore size distribution, and their Hg^{2+} uptake capacities and distribution coefficients K_d values are comparable to those prepared with surfactants.

Granular silica gels **5** (7.0 mol % mercapto ligand) and **3** (17.0 mol % mercapto ligand) were prepared in one step from the co-hydrolysis of HS$(CH_2)_3$Si$(OMe)_3$ and Si$(OMe)_4$ (eq 1). Brunauer-Emmett-Teller (BET) analysis of **5** revealed that it was mesoporous with an average pore diameter: 46 Å, a narrow pore-size distribution, and a large surface area (471 m^2/g).

In the current studies, mesoporous silicas were prepared with surfactants for comparison. Non-ionic Tergitol 15-s-15 (*25*) and Triton x-100 surfactants (*8*) were used in the hydrolysis of Si$(OEt)_4$ and then were removed in the second step by calcinations and EtOH extraction, respectively. Mercapto ligand HS-$(CH_2)_3$Si$(OMe)_3$ was then grafted onto these two mesoporous silica materials to give **6** and **7**, respectively.

In Hg^{2+} solutions (0.5, 5.0, 50, and 500 ppm), **5** was found to have Hg^{2+} uptake capacities comparable to those of sub-micron mesoporous powders (**6** and **7**) prepared with two different surfactants in the current studies (*11*). **5** and **3** were capable of removing Hg^{2+} ions to be below 0.05 ppb (the detection limit of atomic absorption) from 0.368 or 50 ppm solutions, respectively. These studies show that granular **3** and **5** are easier to prepare and handle, and may offer an alternative to sub-micron powders prepared with surfactants for Hg(II) uptake.

Sol-Gels Containing an Encapsulated Crown Ether Ligand for Selective Sr(II) Separation (*10, 13, 14*)

Crown ethers are a class of macrocyclic ligands with high affinity and selectivity for alkali and alkaline earth metal ions, and are widely used in solvent extraction processes (*46–50*). Although these methods are effective at extracting metal ions, they also usually generate secondary organic waste. Crown ether ligands grafted onto organic and inorganic supports have been developed as an alternative in part to reduce organic waste (*46–58*). Grafting ligands onto solid supports usually requires multi-step syntheses, and sometimes additional functional groups are required to achieve selectivity or fast metal intake kinetics (*3*).

The current work followed our earlier work in sol-gel acid and base sensors (*59–66*). In these sensors, organic dyes are encapsulated in sol-gel SiO_2 or SiO_2-ZrO_2-polymer composite thin films. Careful control of porosity yields thin film matrices that encapsulate large organic dye molecules, and yet allow fast diffusion of smaller H^+ and OH^- ions. We reasoned that crown ether molecules are often similar in size to organic dye molecules, and crown ether molecules may be encapsulated in sol-gel silica. The encapsulated crown ethers could bind target metal ions and remove them from aqueous solution. Because SiO_2 sol-gel sorbents are hydrophilic, fast aqueous metal ion diffusion in the gels and thus fast metal ion intake by the ligand-doped gels are likely. Sr^{2+} was chosen as the target metal ion in the current studies to investigate this new approach.

^{90}Sr is a fission product, and it has been generated in large quantities from the production of nuclear power and weapons (*1*). As a relatively long-lived radionuclide (half-life $t_{1/2}$ = 29.1 years) with chemical similarities to Ca, ^{90}Sr is known to incorporate into biological systems and be stored in the bones of vertebrates (*67*). Various forms of cancer including leukemia have been linked to ^{90}Sr exposure (*68*). Leaching of ^{90}Sr into neighboring groundwater from the current storage facilities for high-level radioactive waste is of current concern. It is of intense interest to develop a novel method to separate $^{90}Sr^{2+}$ from contaminated ground water (*68*).

Sol-gel materials containing an encapsulated crown ether were prepared and found to selectively remove 91.4(1.3)% of Sr^{2+} from a solution containing excess of competing ions such as Ca^{2+}. The crown ether ligand, 1,4,10,13-tetraoxa-7,16-diazacyclooctadecane-7,16-bis(malonate) ligand (Na_4oddm, Scheme 3), is known for its high affinity for Sr^{2+} (*69*), and is encapsulated in hydrophilic SiO_2 through a one-step sol-gel process (*10, 13*). Washing the Sr^{2+}-loaded gel with acid or an ethylenediaminetetraacetic acid (EDTA) salt recovered Sr^{2+} from the sol-gel, and regenerated the gel for subsequent Sr^{2+} intake. The approach reported here is a new alternative to the use of crown ethers in metal ion separation through, e.g., solvent extraction or the use of sorbents containing chemically grafted crown ether ligands.

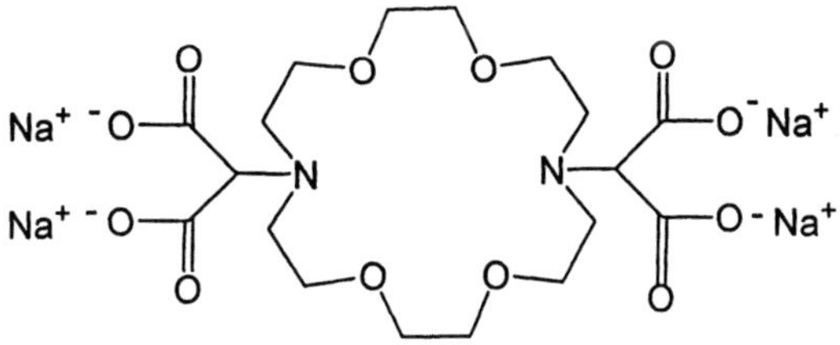

Scheme 3.

The Na_4oddm doped sol-gels (**8**) showed that they effectively removed up to 90(6)% of Sr^{2+} from 25.0 ppm Sr^{2+} solutions. Control blank sol-gel without any ligand was ineffective in Sr^{2+} removal. In a solution containing 31.25 ppm Sr^{2+} and 137.0 ppm Ca^{2+}, **8** removed 91.4(1.3)% of the Sr^{2+} in the presence of the competing Ca^{2+} ions. In these tests, 51(5)% of the Ca^{2+} ions were removed from solution. Up to 65% of the Sr^{2+} in solution was absorbed into the sorbent in the first hour of the extraction, indicating that the Na_4oddm doped sol-gel **8** offers reasonable Sr^{2+} kinetics.

After Sr^{2+} extraction cycles, the metal ions were stripped from the sol-gel with either 6.0-M HCl or 0.1-M EDTA. After washing, the sol-gel sorbents were then dried for the next extraction cycle (Scheme 4). The same sol-gel sample was used to extract Sr^{2+} through four complete cycles. The sol-gel sorbents were proven to be regenerable. Through four extraction cycles, the same sample of sol-gel was found to remove 92(7)% of the Sr^{2+} from solution (Figure 1).

BET gas adsorption experiments indicate that the Na_4oddm doped sol-gel sorbent **8** is a mesoporous material with an average pore diameter of 80 Å and specific surface area of 181.4 m^2/g. The average pore diameter of 80 Å agrees well with pore sizes of 80-100 Å measured by scanning electron microscopy (SEM). A plot of pore volume as a function of pore diameter shows that gel **8** has a narrow pore size distribution (Figure 2). N_2 adsorption isotherm plots indicate that gel **8** is indeed quite porous (Figure 3). The adsorption isotherm is indicative of type Z behavior where the desorption path lies above the adsorption path (*70*). The hysteresis between the adsorption and desorption pathways, and the closed loop are associated with porous materials.

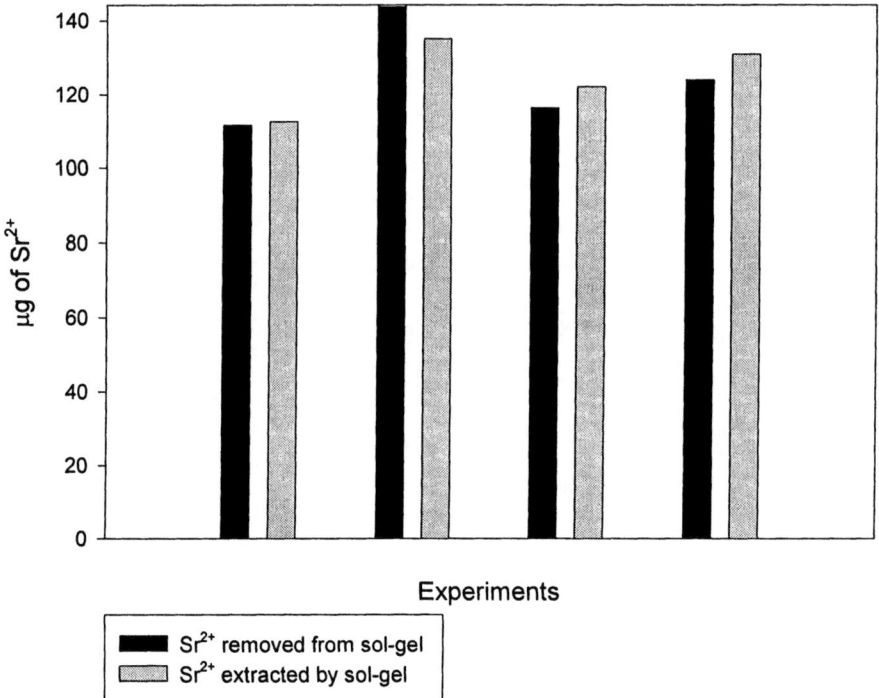

Figure 1. Sr^{2+} stripping efficiency of gel 8 with 6.0 M HCl through four experiments.

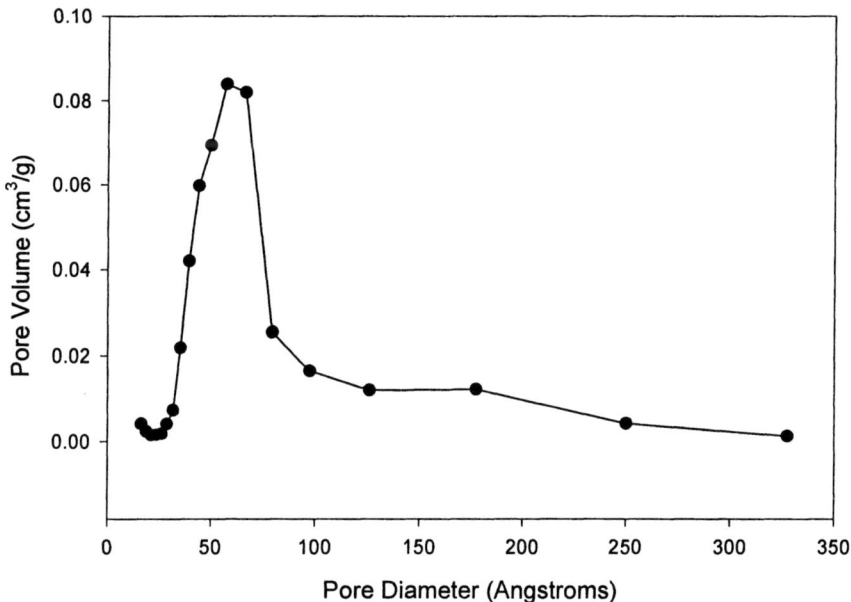

Figure 2. BET pore size distribution for Na_4oddm doped sol-gel 8.

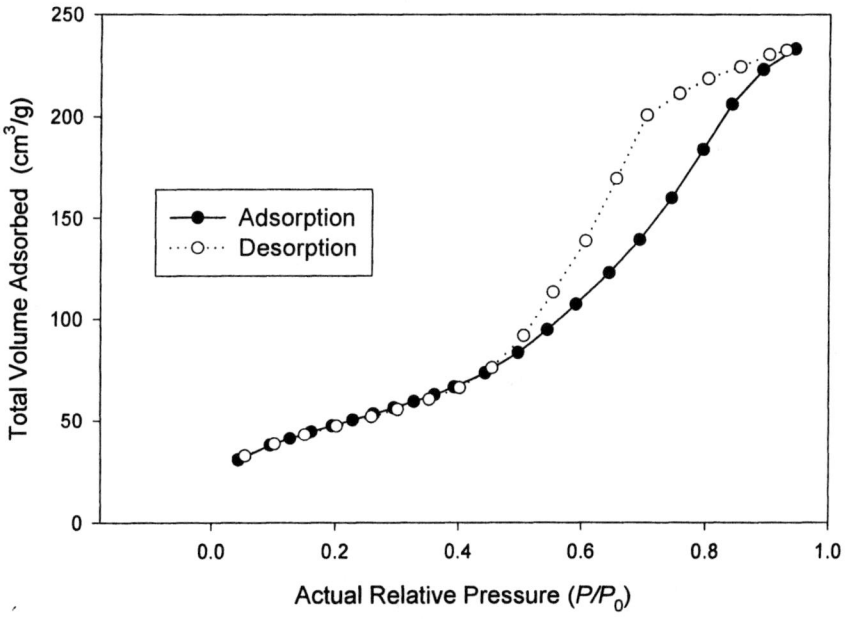

Figure 3. Nitrogen gas adsorption isotherm for Na_4oddm doped sol-gel 8.

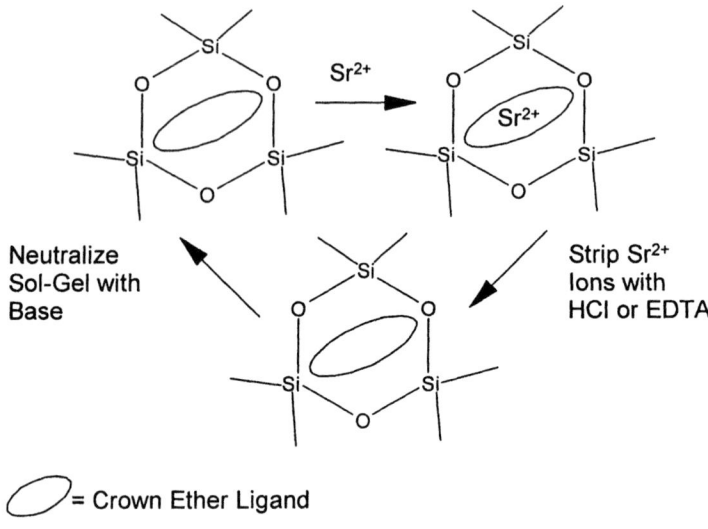

Scheme 4.

Reactions of Thioether Carboxylic Acids with Mercury(II) and Functionalized Silica Gels Grafted with Dithioacetal Derivatives (*18, 19*)

Mercapto (HS-) ligands play a unique role in the Hg^{2+} binding and its separation. Our studies of ligands with high affinity for Hg^{2+} led us to earlier reports of phthalyltetrathioacetic acid $o\text{-}C_6H_4[CH(SCH_2COOH)_2]_2$ (**9**, Scheme 5) (*15, 16*). This dithioacetal reagent was reported to show exceedingly high binding affinity for Hg^{2+} as a multi-dentate ligand. It was also noted in the original report that **9** may have decomposed in its reaction with Hg^{2+} (*15*). This ligand and its high Hg^{2+} binding affinity have attracted much interest, and silica believed to be grafted with similar dithioacetal reagents (**10**, Scheme 5) has been used in Hg^{2+} removal (*17*). Several other dithioacetals $RCH(SR')_2$ have been reported to undergo hydrolysis promoted by Lewis acids such as Hg^{2+} to yield aldehydes RCHO and mercury thiolates $Hg(SR')_2$ (*71*). The earlier report (*15, 16*) of high affinity of **9** for Hg^{2+} thus aroused our interest in this reaction. We have found that the reaction of $o\text{-}C_6H_4[CH(SCH_2COOH)_2]_2$ (**9**) with Hg^{2+} in water indeed leads to its decomposition to $C_6H_4(CH=O)_2$ and $HSCH_2COOH$. Hg^{2+} promotes this hydrolysis reaction, and reacts with $HSCH_2COOH$ to give $Hg(SCH_2COOH)_2$ (Scheme 5) (*18, 19*).

Our studies of the reactions in the preparation of **10** suggest that the dithioacetal group in $(ClCOCH_2S)_2CHPh$ undergoes hydrolysis during its reaction with the tethered amine $(-O)_3Si-(CH_2)_3NH_2$ on silica gel to give benzaldehyde $PhCH=O$ and mercapto ligands $(-O)_3Si-(CH_2)_3NHCOCH_2SH$ (**11**). It is the mercapto ligand **11** rather than dithioacetal **10** that binds Hg^{2+} ions to give the reported high selectivity and capacity (*17*) (Scheme 5).

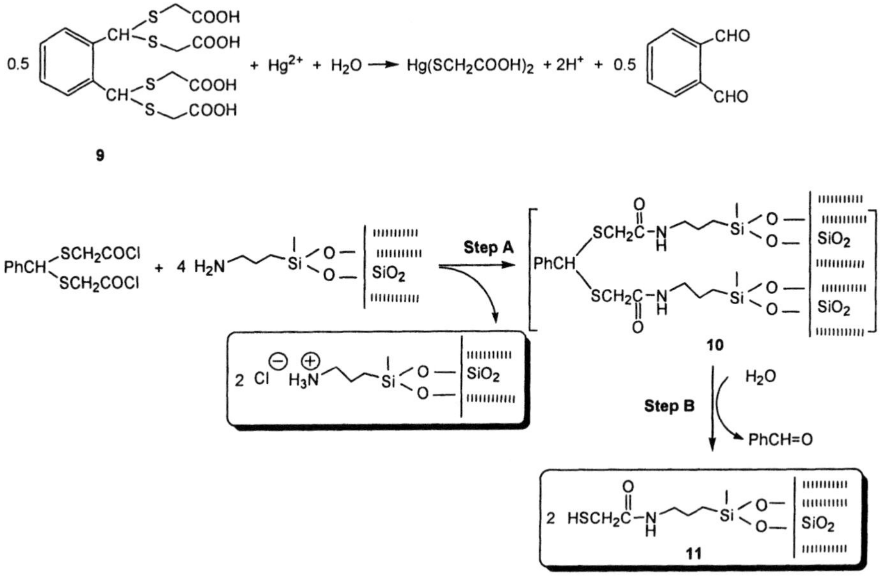

Scheme 5.

Model studies with $PhCH(SCH_2COOH)_2$ and $CH_3(CH_2)_2CH(SCH_2COOH)_2$ indicate they also react with Hg^{2+} to give aldehydes [PhCHO and $CH_3(CH_2)_2CHO$] and $Hg(SCH_2COOH)_2$. The X-ray structure of $Hg(SCH_2COOH)_2$ was determined, and found to have a linear -S-Hg-S- moiety.

Reaction of $CH_3(CH_2)_3NH_2$, a model for tethered amine ligand $(-O)_3Si-(CH_2)_3NH_2$, with $(ClCOCH_2S)_2CHPh$ yields $CH_3(CH_2)_3NH_3^+Cl^-$ and imine $PhCH=N-(CH_2)_3CH_3$ (*19*). These observations support the formation of surface-tethered mercapto $(-O)_3Si-(CH_2)_3NHCOCH_2SH$ (**11**) and imine $(-O)_3Si-(CH_2)_3NH=CHPh$ ligands in the reaction of $(ClCOCH_2S)_2CHPh$ with the tethered amine $(-O)_3Si-(CH_2)_3NH_2$ on silica gel. It is the mercapto ligand **11** that likely binds Hg^{2+} in the Hg^{2+} uptake from aqueous solution, and gives the high Hg^{2+} selectivity and its uptake capacity (94–100%) reported earlier. A trace amount of H_2O in the silica gel or the organic reactants is adequate to catalyze these reactions, driving the formation of **11** and the imine (Scheme 6). The dithioacetal ligand sites that were not hydrolyzed in the reactions to graft the ligands may be hydrolyzed to yield mercapto **11** during Hg^{2+} uptake by a reaction promoted by Hg^{2+} ions.

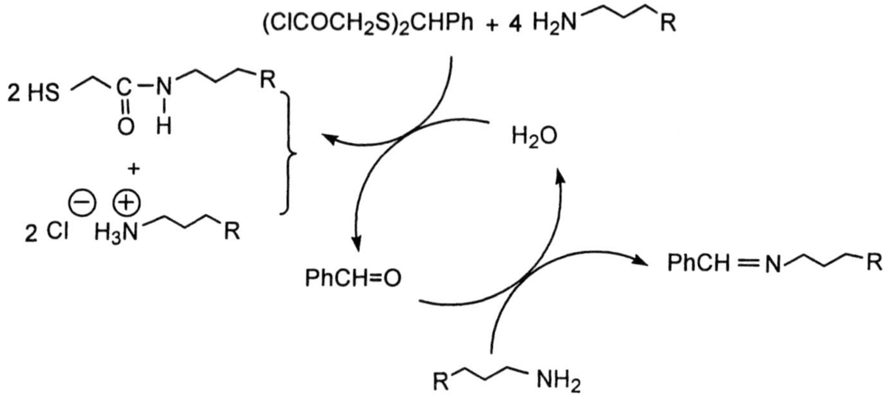

Scheme 6.

Conclusions

Ligands with high selectivity and affinity for target metal ions are a key to the design of organofunctional sol-gel materials for metal separation. Two other important issues are high ligand loading to give high metal uptake capacity, and control of porosity to allow fast diffusion of metal ions to the ligands. The studies here show that the one-step process developed here to graft or encapsulate these ligands leads to granular silica gels with high selectivity and capacity, and fast metal uptake. In addition, the granular form from the current approach is easy to handle, and especially desirable for large-scale separation of toxic metal ions.

Acknowledgments

We thank the DOE EMSP (DE-FG07-97ER14817) and the National Science Foundation, including the Research Sites for Educators in Chemistry program and Camille Dreyfus Teacher-Scholar program, for financial support.

References

1. U.S. Department of Energy. *Linking Legacies;* DOE/EM-0319; Office of Environmental Management; January 1997.
2. Poojary, D. M.; Cahill R. A.; Clearfield, A. *Chem. Mater.* **1994**, *6*, 2364.
3. Alexandratos, S. D.; Crick, D. W. *Ind. Eng. Chem. Res.* **1996**, *35*, 635 and references therein.
4. Feng, X.; Fryxell, G. E.; Wang, L.-Q.; Kim, A. Y.; Liu, J.; Kemner, K. M. *Science* **1997**, *276*, 923.
5. Mercier, L.; Pinnavaia, T. J. *Adv. Mater.* **1997**, *9*, 500.
6. Gash, A. E.; Spain, A. L.; Dysleski, L. M.; Flaschenriem, C. J.; Kalaveshi, A.; Dorhout, P. K.; Strauss, S. H. *Environ. Sci. Technol.* **1998**, *32*, 1007.
7. Lim, M. H.; Blanford, C. F.; Stein, A. *Chem. Mater.* **1998**, *10*, 467.
8. Brown, J.; Richer, R.; Mercier, L. *Microporous Mesoporous Mater.* **2000**, *37*, 41.
9. Im, H.-J.; Yang, Y.-H.; Allain, L. R.; Barnes, C. E.; Dai, S.; Xue, Z.-L. *Environ. Sci. Technol.* **2000**, *34*, 2209.
10. Yost, T. L., Jr.; Fagan, B. C.; Allain, L. R.; Barnes, C. E.; Dai, S.; Sepaniak, M. J.; Xue, Z.-L. *Anal. Chem.* **2000**, *72*, 5516.
11. Im, H.-J.; Barnes, C. E.; Dai, S.; Xue, Z.-L. *Microporous Mesoporous Mater.* **2004**, *70*, 57.
12. Im, H.-J. Ph.D. dissertation, The University of Tennessee, 2002.
13. Yost, T. L., Jr. M.S. thesis, The University of Tennessee, 2000.
14. Fagan, B. C. M.S. thesis, The University of Tennessee, 2000.
15. Jones, M. M.; Banks, A. J.; Brown, C. H. *J. Inorg. Nucl. Chem.* **1975**, *37*, 761.
16. Jones, M. M.; Banks, A. J.; Brown, C. H. *J. Inorg. Nucl. Chem.* **1974**, *36*, 1833.
17. Mahmoud, M. E.; Gohar, G. A. *Talanta*, **2000**, *51*, 77.
18. Bramlett, J. M.; Im, H.-J.; Yu, X.-H.; Chen, T.-N. Roecker, L. E.; Barnes, C. E.; Dai, S.; Xue, Z.-L. *Inorg. Chim. Acta* **2004**, *357*, 243.
19. Im, H.-J.; Barnes, C. E.; Dai, S.; Xue, Z.-L. *Talanta* **2004**, *63*, 259.
20. Dai, S.; Burleigh, M. C.; Shin, Y. S.; Morrow, C. C.; Barnes, C. E.; Xue, Z.-L. *Angew. Chem. Int. Ed. Engl.* **1999**, *38*, 1235.
21. Dai, S.; Shin, Y. S.; Ju, Y. H.; Burleigh, M. C.; Lin, J. S.; Barnes, C. E.; Xue, Z.-L. *Adv. Mater.* **1999**, *11*, 1226.
22. Dai, S.; Burleigh, M. C.; Ju, Y. H.; Gao, H. J.; Lin, J. S.; Pennycook, S. J.; Barnes, C. E.; Xue, Z.-L. *J. Am. Chem. Soc.* **2000**, *122*, 992.
23. Burleigh, M. C.; Dai, S.; Hagaman, E. W.; Barnes, C. E.; Xue, Z.-L. *ACS Symp. Series* **2000**, *778* (Nuclear Site Remediation), 146.
24. Burleigh, M. C.; Dai, S.; Barnes, C. E.; Xue, Z.-L. *Sep. Sci. Technol.* **2001**, *36*, 3395.

25. Clavier, J. W. M.S. thesis, The University of Tennessee, 2000.
26. Loy, D. A.; Shea, K. J. *Chem. Rev.* **1995**, *95*, 1431.
27. Schubert, U.; Hüsing, N.; Lorenz, A. *Chem. Mater.* **1995**, *7*, 2010.
28. Schubert, U. *New J. Chem.* **1994**, *18*, 1049.
29. Wen, J.; Wilkes, G. L. *Chem. Mater.* **1996**, *8*, 1667.
30. Sanchez, C.; Ribot, F. *New J. Chem.* **1994**, *18*, 1007.
31. Cao, W.; Hunt, A. J. *Mat. Res. Soc. Symp. Proc.* **1994**, 346, 631.
32. See, e.g., Prabakar, S.; Assink, R. A. *Mat. Res. Soc. Symp. Proc.* **1996**, *435*, 345.
33. Hüsing, N.; Schubert, U.; Mezei, R.; Fratzl, P.; Riegel, B.; Kiefer, W.; Kohler, D.; Mader, W. *Chem. Mater.* **1999**, *11*, 451.
34. Karakassides, M. A.; Fournaris, K. G.; Travlos, A.; Petridis, D. *Adv. Mater.* **1998**, *10*, 483.
35. See, e.g., Klonkowski, A. M.; Koehler, K.; Widernik, T.; Grobelna, B. *J. Mater. Chem.* **1996**, *6*, 579.
36. Pierce, J. A.; Thorne, K. J. *South. Biomed. Eng. Conf. Proc. 16th*, **1997**, 206.
37. *Stability Constants of Metal Ion Complexes, with Solubility Products of Inorganic Substances. Part 1: Organic Ligands*, The Chemical Society, London, 1957, pp 5, 6.
38. For a recent review of Cu(II)-thiolate complexes, see Mandal, S.; Das, G.; Singh, R.; Shukla, R.; Bharadwaj, P. K. *Coord. Chem. Rev.* **1997**, *160*, 191.
39. Watson, J. S. *Separation Methods for Waste and Environmental Applications*, Marcel Dekker, New York, 1999, p 13.
40. K_d = the amount of adsorbed metal (μg) per gram of gels/metal concentration (μg/mL) remaining in the solution.
41. Valera, N. S.; Hendricker, D. G. *Polymer* **1980**, *21*, 597.
42. Sahni, S. K.; Reedijk, J. *Coord. Chem. Rev.* **1984**, *59*, 1.
43. Beatty, S. T.; Fischer, R. J.; Rosenberg, E.; Pang, D. *Sep. Sci. Technol.* **1999**, *34*, 2723.
44. Fischer, R. J.; Pang D.; Beatty S. T.; Rosenberg, E. *Sep. Sci. Technol.* **1999**, *34*, 3125.
45. Ramsden, H. E. *U.S. Patent* No. 4,540,486, 1985. Assignee: J. T. Baker Co.
46. *Principles and Practices of Solvent Extraction*; Rydberg, J., Musikas, C., Choppin, G. R., Eds.; Dekker: New York, 1992.
47. Bradshaw, J. S.; Izatt, R. M.; Bordunov, A. V.; Zhu, C. Y.; Hathaway, J. K. In *Comprehensive Supramolecular Chemistry*; Atwood, J. L., Davies, J. E. D., Macnicol, D. D., Vogtle, F., Lehn, J. M., Gokel, G. W., Eds.; Elsevier: New York, 1996; Vol. 1, Chapter 2.
48. Moyer, B. A. In *Comprehensive Supramolecular Chemistry*; Atwood, J. L.; Davies, J. E. D.; Macnicol, D. D.; Vogtle, F.; Lehn, J. M.; Gokel, G. W., Eds.; Elsevier: New York, 1996; Vol. 1, Chapter 10.

49. Gokel, G. W. *Crown Ethers and Cryptands*; Royal Society of Chemistry, Cambridge, 1991.
50. Yordanov, A. T.; Roundhill, D. M. *Coord. Chem. Rev.* **1998**, *170*, 93.
51. Bradshaw, J. S.; Izatt, R. M. *Acc. Chem. Res.* **1997**, *30*, 338.
52. Sessler, J. L.; Kral, V.; Genge, J. W.; Thomas, R. E.; Iverson, B. L. *Anal. Chem.* **1998**, *70*, 2516.
53. Hankins, M. G.; Hayashita, T.; Kasprzyk, S. P.; Bartsch, R. A. *Anal. Chem.* **1996**, *68*, 2811.
54. Espinola, J. G. P.; Defreitas, J. M. P.; Deoliveira, S. F.; Airoldi, C. *Colloids Surf.* **1992**, *68*, 261.
55. Dudler, V.; Lindoy, L. F.; Sallin, D.; Schlaepfer, C. W. *Aust. J. Chem.* **1987**, *40*, 1557.
56. Leyden, D. E.; Luttrell, G. H. *Anal. Chem.* **1975**, *47*, 1612.
57. Fischer, R. J.; Pang, D.; Beatty, S. T.; Rosenberg, E. *Separ. Sci. Technol.* **1999**, *34*, 3125.
58. Sharma, C. V. K.; Clearfield, A. *J. Am. Chem. Soc.* **2000**, *122*, 1558.
59. Allain, L. R.; Sorasaenee, K. Xue, Z.-L. *Anal. Chem.* **1997**, *69*, 3076.
60. Allain, L. R.; Xue, Z.-L. *Anal. Chem.* **2000**, *72*, 1078.
61. Allain, L. R.; Canada, T. A.; Xue, Z.-L. *Anal. Chem.* **2001**, *73*, 4592.
62. Allain, L. R.; Xue, Z.-L. *Anal. Chim. Acta* **2001**, *433*, 97.
63. Canada, T. A.; Allain, L. R.; Xue, Z.-L. *Anal. Chem.* **2002**, *74*, 2535.
64. Canada, T. A.; Xue, Z.-L. *Anal. Chem.* **2002**, *74*, 6073.
65. Canada, T. A.; Beach, D. B.; Xue, Z.-L. *Anal. Chem.* **2005**, *77*, 2842.
66. Rodman, D. L.; Pan, H.; Clavier, C. W.; Feng, X.; Xue, Z.-L. *Anal. Chem.* **2005**, *77*, 3231.
67. Thomasset, M. In *Radionuclide: Metab. Toxic. Proc. Symp.* Galle, P., Masse, R., Eds.; 1982, p. 98.
68. Spiers, F. W.; Vaughan, J. *Leuk. Res.* **1989**, *13*, 347.
69. Brucher, E.; Gyori, B.; Emri, J.; Jakab, S.; Kovacs, Z.; Solymosi, P.; Toth, I. *J. Chem. Soc. Dalton Trans.* **1995**, 3353.
70. Dollimore, D.; Spooner, P.; Turner, A. *Surf. Technol.* **1976**, *4*, 121.
71. See, e.g., Satchell, D. P. N.; Satchell, R. S. *Chem. Soc. Rev.* **1990**, *19*, 55 and references therein.

Facility Inspection, Decontamination, and Decommissioning: Materials Science

Chapter 13

Investigation of Nanoparticle Formation During Surface Decontamination and Characterization by Pulsed Laser

Meng-Dawn Cheng[1] and Doh-Won Lee[2]

[1]Environmental Sciences Division, Oak Ridge National Laboratory, Oak Ridge, TN 37831
[2]Oak Ridge Institute for Science and Education, Oak Ridge, TN 37830

The production of ultrafine and nanoparticles from a surface is dependent on the laser energy and laser wavelength used to treat and on the material used to construct the surface. Under dry conditions, the minimal laser fluence (mJ cm^{-2}) required to produce a detectable amount of particles was found to be the greatest for a pure material, alumina, then for a complex mixture, concrete, with the least for a simple mixture, stainless steel, using both visible (532-nm) and UV (266-nm) laser wavelengths. The threshold energy requirement was found to be significantly higher when a shorter laser wavelength was used. The results indicate that for a given amount of laser energy used, there are more than twice the particles produced when a 532-nm wavelength is used than a 266-nm, although a 266-nm photon has 2 times more energy than a 532-nm. For both wavelengths, the total number concentration of produced particles is found to be linearly proportional to laser fluence. The correlation of the log-log linearity is excellent, indicated by a R^2 value close to 1 for all materials. The models were derived, empirically, for predicting the amount of particles that could be removed from the surface of different materials using different lasers operated at low fluence conditions.

Introduction

Decontamination and decommissioning (D&D) of a large number of nuclear facilities is a major effort at the U.S. Department of Energy (DOE) complexes across the country. Use of laser plasma for surface decontamination/cleaning is a new and effective technique. A large quantity of very small particles is produced during the decontamination process. Effective production of particles is critical in determining the surface cleaning efficiency. However, the particles could contain surface contaminants like toxic heavy metals (e.g., Cr, Hg, Pb, and Ni), radionuclides (e.g., Th, Cs, and U), and organic solvents that all might cause health concerns. In this project sponsored by the DOE Environmental Management Science Program (EMSP), we investigated the relationships of nanoparticle formation by the laser decontamination process using laboratory-prepared target surfaces made from Portland cement (concrete), stainless steel 316, and pure alumina. The first two samples are commonly found in DOE installations, while the last sample is used for understanding of fundamental processes. The data were correlated among the particles, surface chemistry, and the laser characteristics. The experiments were conducted to determine the threshold energy needed to remove particles from the surface materials. This is important in understanding the cleaning efficiency and the photon-material interaction at the material surface. The objectives of this study are to determine the minimal laser energy required to decontaminant a surface, which is surface dependent, and to characterize the particles generated during the decontamination processes.

Material and Method

Two targets were selected to represent the range of surfaces commonly found in DOE installations, and one pure substance was selected for contrasting the complex surfaces of the other targets listed. The two complex composition targets were Portland cement or concrete and Stainless Steel 316. The pure substance was alumina. The chemical compositions of the targets are listed in Table I. Shown in the table are the identified compounds and the relative weight proportions of the compounds. Concrete consists of several oxides and some trace elements in which 45% is CaO as the major oxide and SiO_2 16.7%. Portland cement specimens of about 5-cm or 2-in.-diameter circular blocks were prepared initially without aggregate, cured under moist conditions for a minimum of 30 days, and then air dried for a minimum of 1 week before use (*1*). Stainless steel consists of Fe, Cr, and Ni with weight percentages of 70, 19, and 11, respectively.

Figure 1 shows the experimental setup used in this study. An airtight chamber was used to contain a rotating target disc 5-cm in diameter. The disc

Table I. Chemical Composition of Targets Used in This Study

Alumina	Weight %	Stainless Steel	Weight %	Concrete	Weight %
Al_2O_3	100	Fe	70	Al_2O_3	3.35
		Cr	19	CaO	44.79
		Ni	11	Fe_2O_3	2.66
				K_2O	0.49
				MgO	2.01
				MnO_2	0.06
				Na_2O	0.14
				SiO_2	16.65
				SrO	0.04
				TiO_2	0.21
				Loss-on-Ignition (1400 °C)	28.53
				Sum of Oxides	98.95
				Minor Elements	
				Ag	< 0.04
				As	< 0.38
				Ba	0.01
				Be	< 0.002
				Cd	< 0.02
				Co	< 0.04
				Cr	< 0.04
				Cu	< 0.04
				Mo	< 0.01
				Ni	< 0.09
				P	< 0.91
				Pb	< 0.19
				Sb	< 0.38
				Se	< 0.38
				V	< 0.004
				Zn	< 0.09

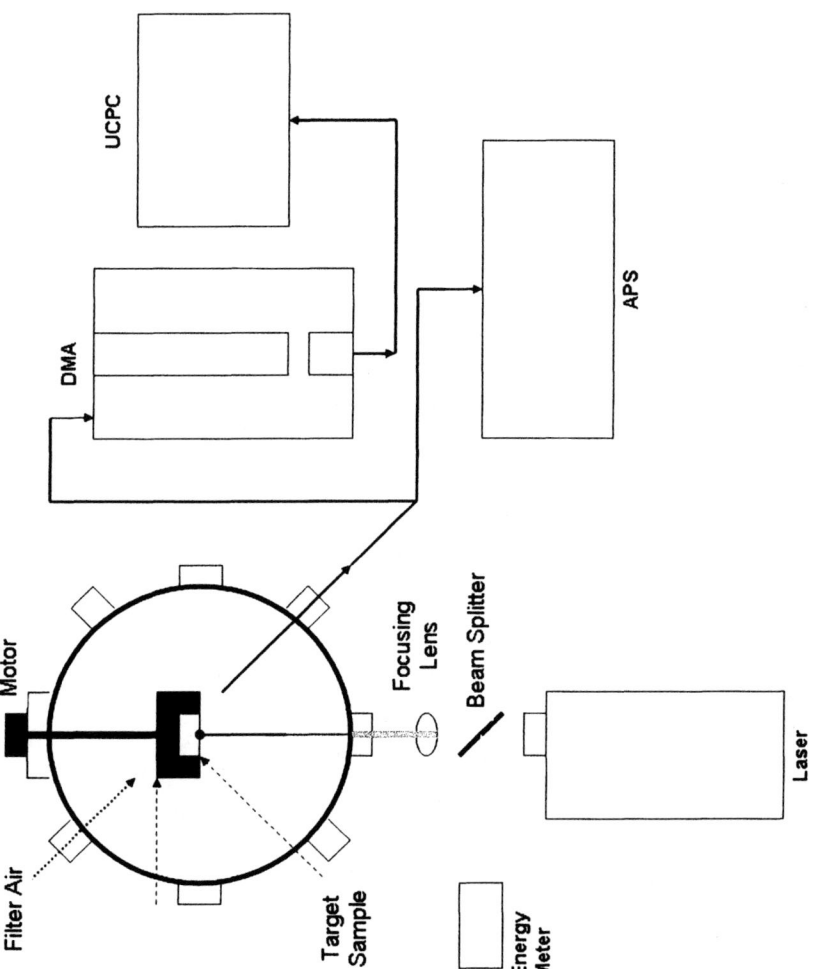

Figure 1. Experimental setup of surface cleaning and particle generation.

was subject to laser treatment through which process particles were generated and characterized in this study. As indicated earlier, the disc material was alumina, concrete, or stainless steel. An Nd: YAG laser was used as the energy source; the laser emits either a 532- or 266-nm wavelength depending on the operational mode for a specific experiment. The maximum energy for the 532-nm wavelength was approximately 100 mJ per pulse, while that for the 266-nm was around 25 mJ. Each pulse is 3–5 nanoseconds. The laser was fired at a 10-Hz repetition rate. The laser energy was monitored by splitting the beam for a small amount to a power meter located a 90° angle location (see Figure 1).

Depending on the number concentration of particles generated and the purpose of a specific experiment, we employed a suite of instruments consisting of the scanning mobility particle sizer (SMPS) and an ultrafine condensation particle counter (UCPC). A diluter was used in some instances, particularly when the UCPC was used. The maximum number concentration that the UCPC can measure is about 10^5 cm^{-3}. The UCPC does not provide the size distribution data, but it yields reliable measurement of total particle counts from 3 nm and larger in diameter (2, 3). The SMPS yields the size distribution of particles, but it would not provide reliable results if the particle number concentration was low, around 10^3 cm^{-3} (or less), for example, because it divides the number into 64 bin sizes, resulting in a small count for each bin and leading to a large Poisson counting uncertainty. The Poisson counting uncertainty is proportional to the square root of the total counts; i.e., for a count of 100, there is a 10% counting uncertainty. The higher the count is, the lower the uncertainty will be.

Results and Discussion

Several experiments were performed; an experiment could run for as long as 8 h or as short as 5 min depending on the objective of the experiment. With a 5-s measurement rate of UCPC, a very large volume of data was generated for each experiment. Under normal circumstances, 120 data points could be generated for analysis in a 10-min experiment, and a total of data points of 5,760 could be produced during an 8-h experiment more than sufficient to calculate the statistics.

Effects of 532-nm Laser Wavelength on Particle Generation from Three Different Substrates

The experimental results obtained with visible (532-nm) wavelength energy are shown in Figure 2. It is noted that both axes are on the log scale, and all three regression lines are virtually parallel to each other. After the data were processed and quality checked, the mean value and standard deviation were computed. Each data point on the plot is a result of at least of 60 data points. The range of data variation is shown by the error bar associated with each data point, which are small indicating high-precision experiments. Also noted is that the log-log plot for alumina, however, was slightly curved upward in appearance in contrast to an observed linear relationship of concrete and stainless steel. However, the goodness-of-fit indicated by the R^2 value was 0.96 for the alumina curve, 0.98, and 0.99 for the concrete and stainless steel curves, respectively. These are indicators of a strong linear relationship between particle concentration and laser fluence on a log-log scale.

Effects of 266-nm Wavelength on Particle Generation

The results obtained with a deep UV (266-nm) wavelength energy are shown in Figure 3, which is also a log-log plot of the total particle concentration vs the laser fluence of 266-nm. The linear relationships for all three materials are strong, with R^2 values of 1.0 for concrete, 0.99 for alumina, and 0.98 for stainless steel. Again, the experimental data variations were extremely small; each point and the error bar were obtained with at least 105 data points. The concrete and stainless steel lines are parallel to each other; however, the alumina line is not parallel, which could be caused by lack of data at larger fluence above 1500 mJ cm^{-2}, for example. Based on the 532-nm results shown previously, we think there is no reason to doubt that the alumina line would not be parallel to the other two lines in Figure 3. Thus, the relationships between particle produced and the laser fluence of 266-nm wavelength for three different materials could be considered as parallel to each other. This result indicates that the total concentration of particle generated is log-log linearly proportional to laser fluence.

Correlation Analysis of Laser Fluence and Concentration of Generated Particles

The numerical values of the regression lines shown in Figures 2 and 3 are reported in Table II. The particle concentrations (Y) are observed to relate to the laser fluence (X) linearly on a log-log scale shown in Figures 2 and 3. We used a log-log linear regression equation to describe the result. Although Liu et al. (4) used log-log linear regression for correlating the electron number density (not particle number concentration) with laser irradiance, we think our

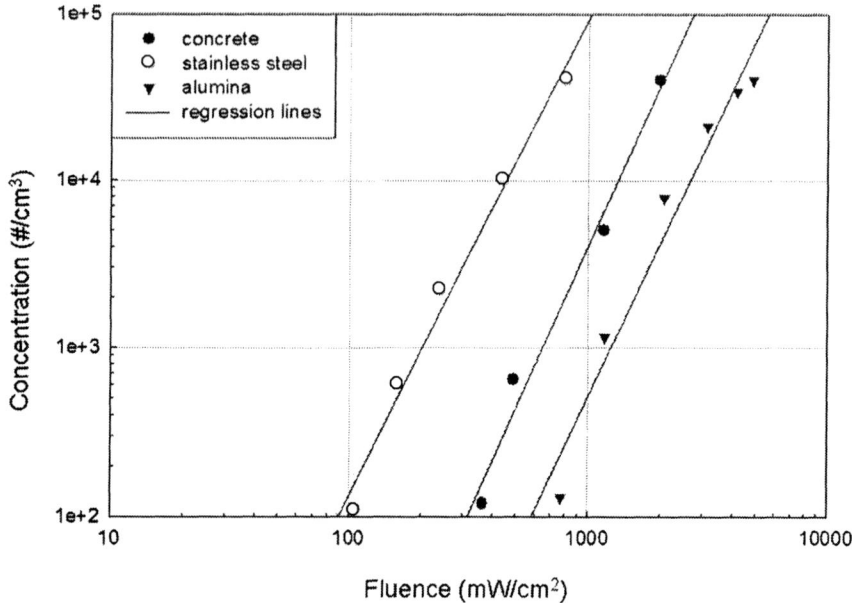

Figure 2. Observed log-log linear relationship between the generated particle concentration (# cm^{-3}) and the 532-nm wavelength laser fluence (mJ cm^{-2}).

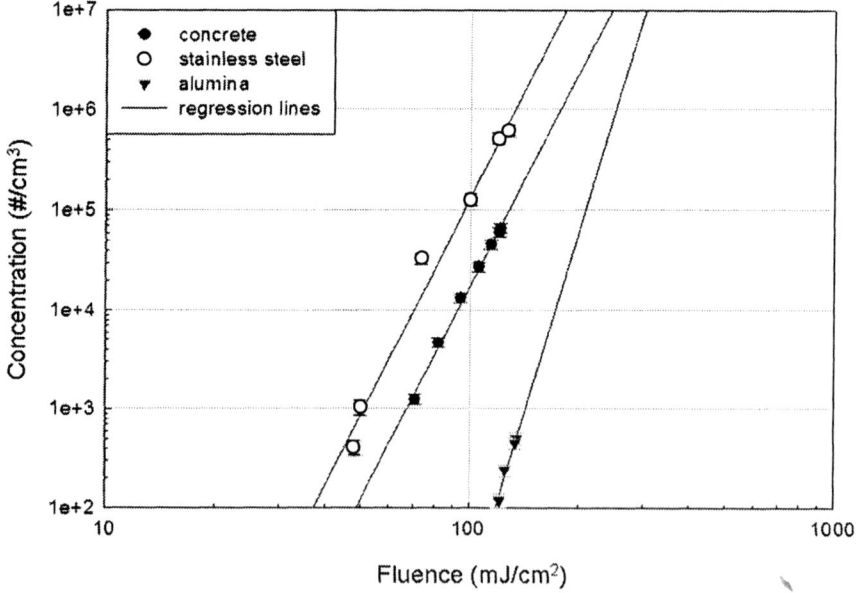

Figure 3. Observed log-log linear relationship between the generated particle concentration (# cm^{-3}) and the 266 nm wavelength laser fluence (mJ cm^{-2}).

results are the first to link the produced particle concentration and laser fluence distinguished by laser wavelength. The log-log function is given as follows:

$$\ln(Y) = \ln(A) + B\ln(X) \qquad (1)$$

The slope B in the regression eq 1 describes the particle generation capacity (ψ) in # of particles cm^{-3} per mJ cm^{-2}, while the parameter A describes the background particle concentration in # of particles cm^{-3}. The particle generation capability (ψ) was found to be 3.33 for concrete, 2.94 for stainless steel, and 3.05 for alumina using the 532-nm wavelength. The calculated ψ values are 7.16 for concrete, 7.43 for stainless steel, and 11.0 for alumina using the 266-nm wavelength. The ratio of ψ values [i.e., ψ(532-nm)/ ψ(266-nm)] is 2.35, but the ratio of single photon energy of the two wavelengths is 0.5 approximately the inverse of the ratio of the ψ values. This derivation leads to a tentative conclusion that, at the same fluence, the 532-nm wavelength was capable of producing 4 times more particles than the 266-nm wavelength.

Table II. Log-log Linear Regression of Total Particle Concentration vs Fluence [ln(Y) = ln(A) + B ln(X)]

	A	B	R^2
532 nm Wavelength			
Concrete	4.70E-7	3.33	0.98
Stainless steel	1.68E-4	2.94	0.99
Alumina	3.57E-7	3.05	0.96
266 nm Wavelength			
Concrete	5.83E-18	7.16	1.0
Stainless steel	6.64E-18	7.43	0.98
Alumina	2.11E-32	11.0	0.99

The background particle concentrations (A) for alumina, stainless steel, and concrete using the 266-nm wavelength were lower than the 532-nm cases by 11 to 25 orders of magnitude. The A value should not be dependent upon laser wavelength nor the material chosen for the target, because theoretically it is obtained as the laser fluence is "extrapolated" to zero in eq 1. In practice, however, the result suggests that a threshold energy level exists for generating particle from a material surface. The threshold of particle generating mechanism is both laser wavelength and target material dependent.

Energy Required to Remove Particles from a Surface

To produce a detectable level (100 particles cm^{-3}) of particles, the threshold energy (ξ) required is obtained by substituting 100 for the particle concentration (Y) to yield the fluence (X). Note that this is not a particle that is deposited on the surface, but a particle that is generated from the target material where there was no particle on the surface. We have also had experimental data (not shown) indicating that the particles are mostly neutral, not carrying surface charges. To perform a reliable assessment of the threshold energy, one has to consider the limitation of current particle technology in counting "one particle." We chose 100 particles as an acceptable counting error that is 10% according to the Poisson counting statistics. The threshold energy is conceptually defined as the minimal energy needed to remove material in the form of ablated particles (with the amount to be reliably detected by the UCPC) from a surface during a laser cleaning process. The results (in Table III) show that ξ is dependent on laser wavelength and material property. For a given material, the 532-nm could remove the minimal detectable amount of particles about twice as efficiently for alumina and concrete, and four times as efficiently for stainless steel as the 266-nm. For the 532-nm wavelength, alumina required 86% higher energy than concrete that required 240% more for stainless steel. This trend is similar when a 266-nm wavelength was used; i.e., 137% higher for alumina than for concrete, and 51% for concrete than for stainless steel. It is interesting to note that ξ is substantially higher for a 266-nm wavelength than for a 532-nm wavelength for a given material, even the energy of a 266-nm photon is 2 times higher than that of a 532-nm photon. This result appears to be consistent with our previous analysis discussed in the Correlation Analysis section. It further indicates that the production of particles at the range of fluence discussed herein may not be a photo-ionization process, as the result is in reverse of the process that one would expect short-wavelength photons to trigger.

Table III. Estimated Threshold Energy (mJ cm^{-2}) of Particle Production for the Three Materials (Estimated Using a Threshold Value of 100 cm^{-3})

	Alumina	*Concrete*	*Stainless Steel*
532 nm	588	317	92
266 nm	1,152	486	381

Conclusions

The threshold energies required to produce particles from alumina, concrete, and stainless steel surface are estimated based on experimental data. The threshold energy is dependent on the laser wavelength, fluence, and the material of the surface. It was found that the threshold energy was the greatest for alumina, then for concrete, and the least for stainless steel for a given laser wavelength used. Use of a shorter wavelength did not improve the production of particles as expected because the threshold energy or energy requirement was substantially higher when a 266-nm laser wavelength was used. Note that the number of particles produced per given volume of air was not twice higher by the 266-nm laser than the 532-nm laser, although the single-photon energy for the former wavelength is two times higher than for that of the later. A log-log linear model was found to provide a good correlation between particle production and laser fluence.

Acknowledgments

The work was performed by researchers at the Oak Ridge National Laboratory (ORNL) for a project (EMSP Project # 82,807) funded by the DOE, Office of Science, Biological and Environmental Research Program, EMSP. ORNL is managed by UT-Battelle, LLC, for the DOE under contract DE-AC05-00OR22725. D.-W. Lee was supported in part by an appointment to the ORNL Postdoctoral Research Associates Program administered jointly by the ORNL and the Oak Ridge Institute for Science and Education. Baohua Gu of the Environmental Sciences Division at the ORNL is acknowledged for providing and preparing the samples used in this study.

References

1. Savina, M. R.; Xu, Z.; Wang, Y.; Leong, K.; Pellin, M. J. Pulsed laser ablation of cement and concrete. *J. Laser Applications* **1999**, *11* (6), 284–287.
2. Stolzenburg, M. R.; McDermott, W. T.; Schwartz, A. Performance studies of ultrafine efficiency membrane filters using a 3.5 nm particle detector. *J. Aerosol Sci.* **1988**, *19*, 1015–1018.
3. Stolzenburg, M. R.; McMurry, P. H. An ultrafine aerosol condensation nucleus counter. *Aerosol Sci. Technol.* **1991**, *14*, 48–65.
4. Liu, H. C.; Mao, X. L.; Yoo, J. H.; Russo, R. E. Early phase laser induced plasma diagnostic and mass removal during single-pulse laser ablation of silicon. *Spectrochim. Acta, Part B* **1999**, *54*, 1607–1624.

Chapter 14

Recent Progress in the Development of Supercritical Carbon Dioxide-Soluble Metal Ion Extractants: Solubility Enhancement through Silicon Functionalization

Mark L. Dietz[1], Daniel R. McAlister[2], Dominique Stepinski[2], Peter R. Zalupski[2], Julie A. Dzielawa[1], Richard E. Barrans, Jr.[1], J. N. Hess[2], Audris V. Rubas[1], Renato Chiarizia[1], Christopher Lubbers[3], Aaron M. Scurto[3], Joan F. Brennecke[3], and Albert W. Herlinger[2]

[1]Chemistry Division, Argonne National Laboratory, 9700 South Cass Avenue, Argonne, IL 60439
[2]Department of Chemistry, Loyola University at Chicago, Chicago, IL 60626
[3]Department of Chemical and Biomolecular Engineering, University of Notre Dame, South Bend, IN 46556

> Efforts to employ partially-esterified alkylenediphosphonic acids (DPAs) in the supercritical fluid extraction of actinides have been hampered by their modest solubility in unmodified supercritical carbon dioxide ($SC-CO_2$). In an effort to design DPAs that are soluble in $SC-CO_2$, various silicon-substituted alkylenediphosphonic acids have been prepared and characterized, and their behavior compared with that of conventional alkyl-substituted reagents. Silicon substitution is shown to enhance the CO_2-philicity of the reagents, while other structural features (e.g., alkylene bridge length) determine their aggregation and extraction properties. The identification of DPAs combining desirable extraction properties with adequate solubility in $SC-CO_2$ is shown to be facilitated by the use of molecular connectivity indices.

Background

In recent years, there has been growing interest in the development of more environmentally benign ("green") approaches to chemical synthesis, catalysis, and separations (*1*). In many instances, a key element in efforts to devise such approaches is the use of so-called "neoteric" solvents (*2–5*). Of this diverse group of reagents, probably none has received more attention than supercritical carbon dioxide (SC-CO_2). Carbon dioxide, in either its liquid or supercritical states, offers many benefits as an environmentally benign solvent, among them low cost, lack of toxicity, and non-flammability. In addition, CO_2 does not contribute to either photochemical smog or ozone depletion. Moreover, in its supercritical state (above its readily accessible critical point of 31 °C and 73.8 atm), its solvating power can be tuned over a wide range by relatively small changes in pressure and temperature (*6*).

Although the solvent power of SC-CO_2 is tunable, it is also modest at best, and although a variety of low molecular weight organic compounds can be dissolved in it, many useful and interesting substances cannot, among them ionic materials (e.g., metal ions) and many metal ion extractants (*7*). The solvent power of SC-CO_2 can, as is well known, be improved by the addition of a co-solvent (*8*). Mixtures of SC-CO_2 and methanol or acetone, for example, can generally dissolve far greater quantities of organic material than can SC-CO_2 alone (*8, 9*). Surfactants containing CO_2-soluble functional groups have also been used to disperse poorly soluble compounds in SC-CO_2 media (*10*). These approaches are not always effective or desirable, however. High concentrations of modifiers can significantly alter the temperature and pressures required for operation, for example (*11*). For this reason, in the case of metal ion extractants, there has been considerable interest in identifying or devising (through appropriate structural modification) reagents that are readily soluble in unmodified SC-CO_2. Most commonly, this structural modification has involved the attachment of fluorinated substituents to a conventional ligand (*12*). Although this is frequently effective, fluorinated compounds as a class are expensive, and therefore, fluorinated extractants are unlikely to be practical for large-scale applications (*13*). As a result, recent efforts to develop CO_2-soluble extractants have focused on ligands bearing branched alkyl substituents (*14, 15*).

One additional option, one that has received little attention to date, involves the use of silicon-based functional groups such as silicones, polymers containing chains of alternating silicon and oxygen atoms (Si-O-Si-O•••). Like fluorinated compounds, silicones are stable under acidic and alkaline conditions, extreme temperatures, and UV light. They are also very hydrophobic (*16*) and soluble in SC-CO_2 (*17, 18*). Unlike fluorinated compounds, however, the preparation of silicones requires only mild conditions (*16, 19*), and they and their precursors are relatively inexpensive. Despite these virtues, the use of silicones or other silicon-based functional groups as a means of improving ligand or extractant compatibility with SC-CO$_2$ has thus far been limited. Rather, efforts have been largely confined to silicon-containing surfactants (20). Moreover, the silicones that have been investigated are polydisperse mixtures (20, 21).

Our work has concerned the possibility of coupling the unique solvent properties of SC-CO$_2$ with the remarkable metal ion complexing properties of diphosphonic acids (22, 23). These compounds have previously been found to form extremely stable complexes with a variety of metal ions, particularly lanthanides and actinides (24–28), and upon appropriate substitution have been shown to provide the basis of powerful metal ion extractants (29–31), novel ion-exchange resins (32–38), and extraction chromatographic materials (39). Their use in SC-CO$_2$ could therefore provide a powerful new means of extracting selected metal ions from a variety of media (e.g., porous solids). Neither unsubstituted diphosphonic acids nor the alkyl-substituted derivatives reported to date, however, are sufficiently soluble in unmodified SC-CO$_2$ to constitute practical metal ion extractants in this medium. We have therefore chosen to prepare and characterize alkylenediphosphonic acids bearing discrete, well-defined, silicon-containing functional groups for possible application in SC-CO$_2$.

Siloxane-Functionalized *Gem*-Diphosphonates

Alkylenediphosphonic acids contain two sites that can be functionalized to adjust their solubility properties, the alkylene bridge separating the two phosphorus atoms (a) and the acidic POH groups (b):

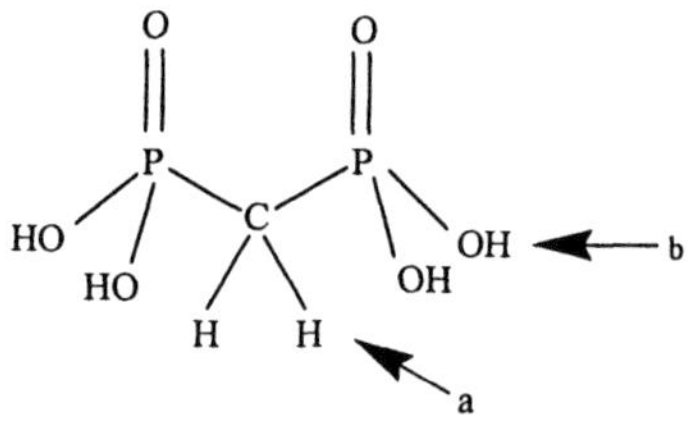

In principle, the availability of two sites for derivatization, the ability to adjust the bridge length, and the multitude of structural variations possible in the functional groups appended offer many opportunities to devise a means of boosting the SC-CO$_2$ solubility of diphosphonic acids. At the same time, however, the situation represents a formidable challenge, namely, how to identify the most appropriate candidates for synthesis and evaluation. To simplify this process, we opted initially to merely demonstrate proof of concept, that is, that appending one or more discrete, silicon-bearing moieties to a ligand could influence its compatibility with carbon dioxide. Furthermore, initial efforts were directed at incorporation into the extractant molecule of dimethylsiloxane oligomers (i.e., methylsilicones), as related polymers have previously been found to be CO_2-philic (20). In addition, to simplify the synthesis as much as possible, geminal diphosphonates (essentially DPAs in which the protons have been replaced by various alkyl groups) were selected for initial investigations (Table I). Finally, the CO_2 compatibility of each extractant was evaluated not by directly measuring its solubility in SC-CO_2, but rather by measuring the uptake of carbon dioxide by each compound under sub-critical conditions. Despite being obtained at lower pressures than would likely be used in an actual extraction, the data are nonetheless indicative of the relative affinity for CO_2 of the various compounds.

Table I. Structures of Oligo(dimethylsiloxane)-Substituted Tetraalkyl *Gem*-Diphosphonates

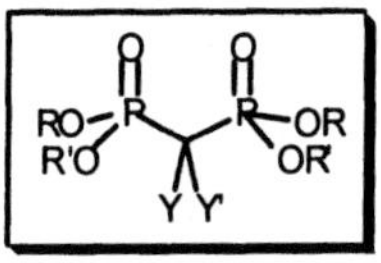

	R	R'	Y	Y'
1	Et	Et	$(CH_2)_3$ SI	H
2	Et	Et	$(CH_2)_3$ SI	$(CH_2)_3$ SI
3	Et	Et	$(CH_2)_3$ SI'	H
4	Et	Et	$(CH_2)_3$ SI'	$(CH_2)_3$ SI'
5	Me	$(CH_2)_3$ SI	H	H
6	Me	$(CH_2)_3$ SI"	H	H
7	Me	Eh	$(CH_2)_3$ SI	$(CH_2)_3$ SI
8	Et	Et	H	H
9	Me	Eh	H	H

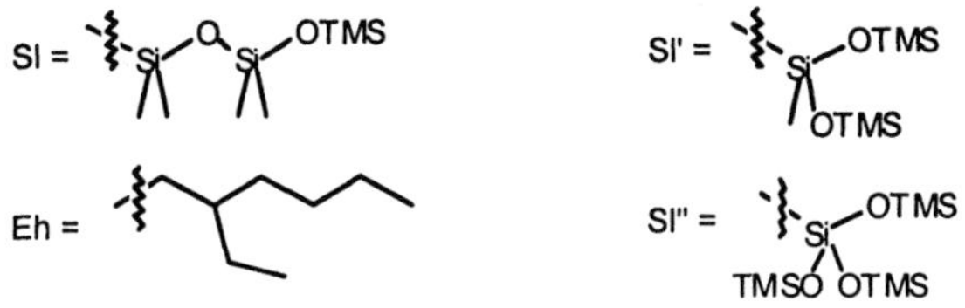

SI = -Si(Me)_2-O-Si(Me)_2-OTMS

SI' = -Si(OTMS)_2-OTMS

Eh = 2-ethylhexyl

SI" = -Si(OTMS)(TMSO)-OTMS

The approaches employed to synthesize the test compounds are summarized in Schemes 1 and 2. Briefly, the *gem*-diphosphonates were functionalized at either the methylene bridge or at the ester groups. Because the phosphonate groups are electron withdrawing, the bridging methylene group can be readily alkylated. Esters of phosphonic acids may be formed by many of the same routes used to esterify carboxylic acids. Here, the bridge-substituted diphosphonates were produced by deprotonation of the appropriate tetraester with potassium *t*–butoxide followed by reaction with allyl iodide. The resulting compounds were hydrosilylated with the appropriate oligo (dimethylsiloxy)silane to yield the desired siloxane. The compounds substituted at the ester groups were prepared by first hydrosilylating an allyl benzyl ether to yield a siloxane-containing benzyl ether. The benzyl protecting group was then removed to yield an alcohol, which was reacted with methylenebis(phosphonic dichloride) in the presence of tetrazole to provide the desired product, which was isolated from by-products by column chromatography.

The CO_2-philicity of each compound was determined by measuring the amount of CO_2 that partitions into each at 40 °C from 30 to 90 bar. Figure 1 summarizes the results as the mole fraction of CO_2 incorporated as a function of

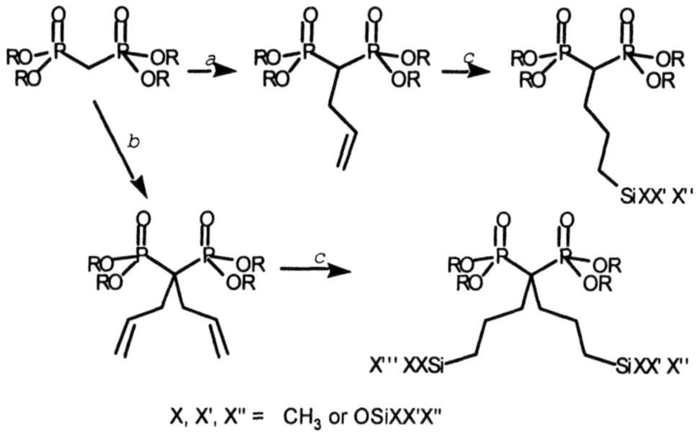

X, X', X" = CH_3 or OSiXX'X"

Key: a. i. KO*t*-Bu (1 eq), benzene, 18 h; ii. allyl iodide (1 eq), 18 h
  b. i. KO*t*-Bu (3 eq), benzene, 18 h; ii. allyl iodide (3 eq), 18 h
  c. HSiXX'X" (1.5 eq/allyl), H_2PtCl_6 (5 x 10^{-5} eq), 100 °C, 2-6 h

Scheme 1. Synthesis of Bridge-Substituted Tetraalkyl Gem-Diphosphonates.

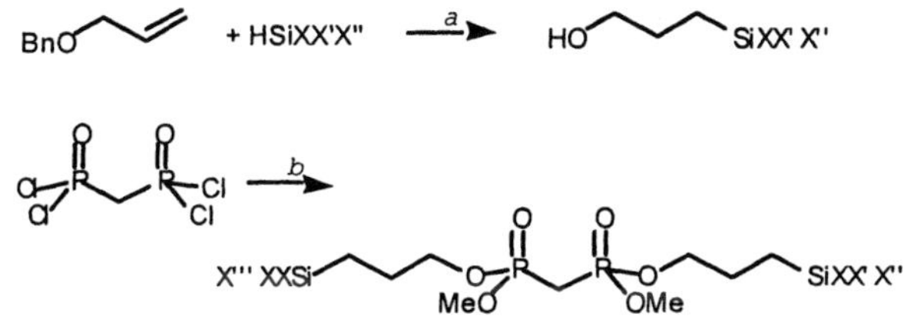

Key: a. i. H_2PtCl_6 (5 x 10^{-5} eq), 100 °C, 2-24 h; ii. H_2, 50 psig, 24 h
b. i. $HO(CH_2)_3SiXX'X''$ (2 eq), benzene, tetrazole (0.1 eq), NPr^i_2Et (4 eq), 18 h, RT; ii. MeOH (excess), 4 h, RT.

Scheme 2. Synthesis of Methylenebis(siloxypropylmethylphosphonate) Esters.

pressure. From these data, several important conclusions can be drawn. First, siloxane substituents clearly enhance the affinity of *gem*-diphosphonates for CO_2. Moreover, the enhancement depends on both the number of siloxane substituents and their structure. For example, in the region below 75 bar, the parent compound, **8**, with no siloxane functionality, absorbs less CO_2 than does compound **1**, which bears one linear trisiloxane substituent on the bridging carbon. This compound, in turn, absorbs less than **2**, which bears two such trisiloxane groups. Also, compounds **3** and **4**, which have their point of attachment as the central, rather than the end, silicon atom, absorb more CO_2 than their straight chain analogs, **1** and **2**. These results demonstrate clearly that the concept of improving the CO_2-philicity of a ligand by incorporation of a simple (i.e., non-polymeric) silicon-bearing substituent has merit. When this approach was extended to the preparation of siloxane-derivatized diphosphonic acids, however, the compounds were found to exhibit signs of deterioration upon s tanding, c learly a n u ndesirable p roperty f or a ny p rospective e xtractant. For this reason, attention was directed to an alternate family of target molecules: di–[3-(trimethylsilyl)-1-propyl]-esterified alkylenediphosphonic acids. (Hereafter, the di-[3-(trimethylsilyl)-1-propyl] functional group is abbreviated as TMSP.) TMSP-esterified DPAs represent a good choice of synthetic target for several reasons. First, the alcohol needed for the esterification of the DPA is available commercially. In addition, the chemistry required for the synthesis is straightforward (*22, 23*). Finally, separation of the trimethylsilyl functionality from the rest of the molecule by a propyl group was expected to provide improved chemical stability.

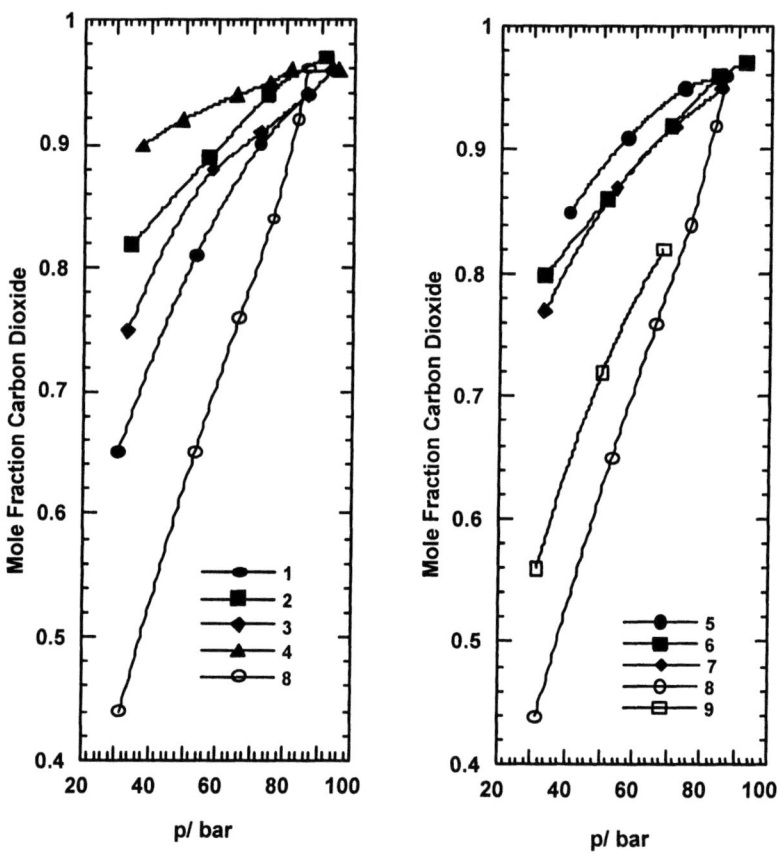

Figure 1. Composition of the ligand-rich phase for siloxane-functionalized gem-diphosphonates in equilibrium with CO_2 at 40 °C and various pressures. Smooth curves are included as a guide to the eye.

TMSP-Esterified Alkylenediphosphonic Acids

Extraction Behavior

Because it was unclear if the introduction of a silicon-based functionality would have an adverse impact on the extraction behavior of diphosphonic acids, initial evaluation of the TMSP-esterified DPAs involved the extraction of Am^{3+} (typically among the most difficult of the actinides to extract with high efficiency) into a conventional solvent, o-xylene. Figure 2 summarizes some of the results obtained as a plot of D_{Am} vs nitric acid concentration. As can be seen, very high D_{Am} values can be obtained using either the TMSP-substituted methylenediphosphonic acid (MDP) or its EDP analog (i.e., compounds containing 1 or 2 bridging methylene groups, respectively), as expected from the behavior of the conventional alkylated analogs (*29, 30*). (In fact, D_{Am} values are actually slightly higher for the TMSP compounds.) Each of the acid dependencies has a slope of -3 over at least part of the range of nitric acid concentrations examined, indicating the loss of 3 protons from the extractant molecules participating in the extraction process, consistent with the +3 charge of the Am ion. The decrease in slope and, in some instances, the fact that the slope turns positive, are indicative of a change in extraction mechanism with acidity, specifically, of the increased importance of extraction by the neutral (i.e., protonated) form of the ligands. All of this is consistent with the behavior of DPAs bearing conventional alkyl substituents.

For the corresponding extractant dependencies (not shown), slopes of ca. 2 are observed for those ligands having an odd number of bridging methylene groups and less than two (or even one) for those bearing an even number of such groups. The latter result seems especially curious. That is, three protons are lost in the extraction process, yet a slope is observed for the extractant dependency that, at first glance, suggests the involvement of only a single extractant molecule, which bears only two protons. This apparent anomaly can be explained if, as is the case for conventional DPAs, extractant aggregation plays a significant role in determining the extraction behavior of these compounds. In fact, using vapor-pressure osmometry (VPO), it has recently been shown that the aggregation behavior of TMSP-substituted DPAs is essentially the same as that seen for the analogous conventional reagents. That is, with an odd number of bridging methylenes, these compounds exist as dimers in toluene, while for those having an even number of bridging -CH_2- units, the species formed are more complex: hexamers in the case of n = 2 and n = 6, and a mixture of trimers and hexamers for n = 4. Thus, as is the case for conventional DPAs, extraction of Am (and by analogy, other metal ions) is the result of the interaction of the cation with extractant aggregates, not individual molecules of extractant. While

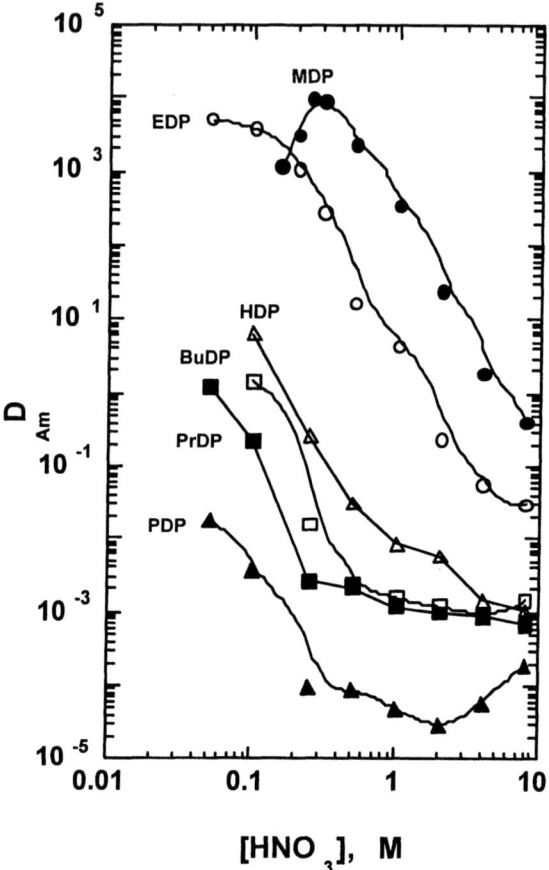

Figure 2. Nitric acid dependency of D_{Am} for various 3-(trimethylsilyl)-1-propyl-substituted alkylenediphosphonic acids. Conditions: T = 23 °C; 0.01 M extractant in o-xylene. (Note that DPAs having 1–6 methylene bridging groups are abbreviated as MDP, EDP, PrDP, BuDP, PDP, and HDP, respectively.)

the details of this are beyond the scope of this paper, these results (together with data for other cations not shown) demonstrate that incorporation of silyl functionalities into a diphosphonic acid has no unexpected or adverse effects on the behavior of these reagents as metal ion extractants (40).

Solubility in SC-CO_2

Because TMSP-derivatization was found not to alter the metal ion complexation or extraction characteristics of diphosphonic acids, the solubility of the various TMSP-substituted DPAs in SC-CO_2 was determined. For these measurements, a dynamic flow method was applied in which a known mass of extractant is briefly equilibrated with CO_2 at the temperature and pressure of interest, then exposed to a flowing stream of SC-CO_2 while the temperature and pressure in the chamber are maintained at the selected values. By determining the mass of extractant transferred as a function of the number of moles of CO_2 required to effect the transfer, the SC-CO_2 solubility of the extractant can be calculated. (Solubility measurements made on several compounds for which literature values are available have previously established the validity of this approach (*41*).

Figure 3 summarizes the results of dynamic transfer experiments for a series of TMSP-substituted DPAs at 60 °C and 250 bar. As can be seen, the compatibility of the extractant with SC-CO_2 is clearly dependent on the length of the bridge connecting the phosphorus atoms. All of the compounds with an odd number of bridging methylene units, in fact, are significantly more soluble than are those with an even number of such units. As noted above, silyl-substituted DPAs with an odd number of bridging groups exist as dimers in toluene, while those with an even number of -CH_2- units are in a more highly aggregated state (e.g., trimers or hexamers). Although no data are available for these compounds in SC-CO_2, studies comparing carboxylic acid association in toluene and SC-CO_2 indicate that the aggregation state of solutes in the two solvents is similar (*42*). Thus, there is reason to believe that the DPAs exist in the same state of aggregation in SC-CO_2 as in toluene. This indicates that less aggregated DPAs exhibit higher solubility, a result consistent with those found for carboxylic acids, for which the monomeric form of the acid has been shown to be more soluble in CO_2 than is the dimer (*42*). Not unexpectedly, for DPAs having an odd number of bridging groups, there is a decrease in SC-CO_2 solubility with increasing bridge length, an observation consistent with the known decrease in SC-CO_2 solubility that accompanies increase in solute molecular weight (*43*).

More important than the variation in the solubility of the silyl-substituted DPAs with bridge length, aggregation, or molecular weight is the solubility of these compounds relative to conventional, alkyl-substituted compounds. As

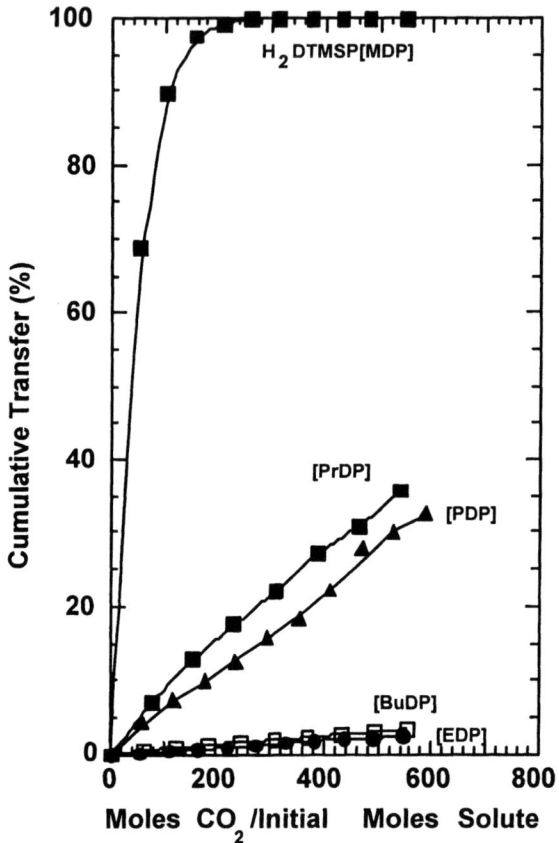

Figure 3. Dynamic transfer measurements for a series of 3-(trimethylsilyl)-1-propyl-substituted alkylenediphosphonic acids of varying bridge length. Conditions: T = 60 °C; 250 bar. (All abbreviations are as noted in Figure 2.)

shown in Figure 4, which compares dynamic transfer results for the di-[3-(trimethylsilyl)-1-propyl]-methylenediphosphonic acid (H$_2$DTMSP[MDP]) with its 2-ethylhexyl analog, H$_2$DEH[MDP], the transfer of the silyl ester is significantly more efficient under the experimental conditions. Although this result suggests that incorporation of a pair of silyl moieties, like incorporation of siloxane units into a DPA tetraester, is sufficient to markedly boost SC-CO$_2$ solubility, this increase may simply arise from differences in the branching pattern for the two compounds. (Increased branching is known to enhance the solubility of solutes in CO$_2$, all else being equal.) To separate the effects of silicon incorporation and extractant branching, a series of methylenediphosphonic acids esterified with a variety of C$_7$ and C$_8$ alkyl groups of various degrees of branching (Table II) were prepared and their solubility in

Table II. Structures of C$_7$ and C$_8$ Methylenediphosphonic Acids (n = 1)

General structure

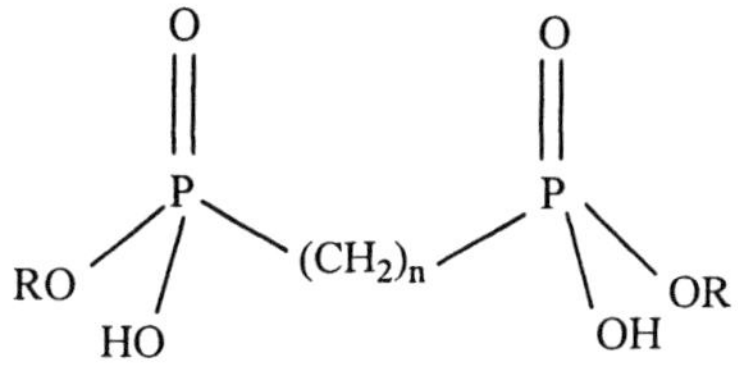

[Structure showing: RO-P(=O)(OH)-(CH$_2$)$_n$-P(=O)(OH)-OR]

R groups

O — (n-octyl)
Hp — (n-heptyl)
EH — (2-ethylhexyl)
MH — (branched heptyl)
PrP — (branched)
iPrBu — (isopropyl-butyl)
cHE — (cyclohexylethyl)
cHM — (cyclohexylmethyl)
TMP — (trimethylpentyl)
TMSP — (trimethylsilylpropyl)

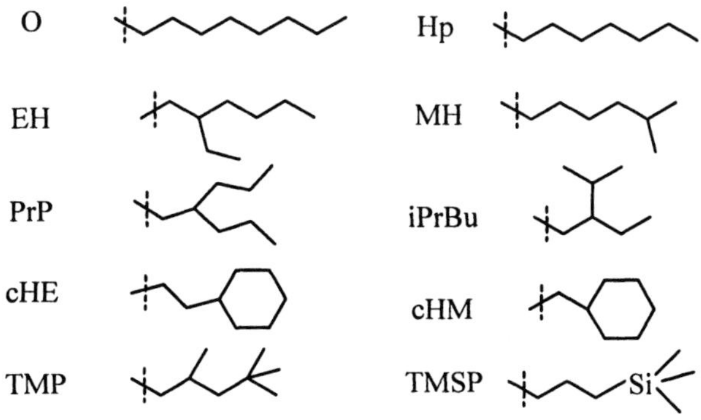

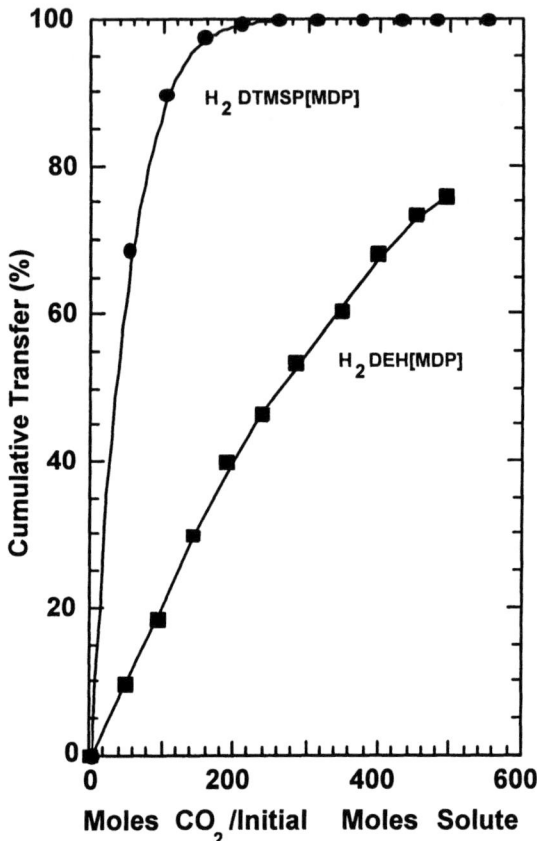

Figure 4. Comparison of dynamic transfer behavior of 3-(trimethylsilyl)-1-propyl- and 2-ethylhexyl-substituted methylenediphosphonic acids. Conditions: T = 60 °C; 250 bar.

SC-CO$_2$ determined. The dynamic transfer results for both the C$_7$ and C$_8$ compounds (not shown) clearly demonstrate that branched isomers are more CO$_2$-soluble than their straight chain or cyclic analogs. If the extent of branching is assigned a numerical value, for example, by application of a molecular connectivity index (*44, 45*), one of a series of topological molecular descriptors that have been found useful in correlating any of a variety of physicochemical properties (*44, 46, 47*), it can be shown (Figure 5) that the SC-CO$_2$ solubility data for both the C$_7$ and C$_8$ compounds can be correlated to a simple linear combination of first- and second-order connectivity indices (*41*). Because the solubility of the TMSP compound lies roughly an order of magnitude above the value expected on the basis of its connectivity index alone (Figure 5), the greater solubilizing power of the TMSP group for the DPAs relative to alkyl substituents arises not only from the introduction of greater branching into the extractant, but also from the presence of silicon atoms. Given that silicon-based functionalities are known to improve the SC-CO$_2$ solubility of polymers and surfactants to which they are appended (*48, 49*), and our own results for the siloxane-derivatized tetraesters (described above), it is not surprising that a silicon-based moiety could boost the solubility of a DPA in SC-CO$_2$. What is somewhat unexpected, however, is the magnitude of the effect, given that each TMSP group contains only one silicon atom.

Previous workers evaluating various organophosphorus extractants for the supercritical fluid extraction of metal ion from various matrices have concluded that both diisodecylphosphoric acid and octyl(phenyl)(*N,N*-diisobutyl carbamoyl)methylphosphine oxide, with SC-CO$_2$ solubilities (at T = 60 °C and P = 200 bar) of 0.041 M and 0.089 M, respectively, are suitable for this application (*50*). Interestingly, the measured solubility of H$_2$DTMSP[MDP], 0.054 M, under the same conditions of temperature and pressure, lies in this range, suggesting that merely by appending a pair of simple, silicon-bearing side arms to a diphosphonic acid, we have taken a very important step toward achieving our stated goal of combining the metal complexing power of this class of extractants with the unique solvent properties of SC-CO$_2$.

Conclusions

Discrete, mono-disperse, silicon-based substituents (either siloxanes or silyl groups) can improve the compatibility of certain organophosphorus extractants with CO$_2$ in its liquid and/or supercritical state. Simple topological indices representing the extent of substituent branching have been shown to be useful in correlating the effect of structural changes in the extractant with changes in extractant solubility in SC-CO$_2$ and in separating the effects of chain branching from the impact of the introduction of silicon atoms. Work addressing the

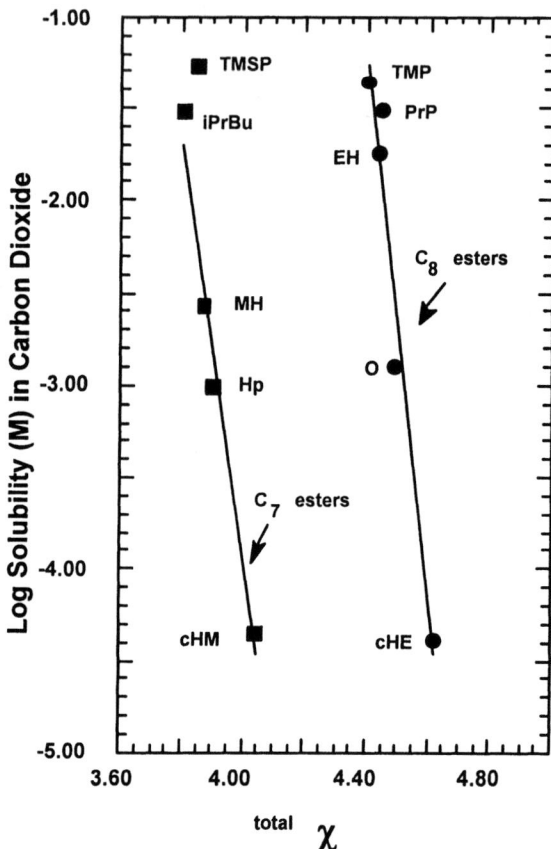

Figure 5. Supercritical carbon dioxide solubility vs $^{total}\chi$ (where $^{total}\chi = {}^1\chi + 0.23\,{}^2\chi$) for methylenediphosphonic acids bearing seven- or eight-carbon ester groups.

opportunities for improved CO_2-based systems for metal ion separation suggested by these results is now underway in our laboratory.

Acknowledgments

This work was performed under the auspices of the Office of Basic Energy Sciences, Division of Chemical Sciences (SC-CO_2 studies) and the Environmental Management Science Program of the Offices of Science and Environmental Management (extractant syntheses), U.S. Department of Energy, under grant numbers W-31-109-ENG-38 (ANL), DE-FG-07-98ER14928 (LUC), and DE-FG-07-98ER14924 (UND). The authors thank Loyola University Chicago for Dissertation Fellowship support (JAD and DCS) and the Department of Education for support through a GAANN Fellowship (DRM and PRZ).

References

1. Anastas, P. T.; Warner, J. C. *Green Chemistry: Theory and Practice*; Oxford University Press: New York, 1998.
2. Holbrey, J. D.; Seddon, K. R. *Clean Prod. Processes* **1999**, *1*, 223–236.
3. Earle, M. J.; Seddon, K. R. *Pure Appl. Chem.* **2000**, *72*, 1391–1398.
4. Welton, T. *Chem. Rev.* **1999**, *99*, 2071–2083.
5. Brennecke, J. F.; Maginn, E. J. *AIChE Journal* **2001**, *47*, 2384–2389.
6. Jessop, P. G.; Leitner, W. In *Chemical Synthesis Using Supercritical Fluids*; Jessop, P. G., Leitner, W., Eds.; Wiley-VCH: New York, 1999; pp 9–13.
7. Laintz, K. E.; Wai, C. M.; Yonker, C. R.; Smith, R. D. *Anal. Chem.* **1992**, *64*, 2875–2878.
8. Dobbs, J. M.; Wong, J. M.; Lahiere, R. J.; Johnston, K. P. *Ind. Eng. Chem. Res.* **1987**, *26*, 1476–1482.
9. Zhang, X.; Han, B.; Hou, Z.; Zhang, J.; Liu, Z.; Jiang, T.; He, J.; Li, H. *Chem. Eur. J.* **2002**, *8*, 5107–5111.
10. Hoefling, T. A.; Beitle, R. R.; Enick, R. M.; Beckman, E. J. *Fluid Phase Equilibria* **1993**, *83*, 203–212.
11. Beckman, E. J.; Russell, A. J. US Patent 5 641 887, 1997.
12. Laintz, K. E.; Wai, C. M.; Yonker, C. R.; Smith, R. D. *J. Supercrit. Fluids* **1991**, *4*, 194–198.
13. Smart, N. G.; Carleson, T.; Kast, T.; Clifford, A. A.; Burford, M. A.; Wai, C. M. *Talanta* **1997**, *44*, 137–150.

14. Smart, N. G.; Carleson, T. E.; Elshani, S.; Wang, S.; Wai, C. M. *Ind. Eng. Chem. Res.* **1997**, *36*, 1819–1826.
15. Meguro, Y.; Iso, S.; Sasaki, T.; Yoshida, Z. *Anal. Chem.* **1998**, *70*, 774–779.
16. Noll, W. *Chemistry and Technology of Silicones*; 2nd ed.; Academic Press: New York, 1968.
17. DeSimone, J. M.; Maury, E. E.; Menceloglu, Y. Z.; McClain, J. B.; Romack, T. J.; Combes, J. R. *Science* **1994**, *265*, 356–359.
18. Hoefling, T. A.; Enick, R. M.; Beckman, E. J. *J. Phys. Chem.* **1991**, *95*, 7127–7129.
19. Kendrick, T. C.; Parbhoo, B.; White, J. W. In *The Chemistry of Organic Silicon Compounds*; Patai, S., Rappoport, Z., Eds.; John Wiley & Sons: Chichester, 1989; Vol. 2, pp 1289–1361.
20. Hoefling, T. A.; Newman, D. A.; Enick, R. M.; Beckman, E. J. *J. Supercrit. Fluids* **1993**, *6*, 165–171.
21. Yazdi, A. V.; Beckman, E. J. *J. Mater. Res.* **1995**, *10*, 530–537.
22. Stepinski, D. C.; Herlinger, A. W. *Synth. Commun.* **2002**, *32*, 2683–2690.
23. Stepinski, D. C.; Nelson, D. W.; Zalupski, P. R.; Herlinger, A. W. *Tetrahedron* **2001**, *57*, 8637–8645.
24. Gatrone, R. C.; Horwitz, E. P.; Rickert, P. G.; Nash, K. L. *Sep. Sci. Technol.* **1990**, *25*, 1607–1627.
25. Nash, K. L.; Rickert, P. G. *Sep. Sci. Technol.* **1993**, *28*, 25–41.
26. Nash, K. L. *J. Alloys Compd.* **1997**, *249*, 33–40.
27. Fugate, G. A.; Nash, K. L.; Sullivan, J. C. *Radiochim. Acta.* **1997**, *79*, 161–166.
28. Nash, K. L. *Radiochim. Acta.* **1991**, *54*, 171–179.
29. Chiarizia, R.; Horwitz, E. P.; Rickert, P. G.; Herlinger, A. W. *Solvent Extr. Ion Exch.* **1996**, *14*, 773–792.
30. Chiarizia, R.; Herlinger, A. W.; Horwitz, E. P. *Solvent Extr. Ion Exch.* **1997**, *15*, 417–431.
31. Chiarizia, R.; Herlinger, A. W.; Chang, Y. D.; Ferraro, J. R.; Rickert, P. G.; Horwitz, E. P. *Solvent Extr. Ion Exch.* **1998**, *16*, 505–526.
32. Horwitz, E. P.; Chiarizia, R.; Diamond, H.; Gatrone, R. C.; Alexandratos, S. D.; Trochimczuk, A. W.; Crick, D. W. *Solvent Extr. Ion Exch.* **1993**, *11*, 943–966.
33. Alexandratos, S. D.; Trochimczuk, A. W.; Crick, D. W.; Horwitz, E. P.; Gatrone, R. C.; Chiarizia, R. *Macromolecules* **1996**, *29*, 1021–1026.
34. Chiarizia, R.; D'Arcy, K. A.; Horwitz, E. P.; Alexandratos, S. D.; Trochimczuk, A. W. *Solvent Extr. Ion Exch.* **1996**, *14*, 519–542.
35. Chiarizia, R.; Horwitz, E. P.; D'Arcy, K. A.; Alexandratos, S. D. Trochimczuk, A. W. *Spec. Publ. - R. Soc. Chem.* **1996**, *182*, 321–328.
36. Chiarizia, R.; Horwitz, E. P.; D'Arcy, K. A.; Alexandratos, S. D.; Trochimczuk, A. W. *Solvent Extr. Ion Exch.* **1996**, *14*, 1077–1100.

37. Chiarizia, R.; Horwitz, E. P.; Alexandratos, S. D.; Gula, M. J. *Sep. Sci. Technol.* **1997**, *32*, 1–35.
38. Chiarizia, R.; Horwitz, E. P.; Beauvais, R. A.; Alexandratos, S. D. *Solvent Extr. Ion Exch.* **1998**, *16*, 875–898.
39. Horwitz, E. P.; Chiarizia, R.; Dietz, M. L. *React. Funct. Polym.* **1997**, *33*, 25–36.
40. Herlinger, A. W.; McAlister, D. R.; Chiarizia, R.; Dietz, M. L. *Sep. Sci. Technol.* **2003**, *38*, 2741-2762.
41. McAlister, D. R.; Dietz, M. L.; Stepinski, D.; Zalupski, P. R.; Dzielawa, J. A.; Barrans, R. E., Jr.; Hess, J. N.; Herlinger, A. W. *Sep. Sci. Technol.* **2004**, *39*, 761-780.
42. Yamamoto, M.; Iwai, Y.; Nakajama, T.; Arai, Y. *J. Phys. Chem. A.* **1999**, *103*, 3525–3529.
43. Dandge, D. K.; Heller, J. P.; Wilson, K. V. *Ind. Eng. Chem. Prod. Res. Dev.* **1985**, *24*, 162–166.
44. Kupchik, K. J. *J. Chromatog.* **1993**, *630*, 223–230.
45. Kier, L. B.; Hall, L. H. *Molecular Connectivity in Structure-Activity Analysis*; John Wiley and Sons, Ltd.: New York, 1986; pp 1–23, 43–66.
46. White, C. M. *J. Chem. Eng. Data* **1986**, *31*, 198–293.
47. Hong, H.; Shuokui, H.; Xiaorong, W.; Liansheng, W. *Environ. Sci. Technol.* **1995**, *29*, 3044–3048.
48. Fink, R.; Hancu, D.; Valentine, R.; Beckman, E. *J. Phys. Chem. B* **1999**, *103*, 6441–6444.
49. Fink, R.; Beckman, E. J. *J. Supercrit. Fluids* **2000**, *18*, 101–110.
50. Meguro, Y.; Iso, S.; Sasaki, T.; Yoshida, Z. *Anal. Chem.* **1998**, *70*, 774–779.

Chapter 15

Investigation of SOMS and Their Related Perovskites

Yali Su[1,4], Liyu Li[1], Tina M. Nenoff[2], May D. Nyman[2],
Alexandra Navrotsky[3], and Hongwu Xu[3]

[1]Pacific Northwest National Laboratory, P.O. Box 999,
Richland, WA 99352
[2]Sandia National Laboratories, P.O. Box 5800, Albuquerque, NM 87185
[3]University of California, One Shields Avenue, Davis, CA 95616
[4]Current address: Arizona State University, University Drive and Mill
Avenue, Tempe, AZ 85287

We are evaluating new metal niobate and silicotitanate ion exchangers for Cs and Sr removal and their related condensed phases as potential ceramic waste forms. The goal of the program is to provide the U.S. Department of Energy (DOE) with alternative materials that can exceed the solvent extraction process for removing Cs and Sr from high-level wastes and with technical alternatives for disposal of silicotitanate and niobate based ion exchange materials. One class of the new phases, which are $Na_2Nb_{2-x}M^{IV}_xO_{6-x}(OH)_x \cdot H_2O$ (M^{IV} = Ti, Zr, x = 0.04 ~ 0.4, SOMS) and their thermally converted Sr loaded perovskites $Na_{2-x}Sr_xNb_{1.6}Ti_{0.4}O_{5.8+0.5x}$ ($0 \leq x \leq 0.4$), have been synthesized and characterized. SOMS exhibits very high Sr selectivity with K_d larger than 99,800 mL/g in the absence of competitive ions such as Na in solution, and about 10^3 mL/g when 0.1M Na presented in solution. This class of SOMS is easily converted to perovskites through low-temperature heat treatment (500 to 600 °C). The thermally converted perovskites exhibit extremely low Sr leach rates, ranging from 2.4×10^{-7} to 1×10^{-6} g/m^{2}day for 5% to 20% Sr loading over a 7-day leaching period. Fractional Sr release FR% is

0.001% to 0.02% for 5% to 20% Sr loading over the 7-day leaching period. These results indicate SOMS could potentially be used for separating Sr from high-level waste and that their related perovskites could be used as potential ceramic waste forms.

Introduction

Millions of gallons of radioactive waste contained in tanks at U.S. Department of Energy (DOE) sites require immobilization in stable waste forms. Over 99% of the radioactivity in tank waste is due to ^{137}Cs and ^{90}Sr that will be separated in order to concentrate and minimize final volumes of expensive high-level waste (HLW). Separations research in the last decade has resulted in the discovery of a number of new ion exchangers. In particular, Crystalline Silicotitanate (CST), which was discovered by Sandia National Laboratories (SNL), Texas A&M, and Universal Oil Products (UOP), is one of the best separation agents available for cesium (*1*). It has a high selectivity for cesium over other alkali metals over a broad pH range. Additionally, it is stable in extreme radioactive and chemical environments (*1, 2*). These properties make the CST ion exchanger useful for removal of ^{137}Cs from defense wastes such as those stored at the Hanford Site. However, different from other ion exchanger technologies such as solvent extraction process, Cs is retained in CST and is nonelutable from CST. Processing the Cs-loaded CST requires the direct use of the loaded exchanger as a feed to make the final waste form.

Borosilicate glass has been chosen as the baseline host for immobilization of HLW present at the Hanford site. Borosilicate glass exhibits a high resistance to aqueous leaching, good glass forming ability or low crystallization tendency, and relatively low melting temperatures. However, many waste feeds are not completely soluble in or are incompatible with this baseline borosilicate glass host. For example TiO$_2$, which is present in this CST waste stream, has been identified as a risk to the vitrification process. High TiO$_2$ concentrations promote crystallization and immiscible phase separation, and affect the redox state and solubility of uranium in glass (*3–6*). Because of this, a TiO$_2$ limit of 1 wt % was set for borosilicate waste glass at the Savannah River Defense Waste Processing Facility (DWPF) (*6*). If high levels of waste dilution are required to stabilize CST waste, the volume of expensive high activity borosilicate waste glass produced for subsequent storage will be substantially increased.

Therefore, an alternate waste form based on the composition of the CST and direct, in situ thermal conversion can eliminate risks associated with borosilicate glass dissolution, simplify processing, expand the suite of available waste forms, and significantly reduce the high activity waste volume and cost. Our previous studies demonstrated that direct thermal conversion of the Cs-loaded CST ion exchanger can produce thermally stable and chemically durable ceramic waste forms (7–12). The structure/property relationship of silicotitanates, Cs-containing phases, and other related phases in CST ceramic waste forms has been studied extensively (13–26). All results indicate that this viable ceramic matrix can be used as both a long-term and short-term storage form. While CST is a good for Cs removal, however, it may not be the best material for separating Sr. In addition, issues with the durability and stability of the CST ion exchanger at elevated temperature due to radioactive decay of ^{137}Cs if the process waste stream removing this heat were interrupted have recently been encountered (27–29). Therefore, developing new ion exchangers for Sr and Cs removal is essential.

As we described above, a full description of the separation science and technology of Cs or Sr removal must always include the element of waste separation and disposal. This is particularly true for using CST as a separation media. Hence, our studies not only focus on developing new silicotitanate and niobate ion exchanger materials, but also their direct thermal converted phases as potential ceramic waste forms. The goal of our studies is to provide the DOE with alternative materials that can exceed the solvent extraction process for removing Cs and Sr from high-level wastes at Hanford and other DOE sites and with technical alternatives for disposal of silicotitanate and niobate based ion exchange materials.

With the common interest of developing new ion exchanger materials and their ceramic waste forms, a collaborative effort between Pacific Northwest National Laboratory (PNNL), Sandia National Laboratories (SNL), and University of California – Davis has been carried out involving phase search using component oxides of Cs-loaded CST material. Our investigation has resulted in several new silicotitanate and niobate based ion exchanger materials for Sr or Cs removal and their related condensed phases as potential ceramic waste forms. In this paper, we report one class of niobate based ion exchangers, $Na_2Nb_{2-x}M^{IV}_xO_{6-x}(OH)_x \cdot H_2O$ (M^{IV} = Ti, Zr, x = 0.04 ~ 0.4, SOMS), and their related perovskites $Na_{2-x}Sr_xNb_{1.6}Ti_{0.4}O_{5.8+0.5x}$ ($0 \leq x \leq 0.4$) only to demonstrate the Sr removal and disposal. The other silicotitanates related to Cs and Sr removal and disposal will be reported elsewhere. SOMS is among the ion exchangers with extremely high selectivity for divalent cations such as Sr, and thus can potentially be used for ^{90}Sr extraction (30–33). This class of SOMS can easily be converted to perovskite through low-temperature heat treatment (500–600 °C). To assess the feasibility of a disposal plan, a thorough understanding of the structures, and the chemical and thermodynamic stability for SOMS and their thermally converted perovskite phases (especially Sr loaded perovskite phases), is essential. This paper reports Sr selectivity and thermodynamic stability of SOMS, along with the chemical durability of Sr loaded perovskites and their viability as a potential ceramic waste form.

Experimental Section

Synthesis of SOMS

The detailed synthesis method of SOMS has been described previously (*33*). Briefly, the SOMS phases were synthesized by hydrothermal treatment of sol mixtures containing water, sodium hydroxide, and hydrolyzed metal (Nb, Ti, Zr) alkoxides. The M:Nb ratio (M = Ti, Zr) in the resultant SOMS phase was directly correlated with the precursor ratio for the range of 1:50–1:4 M:Nb. Within this composition range, pure phase, isostructural materials were formed, based on powder X-ray diffractions analysis.

Synthesis of $Na_{2-x}Sr_xNb_{1.6}Ti_{0.4}O_{5.8+0.5x}$: ($0 \leq x \leq 0.4$)

Five samples with compositions $Na_{2-x}Sr_xNb_{1.6}Ti_{0.4}O_{5.8+0.5x}$ compositions (x = 0, 0.1, 0.2, 0.3, and 0.4, corresponding to 0%, 5%, 10%, 15%, and 20% Sr loading) were synthesized via a sol-gel processing route. First, an amorphous, homogeneous precursor for each composition was prepared using niobium ethoxide, titanium isopropoxide (TIP), strontium nitrate, and sodium hydroxide. The alkoxides (niobium ethoxide and TIP) were mixed in the stoichiometric ratio in a glove box under Ar atmosphere, and then a mixture of NaOH, $Sr(NO_3)_2$, water, and ethanol was added. After gelation, additional water and ethanol were added to dissolve the gel, and the liquid was then stirred and dried overnight. Second, ~ 0.5 g of the obtained precursor was heat-treated at 1173 K for ~ 15 h. The resulting product is a white, monophasic crystalline material, as revealed by XRD.

Distribution Coefficient (K_d) Measurements

Detailed procedure has been described previously (*33*). Briefly, SOMS selectivity for Sr was measured as a function of pH and Na-concentration. In these selectivity experiments, $NaNO_3$, HNO_3, and NaOH were used to produce Na-containing solutions, acidic solutions, and basic solutions, respectively. Distribution coefficient was calculated by the following relationship:

$$K_d(mL/g) = ([Sr_{ix}]/g_{ix})/([Sr_{sln}]/mL\ sln) \quad (1)$$

where K_d is the distribution coefficient, $_{ix}$ is ion exchanger, $[Sr_{ix}]$ is the concentration of Sr adsorbed by the ion exchanger, g_{ix} is the weight of the SOMS ion exchanger, $[Sr_{sln}]$ is the concentration of the Sr remaining in solution after contacting SOMS, and mL sln is milliliters of solution.

Chemical Durability Measurement

Chemical durability was measured using the American Society for Testing and Materials (ASTM) product consistency test (PCT) (*34*). Powder specimens, the surface areas of which were measured by nitrogen adsorption and the Brunauer–Emmett–Teller (BET) method, were used for the PCT test. Leachant volume to surface area of the samples was 1/20 cm. The samples were placed into deionized water in a Teflon container. The sealed Teflon containers were placed into an oven at 90 °C. Samples of the leachate were removed at intervals of 1, 2, 4, and 7 days. The leachant was analyzed for Sr using an Optima 3000 DV Perkin Elmer ICP-AES.

Powder X-ray Diffraction

Powder XRD experiments were conducted with a Philips PW3050 diffractometer using Cu Kα radiation. Sample powders were mounted in a front-loading, shallow-cavity zero-background quartz holder, and the data were collected from 5° to 120° 2θ in step scan mode using steps of 0.02° with a counting time of 10 s.

TGA/DTA measurements

Thermogravimetric analysis-differential thermal analysis (TGA-DTA) experiments were performed on a STD 2960 TA DTA-TGA instrument with alumina as a standard for DTA. Samples (10–15 mg) were heated at 10 °C/min to 900 °C, and argon was used as a sweep gas with a flow of 20 cm^3/min.

Results and Discussion

SOMS

SOMS is an isostructural, variable composition class of ion exchangers with the general formula $Na_2Nb_{2-x}M^{IV}{}_xO_{6-x}(OH)_x \cdot H_2O$ (M^{IV} = Ti, Zr; x = 0.04–0.40). Up to 20% of the framework Nb^V in SOMS can be substituted with Ti^{IV} or Zr^{IV}. This paper focuses on one member of this family, SOMS-1 ($Na_2Nb_{2-x}Ti_xO_{6-x}(OH)_x \cdot H_2O$ (x = 0.4), for our discussion. The structure of the SOMS-1 was determined from single-crystal X-ray diffraction data and was described in detail previously (*32, 33*). Briefly, the overall architecture of SOMS-1 is a three-dimensional framework with one-dimensional channels oriented parallel to the *b*-direction, and three distinct structural units. The first unit is edge-sharing double chains of Nb/Ti octahedra containing off-center atoms common to Ti/Nb chemistry (*35*) which run parallel to [010]. The second building unit is a layer of edge-linked, six-coordinated Na1 and Na2 polyhedra. The framework then consists of sheets of these Na-layers alternating with layers containing the double chains of Ti/Nb octahedra. The third structural unit, the Na3 site, resides between these double chains.

The SOMS phases exhibit ion-exchange selectivity for divalent cations over monovalent cations. For example, without competing ions present in solution, K_d of SOMS-1 is as follows: $K_d(Ba^{2+})$ > 99,800 mL/g, $K_d(Sr^{2+})$ > 99,800 mL/g, $K_d(Cs^+)$ = 150 mL/g, and $K_d(K^+)$ = 95 mL/g. The divalent transition metals, Ba^{2+} and Sr^{2+}, were completely removed from solution by SOMS-1. Selectivity of SOMS-1 for the alkali metals is much lower than that for alkaline earth metals. The selectivity of SOMS-1 for Sr^{2+} as a function of [Na] concentration and pH was also investigated. Distribution coefficients of Sr on the SOMS-1 as a function of pH and concentration of Na as a competing cation are plotted in Figure 1. The purpose of using Na as a competing ion for these studies is twofold: (1) to suppress the Sr selectivity so that K_d values may be obtained, and (2) adding NaOH is necessary for the higher pH experiments so that, to the extent possible, consistent concentration of sodium (a combination of $NaNO_3$ and NaOH) is maintained in all of the solution matrices. In the absence of Na, (0 M Na), the K_d values approach 10^6 mL/g, which is the value obtained for approximately 0.1 ppm Sr remaining in solution and the detection limit of the Sr analytical technique (atomic absorption spectroscopy, AAS). As [Na] increases to 0.01M, the K_ds decrease to the range of 10^3 to 10^5 mL/g. As [Na] increases to 0.1 M, the K_ds further decrease to the range of 10^2 to 10^3 mL/g, in the pH range of 4 to 12.

In general, there is an increase in Sr selectivity with increasing basicity, as shown in Figure 1. The increase in Sr selectivity with increased pH is consistent with the following exchange mechanism:

$$Na^+_{SOMS} + H^+_{SOMS} + Sr^{2+}_{solution} \rightarrow Sr^{2+}_{SOMS} + Na^+_{solution} + H^+_{solution} \quad (2)$$

A basic aqueous medium (high pH) removes framework protons more easily than a lower pH solution. The selectivity experiments with 0 M Na were carried out only in the acidic range because at pH > 6, all Sr is removed from solution. With lower concentrations of Na (0.01 M, 0 M), the selectivity for Sr remains surprisingly high, $10^4 \sim 10^6$ mL/g, which is several orders of magnitude higher than other Sr selective phases (*1, 36*). On the other hand, high concentrations of Na inhibit Sr selectivity considerably. However, to compare this effect directly to the performance of other Sr selective ion exchangers such as layered titanates and silicotitanates, experiments using very high hydroxyl concentrations need to be executed (*1*).

The DTA-TGA analysis of SOMS-1 with 10% of Na^+ exchanged for Sr^{2+} is shown in Figure 2. The weight loss between 100 and 300 °C corresponds to dehydration (*calcd 7.6 wt %, obsd 7.5 wt %*) followed by structure change to a new crystalline phase, as observed by X-ray diffraction. The exothermic transition at 550 °C is associated with conversion to a perovskite form. Perovskite (titanate-based) is a major component in the well known SYNROC ceramic waste form for high-level radioactive waste storage, and thus a reliable commodity for stability in radioactive fields and in repository conditions (*37–40*). Micrographs of the Sr^{2+}-loaded SOMS-1 (Figure 3a) and the perovskite (Figure 3b) reveal that this phase change takes place with remarkable morphology preservation, which indicates that remobilization of the strontium during heating is improbable. The chemical durabilities of Sr-loaded perovskite phases are discussed in the next section in detail.

$Na_{2-x}Sr_xNb_{1.6}Ti_{0.4}O_{5.8+0.5x}$ ($0 \leq x \leq 0.4$)

Because of the unique properties of thermal conversion of SOMS to perovskite, we further investigated the chemical and thermal durability of thermally converted Sr loaded perovskite phases. A suite of perovskite phases with the compositions $Na_{2-x}Sr_xNb_{1.6}Ti_{0.4}O_{5.8+0.5x}$ has been investigated. $Na_{2-x}Sr_xNb_{1.6}Ti_{0.4}O_{5.8+0.5x}$ phases are the thermally converted ceramic waste forms from Sr loaded SOMS-1. Sr loadings in these phases range from 0, 5%, 10%, 15%, to 20% for x = 0, 0.1, 0.2, 0.3, and 0.4. In these series, a portion of Na^+ is replaced by Sr^{2+}, and the charge is balanced by incorporation of additional O^{2-} anions. Figure 4 shows XRD pattern of $Na_{2-x}Sr_xNb_{1.6}Ti_{0.4}O_{5.8+0.5x}$ ($0 \leq x \leq 0.4$). All the XRD patterns can be indexed based on the structural model of $NaNbO_3$. Unit-cell volume determined from Rietveld analysis of the

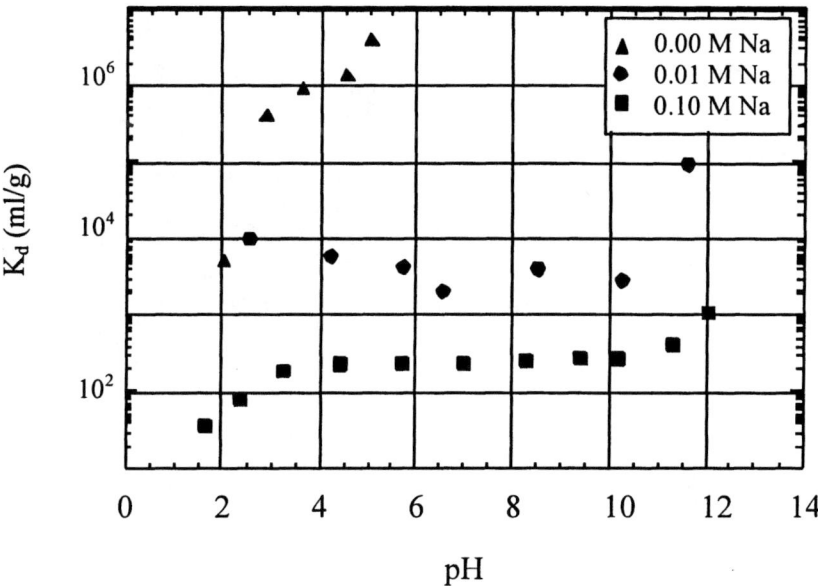

Figure 1. Selectivity of Sr (50 ppm) as a function of pH on SOMS-1.

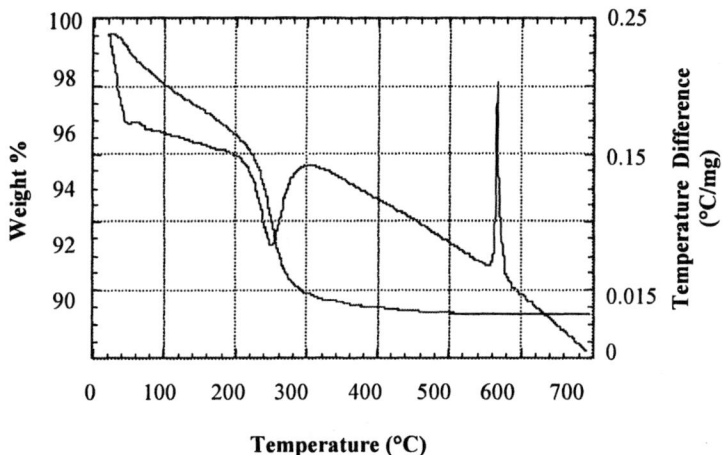

Figure 2. DTA-TGA of Sr^{2+}-exchanged SOMS-1.

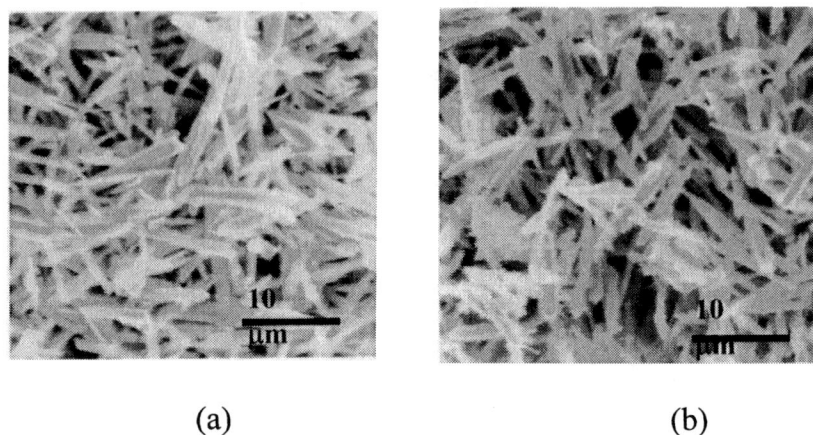

Figure 3. (a) SEM micrograph of Sr^{2+}-exchanged SOMS-1; (b) SEM micrograph of post thermal treatment of Sr^{2+} exchanged SOMS-1 (perovskite).

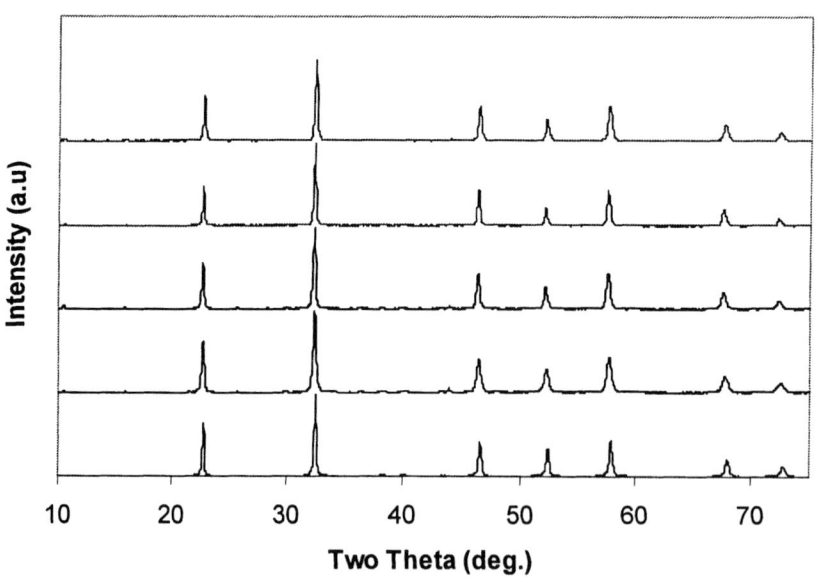

Figure 4. XRD Patterns of $Na_{2-x}Sr_xNb_{1.6}Ti_{0.4}O_{5.8+0.5x}$ with $x = 0, 0.1, 0.2, 0.3,$ and 0.4.

data are plotted as a function of composition in Figure 5. The result reveals that with increasing Sr content, cell volume increases. This trend is consistent with the replacement of Na^+ by the larger Sr^{2+} and the occurrence of additional O^{2-} in the structure.

Aqueous durability of $Na_{2-x}Sr_xNb_{1.6}Ti_{0.4}O_{5.8+0.5x}$ ($0 \leq x \leq 0.4$) series was measured by Sr release rate in ASTM-PCT test. Sr release rates from the PCT test are reported in two ways, to provide a better means of comparison with literature data. These methods involve the fraction of Sr released and the normalized Sr mass loss.

The fraction released (FR%) is given to illustrate the percentage of Sr lost during the leach period and is expressed as follows:

$$FR\% = 100 \frac{mass\ of\ Sr\ released}{total\ mass\ of\ Sr\ in\ the\ sample} \qquad (3)$$

This expression provides a correlation between the Sr released and the total amount of Sr in a sample without explicitly considering the effect of geometric surface area.

The normalized Sr mass loss (*NL*) is also used to reflect the Sr loss:

$$NL = \frac{C\ V}{F\ At}(g/m^2 day) \qquad (4)$$

where C is the concentration of Sr released to solution, V is the volume of leachant, F is the fraction of the Sr in the sample, A is the surface area of the sample, and t is leaching time. The *NL* is usually associated with dissolution of the material and accounts for the surface area effect. For the purposes of this study, leach rates expressed by normalized loss use BET surface area and are denoted as *NL*.

The PCT leach test results for $Na_{2-x}Sr_xNb_{1.6}Ti_{0.4}O_{5.8+0.5x}$ are shown in Table I and illustrated in Figure 6 (for Sr fractional release) and in Figure 7 (for normalized Sr leach rates). Sr substitution is from 5% to 20% (x = 0.1 to 0.4). While the maximum Sr loading on the ion exchanger is likely to be around 5%, samples were loaded to 20% Sr in order to exaggerate potential Sr loss during leaching. The results showed that the Sr release rate decreased as time increased. The normalized 7-day Sr leach rate is 2.37×10^{-7} g/m²day for 5% Sr loading. The Sr leach rates remained nearly unchanged for Sr loading up to 15%, and increased to 1.14×10^{-6} g/m²day when 20% Sr was loaded. The cumulative fraction of Sr lost after 7 days (Figure 6) was 0.0013%, 0.016%, 0.0060%, and 0.023% for 5%, 10%, 15%, and 20% Sr loading, respectively. These results show that percentage of Sr loss is extremely low, even for 20% Sr loading.

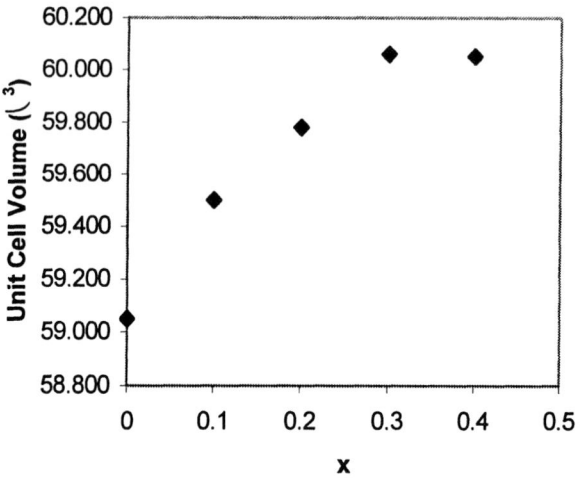

Figure 5. Variation of Cell volume as a function of composition for $Na_{2-x}Sr_xNb_{1.6}Ti_{0.4}O_{5.8+0.5x}$.

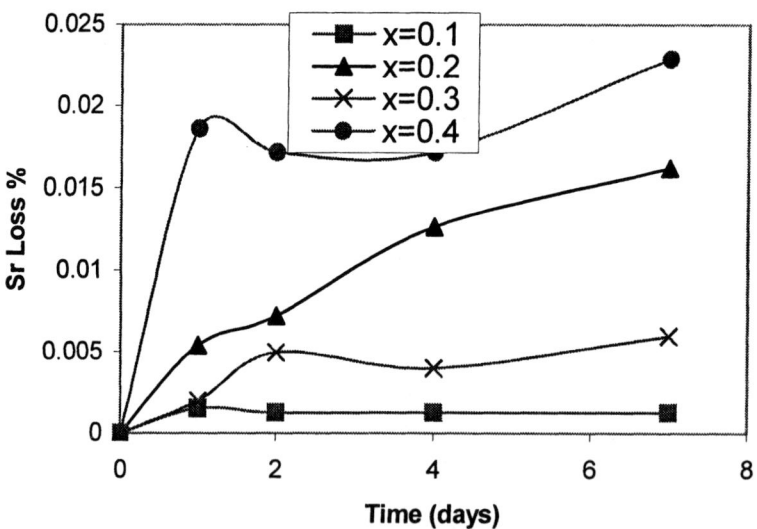

Figure 6. Fraction (FR%) of Sr loss for $Na_{2-x}Sr_xNb_{1.6}Ti_{0.4}O_{5.8+0.5x}$ as a function of time.

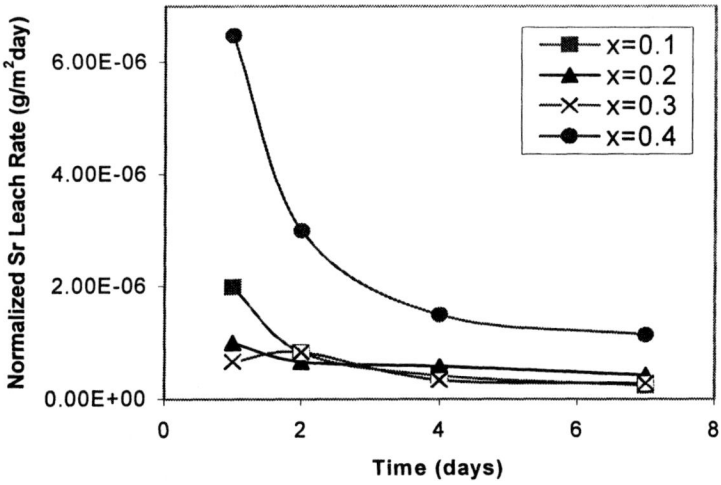

Figure 7. Normalized Sr leach rates for $Na_{2-x}Sr_xNb_{1.6}Ti_{0.4}O_{5.8+0.5x}$.

Several matrices have been studied to immobilize Sr. While it is difficult to compare Sr leach rates to those of previous reports in that Sr release rate experiments were rarely done under identical conditions, several reports (*41–45*) with Sr leach rates are nonetheless summarized in Table II for comparison. Raman reported glass-ceramic waste forms in which the Sr leaching rate in deionized water at 90 °C was 0.3 ~ 3.7g/m^{2}day in the 14-day leach test (*41*). He et al. reported apatite glass-ceramics with Sr leaching rates at 90 °C of 1 x 10^{-2} g/m^{2}day for 3 days and 5.6 x 10^{-3} g/m^{2}day for 7 days (*42*). The lowest leach rates were reported by Nakayama et al., which ranged from 1 x 10^{-4} g/m^{2}day to 1x10^{-3} g/m^{2}day at 160 °C for a 1-day leach test (*44*). The Sr leach rates of the perovskite phases in our studies range from 10^{-6} to 10^{-7} g/m^{2}day, which is several orders of magnitude lower than those presented in Table II. Based on the comparison, it is clear that the perovskites have effectively immobilized Sr. The low normalized leach rate indicates the pevroskite would be a workable host candidate for immobilizing Sr-rich HLW.

Conclusions

$Na_2Nb_{2-x}M^{IV}_xO_{6-x}(OH)_x \cdot H_2O$ (M^{IV} = Ti, Zr, x = 0.04 ~ 0.4, SOMS) and their thermally converted Sr loaded perovskites $Na_{2-x}Sr_xNb_{1.6}Ti_{0.4}O_{5.8+0.5x}$ (0 ≤ x ≤ 0.4) have been synthesized and characterized. Our results suggest that SOMS is a variable-composition class of niobate-based ion exchangers that exhibit extremely high slectivity for divalent cation (such as Sr^{2+}, Ba^{2+}) over monovalent cations (such as K^+, Cs^+). Ion exchange occurs by exchange of a sodium plus a proton for a divalent cation, such as strontium. This class of SOMS can be easily converted to perovskites through low-temperature heat treatment (500–600 °C). The thermally converted perovskites exhibit extremely low Sr leach rate, ranging from 2.4 x 10^{-7} to 1 x 10^{-6} g/m^{2}day for 5% to 20% Sr loading over the 7-day leaching period. Fractional Sr release FR% ranges only from 0.001% to 0.02% for 5% to 20% Sr loading over the 7-day leaching period. These results indicate SOMS could potentially be used for separating Sr and their related perovskites could be used as potential ceramic waste forms.

Acknowledgments

This research was supported by the DOE Environmental Management Science Program. Some of the experiments were performed at the Environmental Molecular Sciences Laboratory, a national scientific user facility sponsored by the DOE Office of Biological and Environmental Research and

Table I. PCT Sr Leach Rates for $Na_{2-x}Sr_xNb_{1.6}Ti_{0.4}O_{5.8+0.5x}$ $(0 < x \leq 0.4)$

X	X=0.1		X=0.2		X=0.3		X=0.4	
Time (days)	NL (g.m^{2}day)	FR%	NL (g.m^{2}day)	FR%	NL (g.m^{2}day)	FR%	NL (g.m^{2}day)	FR%
1	1.99×10^{-6}	0.0015	9.97×10^{-7}	0.0054	6.65×10^{-7}	0.0020	6.48×10^{-6}	0.018
2	8.29×10^{-7}	0.0013	6.65×10^{-7}	0.0072	8.31×10^{-7}	0.0050	2.99×10^{-6}	0.017
4	4.14×10^{-7}	0.0013	5.82×10^{-7}	0.013	3.32×10^{-7}	0.0040	1.49×10^{-6}	0.017
7	2.37×10^{-7}	0.0013	4.27×10^{-7}	0.016	2.85×10^{-7}	0.0060	1.14×10^{-6}	0.023

Table II. Comparison of Normalized Sr Leach Rates among Perovskites and Several Waste Forms

Waste Form	Leach Test	Leach Days	Leach media	Leach Rate(g/m^{2}day)	Reference
Glass-Ceramic	MCC-1	14	90 °C, H_2O	0.3 ~ 3.7	41
Apatite Glass-Ceramic	MCC-1	7	90 °C, H_2O	1.56×10^{-3} ~ 5.6×10^{-3}	42
Geoceramic	Modified MCC-1	7	90 °C, H_2O	0.18 ~ 4.7	43
Zirconium Phosphate	Modified PCT	24 h	160 °C, H_2O	1×10^{-4} ~ 1×10^{-3}	44
Slag matrix	GB7023-86	7	25 °C, H_2O	6×10^{-4} ~ 6×10^{-3}	45
Perovskites	PCT	7	90 °C, H_2O	2.4×10^{-7} ~ 1×10^{-6}	This work

located at PNNL. PNNL is operated for the DOE by Battelle Memorial Institute under contract DE-AC06-76RLO (1830). Sandia is a multiprogram laboratory operated by Sandia Corporation, a Lockheed Martin Company, for the DOE National Nuclear Security Administration under contract DE-AC04-AL85000.

References

1. Anthony, R. G.; Phillip, C. V.; Dosch, R. G. *Waste Management*, **1993**, *13*, 503.
2. Bunker, B. C. *Evaluation of Inorganic Ion Exchangers for Removal of Cs from Tank Wastes*; TWRSPP-94-085; Pacific Northwest Laboratory: Richland, WA, 1994.
3. Ewest, E. E.; Wiese, H. *High Level Liquid Waste Vitrification with the Pamela Plant in Belgium*; IAEA-CN-48/177; Vienna, Austria, 1987, pp 269–280.
4. Galakhov, F. Y.; Vavilonova, V. T.; Aver'yanov, V. I.; Slyshkina, T. V. *Fiz. khim. Stekla* **1988**, *14*(1), 38–46.
5. Bickford, D. F.; Applewhite-Ramsey, A.; Jantzen, C. M.; Brown, K. G. *J. Am. Ceram. Soc.* **1990**, *73*, 2896–2902.
6. Plodinec, M. J. In *Scientific Bases for Nuclear Waste Management*; Northrup, C. J. M., Jr., Ed.; Materials Research Society, 1980; Vol. 2, pp 223–229.
7. Su, Y.; Balmer, M. L.; Bunker, B. C. In *Scientific Basis for Nuclear Waste Management XX;* Gray, W. J., Triay, I. R., Eds.; Vol. 465; Materials Research Society, 1996; p 457.
8. Balmer, M. L.; Su, Y.; Grey, I. E.; Santoro, A.; Roth, R. S.; Huang, Q.; Hess, N.; Bunker, B. C. In *Scientific Basis for Nuclear Waste Management XX*; Gray, W. J., Triay, I. R., Eds.; Vol. 465; Materials Research Society, 1996.; p 449.
9. Su, Y.; Balmer, M. L.; Wang, L.; Bunker, B. C.; Nyman, M. D.; Nenoff, T. M.; Navrotsky, A. In *Scientific Basis for Nuclear Waste Management XXII;* Wronkiewicz, D. J., Lee, J. H., Eds.; Vol. 556; Materials Research Society, 1998; pp 77–84.
10. Nyman, M. D.; Nenoff, T. M.; Su, Y.; Balmer, M. L.; Navrotsky, A.; Xu, H. In *Scientific Basis for Nuclear Waste Management XXII;* Wronkiewicz, D. J., Lee, J. H., Eds.; Vol. 556; Materials Research Society, 1998; pp 71–76.
11. Balmer, M. L.; Su, Y.; Xu, H.; Bitten, E.; McCready, D.; Navrotsky, A. *J. Am. Ceram. Soc.* **2001**, *84*(1), 153–160.
12. Balmer, M. L.; Su, Y.; Nenoff, T. M.; Nyman, M. D.; Navrotsky, A.; Xu, H. *New Silicotitanate Waste Forms: Development and Characterization;* Science to Support DOE Site Cleanup: The Pacific Northwest National

Laboratory Environmental Management Science Program Award, Fiscal Year 1999 and 2000 Mid-Year Progress Report, PNNL-12208 UC-2000 and PNNL-13262; Pacific Northwest National Laboratory: Richland, WA.
13. Balmer, M. L.; Huang, Q.; Santoro, A.; Roth, R. *J. Solid State Chem.* **1997**, *130*, 97–102.
14. Hess, N. J.; Balmer, M. L.; Bunker, B. C.; Conradson, S. D. *J. Solid State Chem.* **1997**, *129*, 206–213.
15. McCready, D. E.; Balmer, M. L.; Keefer, K. D. *Powder Diffraction* **1997**, *12*(1), 40–46.
16. Grey, I. E.; Roth, R. S.; Balmer, M. L. *J. of Solid State Chem.* **1997**, *131*, 38–42.
17. Balmer, M. L.; Bunker, B. C.; Wang, L. Q.; Peden, C. H. F.; Su, Y. *J. Phys. Chem.* **1997**, *101*, 9170–9179.
18. Su, Y.; Balmer, M. L.; Bunker, B. C. *J. Phys. Chem. B.* **2000**, *104*, 8160–8169.
19. Hess, N.; Su, Y.; Balmer, M. L. *J. Phys. Chem. B* **2001**, *105*, 6805–6811.
20. Xu, H.; Navrotsky, A.; Balmer, M. L.; Su, Y.; Bitten, E. R. *J. Am. Ceram. Soc.* **2001**, *84*(3), 555–560.
21. Xu, H.; Su, Y.; Balmer, M. L; Navrotsky, A. *Chemistry Mater.* **2003**, *15*(9), 1872–1878.
22. Xu, H.; Navrotsky, A.; Balmer, M. L.; Su, Y. *Mater. Res. Soc. Symp. Proc.*, Boston, MA, Dec 2, 2002; Wentzcovitch, R., Navrotsky, A., Poeppelmeier, K., Eds.; 718 (Perovskite Materials); Warrendale, PA: Materials Research Society, 2003; pp 65–70.
23. Xu, H.; Navrotsky, A.; Balmer, M. L.; Su, Y. *J. Am Ceram. Soc.* **2002**, *85*, 1235–1242.
24. Nyman, M.; Bonhomme, F.; Tripathi, A.; Parise, J.; Nenoff, T. M. *Chem. Mater.* **2002**, *13*(12), 4603–4611.
25. Nyman, M.; Bonhomme, F.; Teter, D. M.; Maxwell, R. S.; Gu, B. X.; Wang, L. M.; Ewing, R. C.; Nenoff T. M. *Chem. Mater.* **2000**, *12*, 3449.
26. Xu, H.; Navrotsky, A.; Nyman, M. D.; Nenoff, T. M. *J. Mater. Res.* **2000**, *15*(3), 815.
27. Su, Y.; Li, L.; Young, J. S.; Balmer, M. L. *Investigation of Chemical and Thermal Stabilities of Cs-Loaded UOP IONSIV IE-911 Ion Exchanger*; PNNL-13392-2; Pacific Northwest National Laboratory: Richland, WA, 2001; pp 1–36.
28. Nyman, M.; Krumhansl, J. L.; Jove-Colon, C.; Zhang, P.; Nenoff, T. M.; Headley, T. J.; Su, Y.; Li, L. In *Mater. Res. Soc. Symp. Proc.*, Boston, MA, Dec 4, 2002; McGrail, B. P., Cragnolino, G. A., Eds.; 713 (Scientific Basis for Nuclear Waste Management XXV); Warrendale, PA: Materials Research Society, 2003; pp 885–891.
29. Young, J. S.; Su, Y.; Li, L.; Balmer M. L. *Microscopy and Microanalysis* **2001**, *7*(2), 498.

30. Nenoff, T. M.; Nyman, M.; Su, Y.; Balmer, M. L.; Navrotsky, A.; Xu, H. In *Nuclear Site Remediation, First Accomplishments of Environmental Management Science Program*; Eller, P. G., Heineman, W. R., Eds.; ACS Symposium Series 778; American Chemical Society: Washington, DC, 1999; pp 175–186.
31. Nyman, M.; Gu, B. X.; Wang, L. M.; Ewing, R. C.; Nenoff T. M. *Micro. and Meso. Materials* **2000**, *40(*1-3*)*, 115.
32. Nyman, M.; Tripathi, A.; Parise, J. B.; Maxwell, R. S.; Harrison, W. T. A.; Nenoff, T. M. *J. Amer. Chem. Soc.* **2001**, *123*(7), 1529–1530.
33. Nyman, M.; Tripathi, A; Parise, J. B.; Maxwell, R. S.; Nenoff, T. M. *J. Amer. Chem. Soc.* **2002**, *124*(3), 1704–1713.
34. *Standard Test Methods for Determining Chemical Durability of Nuclear, Hazardous, and Mixed Waste Glasses: The Product Consistency Test*; Designation ASTM C1285-94, 1997 Annual Book of ASTM Standards; American Society for Testing and Materials Standards: West Conshohocken, PA, 1997.
35. Kunz, M.; Brown, I. D. *J. Solid State Chem.* **1995**, *115*, 395–406.
36. Nenoff, T. M.; Miller, J. E.; Thoma, S. G.; Trudell, D. E. *Environ. Sci. Technol.* **1996**, *30*, 3630.
37. Dosch, R. G.; Northrup, C. J.; Headley, T. J. *J. Am. Ceram. Soc.* **1985**, *68*, 330–337.
38. Li, L.; Luo, S.; Tang, B.; Wang, D. *J. Am. Ceram. Soc.* **1997**, *80*, 250–252.
39. Luo, S.; Li, L.; Tang, B.; Wang, D. *Waste Management* **1998**, *18*, 55–59.
40. Donze, S.; Montagne, L.; Palavit, G. *Chem. Mater.* **2000,** *12*, 1921–1925.
41. Raman, S. V. *J. Non-Cryst. Solids* **2000**, *263&264*, 395–408.
42. Ye, Y.; Bao, W.; Song, C. *J. Nucl. Mater.* **2002**, *305*, 202–208.
43. Arancibia, N. E.; Bogdanov, R. V.; Kuznetsov, R. A.; Sergeev, A. S.; Glushkova, V. B.; Panova, T. I. *Radiochemistry* **2002***, 44*, 508–517.
44. Nakayama, S.; Itoh, K. *J. European Ceramic Society* **2003**, *23*, 1047–1052.
45. Qin, G.; Li, Y.; Yi, F.; Shi, R. *J. Hazard. Mater.* **2002**, *B29*, 289–300.

Modeling and Waste Treatment Chemistries

Chapter 16

Solubility of $TcO_2 \cdot xH_2O$(am) in the Presence of Gluconate in Aqueous Solution

Nancy J. Hess, Yuanxian Xia, and Andrew R. Felmy

Environmental Simulations and Dynamics, Pacific Northwest National Laboratory, P.O. Box 999, Richland, WA 99352

The solubility of $TcO_2 \cdot xH_2O$(am) in the presence of 0.01 M of gluconate was measured over a wide range of hydroxide concentrations from pH = 3.5 to 5 M of NaOH and as a function of gluconate concentration from 0.01 M to 0.5 M at a fixed hydroxide concentration of pH = 10.5. The presence of gluconate increased the solubility of $TcO_2 \cdot xH_2O$(am) by approximately two orders of magnitude over most of the pH range compared to the solubility of $TcO_2 \cdot xH_2O$(s) when no complexant is present. At extremely high hydroxide concentrations, above 3 M of NaOH, the Tc-gluconate solution species appears to polymerize and then precipitate as a white solid. At a fixed pH of 10.5, increasing the gluconate concentration by two orders of magnitude resulted in only a modest increase in measured Tc concentration. These initial results suggest the Tc-gluconate species may form polymerized species under basic conditions.

Introduction

Gluconate was used at the U.S. Department of Energy (DOE) Hanford Site during the treatment of waste streams generated from the processing spent nuclear fuel in two applications. One application was as a denitration agent to reduce the acidity of cladding waste prior to storage in underground tanks. In this application, virtually all the gluconate, added as sucrose, is consumed by nitric acid and as a result very little, if any, would be available to complex radionuclides. However, in 1968 a separations processing campaign was initiated to recover ^{90}Sr and ^{137}Cs from redox and purex waste that was retrieved from the storage tanks. The redox wastes in particular, which were generated from 1951 to 1967, produced enough heat from the decay of short-lived radionuclides so that the waste self-boiled in the tanks. The separations processing was designed to reduce the thermal load by removing Cs and Sr. In the separations process, large amounts of complexants, EDTA, HEDTA, and citrate, were used to keep Fe, Al, and Mn in the aqueous phase during extraction of Sr into the organic phase. A Na-gluconate-NaOH solution was used to wash the organic complexants under alkaline conditions of tightly bound "impurities" (Fe, Al, Y, Ce, U) prior to recycling the organic constituents. All of the separations process took place under acidic conditions, but the pH of the generated waste streams were adjusted to pH = 11 prior to returning the wastes to the tanks. Many of the organic complexants used in the Sr separations processing have been identified in Hanford wastes. Although these analyses did not look for gluconate specifically, it is likely that gluconate is also present. Our work on carboxylic acid based ligands indicates that their ability to complex tetravalent radionuclides decreases above pH 10. So far, gluconate is a notable exception to this observation. Stability of the Tc(V)-gluconate complex at 2 M NaOH was also observed by Schroeder et al. (1). For this reason, the presence of gluconate remains an issue for pretreatment strategies that aim to remove radionuclides from the waste stream prior to immobilization.

The nature and stability of metal complexes with d-gluconic acid ($C_6H_7O_{12}$) as a function of pH has been studied by a variety of techniques including polarography, pH titration, UV-Vis spectrophotometry, and potentiometry. The metal-gluconic acid complexes studied include the ferric-gluconate system (2) and the Mn(II, III, IV) system (3, 4). In these early studies, various bonding schemes were suggested to explain stability of the metal-gluconate complexes over a wide range of pH and oxidative conditions including bidentate bonding involving deprotonation of the carboxyl proton and the adjacent carbon, oxidative deprotonation of the terminal primary alcohol group (4), and deprotonation of the mid-chain carbons leaving the carboxyl proton intact (2). The gluconate ligand and the various bonding sites are shown schematically in

Figure 1. The crystal structure of Mn(II)-gluconate (5) confirmed both the bidentate bonding and bonding involving the terminal alcohol. Recently, gluconate complexes with hexavalent Mo and W have been investigated as a function of pH determined by NMR (6). These investigations revealed a large number of complexes with a range of the metal:ligand ratios. High metal:ligand ratios involved extensive deprotonation along mid-chain carbons. Although gluconate complexes with Tc(IV), per se, have not been studied previously, Tc(V)-gluconate (7) and analogous Tc(V)-glucoheptonate ($C_7H_{13}O_8$) complexes have been used as kidney and brain imaging agents (7, 8) and are commonly used as a derivative for more targeted Tc-bearing imaging agents and radiopharmaceuticals. Two different structures have been proposed for Tc(V)-glucoheptonate. One structure has a 1:2 metal to ligand stoichiometry and the gluconate bonds in bidentate fashion through the carboxyl proton and the adjacent carbon resulting in a five member ring consisting of Tc-O-C1-C2-O (8). The second structure is a dimer with a 1:1 metal to ligand stoichiometry and a linear Tc-O-Tc bridge. In this second structure bonding that does not involve the carboxylate group (7).

We present initial results from $TcO_2 \cdot xH_2O$(am) solubility experiments in the presence of gluconate over a wide range of pH and gluconate concentrations. Our initial results indicate that Tc(IV) solution species may be stabilized by a series of polymerized gluconate species under alkaline conditions.

Methods and Materials

All experiments and sample preparations were conducted in an atmosphere-controlled chamber under an Ar atmosphere. Deionized distilled water, degassed by boiling and cooling in the Ar atmosphere, was used in all cases. The solubility experiments were conducted in the "under-saturation direction" meaning that the concentration of Tc(IV) in solution was measured after the addition of gluconate to an initial solution of Tc(IV) aqueous species in the presence of $TcO_2 \cdot xH_2O$(am) solid phase at a fixed pH. The concentration of aqueous Tc(IV) species was observed to increase due to the formation of the Tc(IV)-gluconate aqueous complex and the increased solubility of $TcO_2 \cdot xH_2O$(am). Repeated measurement of the Tc concentration in solution was found not to change or to reach a "steady-state" level after a period of time. Solubility experiments were also attempted from the "over-saturation direction" meaning that the pH of a solution initially containing a known, steady-state concentration of Tc(IV)-gluconate complex in the presence of $TcO_2 \cdot xH_2O$(am) solid phase is adjusted to a different pH where the solubility of $TcO_2 \cdot xH_2O$(am) is believed to be lower resulting in a solution that is over-saturated with respect to the Tc(IV)-gluconate complex. In theory, the observed Tc(IV) concentration in solution should decrease to the new steady-state levels for the new pH.

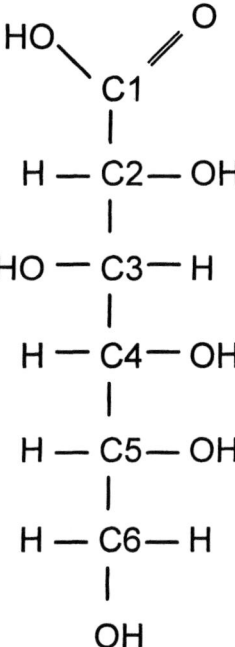

Figure 1. Schematic of gluconate ligand. Different bonding schemes are possible through a variety of deprotonation reactions. Bidentate bonding is possible through deprotonation of the carboxyl group (C1) and the adjacent carbon (C2). Oxidative deprotonation involves removal of hydrogen from the terminal alcoholic carbon (C6). Other bonding schemes involve progressive deprotonation of the C2 through C5 along the carbon backbone.

However these experiments resulted in the destruction of the Tc(IV)-gluconate complex, resulting in the formation of Tc-colloids under some pH conditions. The experimental method is described in more detail below.

A 0.29 M stock solution of NH_4TcO_4 was prepared, following purification, from irradiated MoO_3 obtained from Oak Ridge National Laboratory. The purity of ammonium pertechnetate was checked spectrophotometrically, and its content was established by comparative measurement of the β-activity of actual and standard ^{99}Tc solutions. $TcO_2 \cdot xH_2O$(am) precipitate was prepared individually for each sample starting from the stock solution and under basic conditions. Specifically, a small amount (0.18 ml) of the Tc stock solution was added to a 25-ml glass centrifuge tube that contained 4.12 ml of water and 1.0 ml of 1.0 M freshly prepared $Na_2S_2O_4$ aqueous solution and a magnetic stir bar. The pH was then adjusted to about 12 using NaOH. A black precipitate formed quickly and was allowed to mature in the mother liquid for 72 h. The precipitate was then washed three times using 20 ml of freshly prepared 0.01 M $Na_2S_2O_4$ in near neutral aqueous solution. The amorphous nature of the precipitate was confirmed by powder X-ray diffraction. A 1.0-M gluconate stock solution was prepared by dissolving sodium gluconate. For the preparation of Tc-gluconate sample solution, 0.2-ml of 1.0-M gluconate stock solution was added into the sample tube that contained the washed precipitates. 19.8 ml of pH adjusted solution, which contained 0.02 M of hydrazine to maintain reducing conditions, was then added resulting in a 0.01-M gluconate Tc sample.

Three sets of experiments were conducted using the washed precipitates. These sets were conducted (1) over the pH range from 3 to 12 in the presence of 0.01 M of gluconate, (2) at high hydroxide concentrations (from 0.01 M to 5 M NaOH) in the presence of 0.01 M of gluconate, and (3) at fixed pH = 10.5 as a function of gluconate concentrations from 0.01 M to 0.5 M gluconate. The pH values of the samples were readjusted using NaOH and HCl. The tubes were tightly capped and placed on a shaker in an Argon atmospheric chamber.

After 12 to 22 days of equilibration, the solution Tc concentrations and solution pH were measured. The pH was measured using an Orion-Ross combination glass electrode calibrated against pH buffers. The sampling procedure involved centrifuging a 1.5-ml aliquot of solution at 2000 g for 10 min followed by filtering through a Centricon-30 filter (Amicon, Inc.) with an approximate 0.0036-micron pore size. The filtrate was acidified with HNO_3 prior to the determination of the Tc concentration and oxidation state in Sets 1 and 3. In Set 2, the filtrate was diluted to reduce the hydroxide concentration to below 0.5 M. The total amount of Tc in solution was determined by beta-scintillation counting using a 0.5-ml subsample of the aliquot and 10 ml of scintillation cocktail. The detection limit for this technique is approximately 10^{-8} M Tc. Oxidation state analysis was determined using solvent extraction techniques where tetraphenyl-phosphoniumchloride (TPPC) in chloroform extracts pertechnetate (TcVII) into the organic phase leaving reduced Tc, assumed to be Tc(IV), in the aqueous phase. The extraction equilibration time was 5 min. The concentration of Tc in the organic and aqueous fractions was subsequently determined by beta-scintillation counting.

Raman spectra were collected on three 3.0 ml of centrifuged and filtered Tc-gluconate aqueous solutions which were sealed in Ar atmosphere in a quartz cuvette. The Raman spectra were measured in a backscattering geometry using approximately 60 mW of 532.0 nm excitation from a CW diode Nd:YAG laser. The Raman scattered light was focused on the entrance slit of an XY triple spectrometer. The scattered light was dispersed by a 1600 grooves/mm grating and then dispersed onto a liquid nitrogen cooled, charged coupled device Ge detector. The exit slit was maintained at 100 microns. The Raman signal was collected for 1800 seconds for all samples. The Raman spectra were collected in four overlapping regions, three of which were then combined to yield a spectral range from 150 to 1750 cm^{-1}. The fourth region was collected to observe the ethylenic stretching modes from 2700 to 3100 cm^{-1}. The Raman spectra were collected, processed, and analyzed using the software package, LABSPEC v.2.08.

Results and Discussion

The results of $TcO_2 \cdot xH_2O$(am) solubility experiments in the presence of 0.01 M of gluconate over the pH range from 3 to 12.5 are presented in Figure 2. There is a slight increase in the Tc concentrations measured between 12 and 138 days which would indicate that steady state concentrations attained relatively slowly. The high ratio of Tc(IV) to Tc(VII) and the near equivalence of the measured total Tc and Tc(IV) concentrations indicate that we have successfully maintained reducing conditions in these experiments. Compared to the solubility of $TcO_2 \cdot xH_2O$(am) with no complexant present (9, 10), the measured concentration of Tc increases by nearly two orders of magnitude in the presence of 0.01 M of gluconate. In addition, the measured Tc concentration dependence on pH can be divided into at least two distinct regions. Below pH = 9, the measured Tc concentrations displays at most a weak dependence on pH, especially if the data at pH = 3 is ignored. Above pH = 9, the measured Tc concentrations increase with increasing pH with a slope of approximately 0.3. The presence of distinct pH dependencies indicates the existence of at least two Tc-gluconate species in this pH region. Furthermore, the lack of an observed pH dependence in the pH range between 6 and 9 suggests that this species may be a neutral complex.

The $TcO_2 \cdot xH_2O$(am) solubility in the presence of 0.01 M of gluconate at higher hydroxide concentrations is shown in Figure 3. At high hydroxide concentrations, the solubility of $TcO_2 \cdot xH_2O$(am) continues to increase with

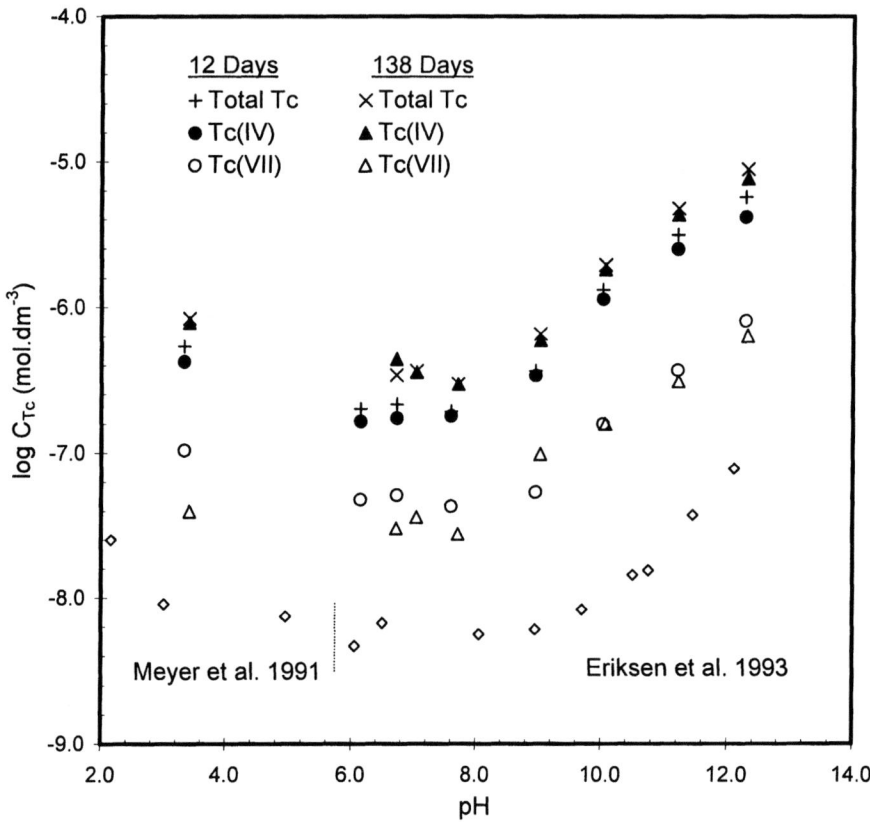

Figure 2. $TcO_2 \cdot xH_2O(am)$ solubility in the presence of 0.01 M of gluconate in the pH range from 3 to 12.5. The measured solubility in the presence of gluconate is approximately two orders of magnitude greater than the solubility reported by Meyer et al. (9) and Eriksen et al. (10) in the absence of complexants.

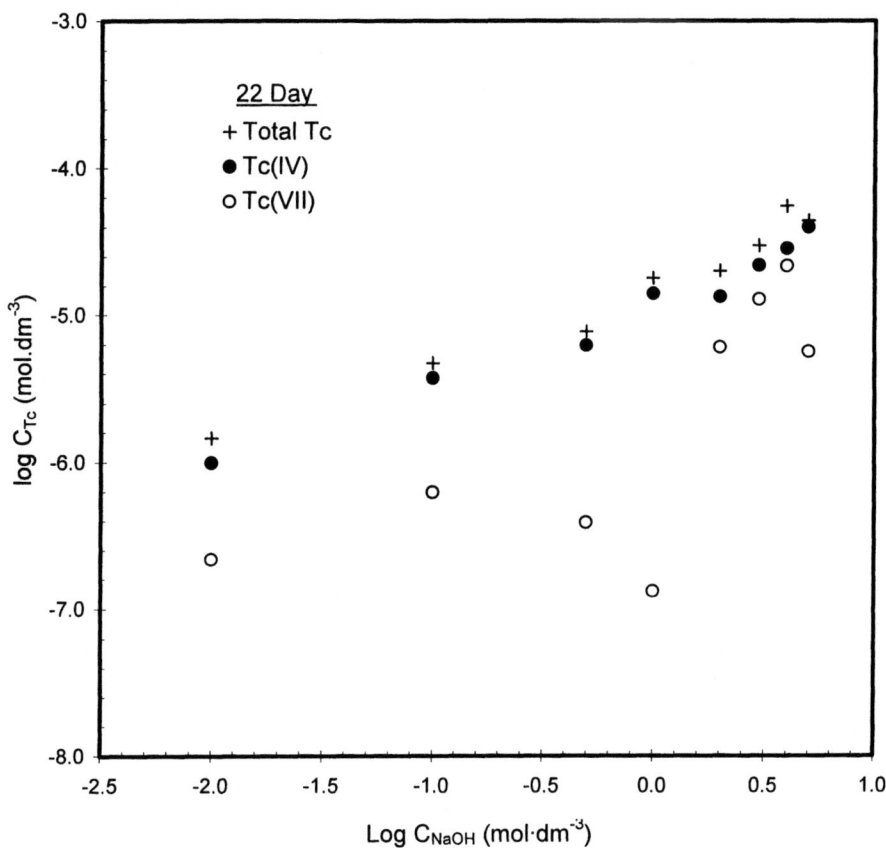

Figure 3. $TcO_2 \cdot xH_2O(am)$ solubility in the presence of 0.01 M of gluconate at high hydroxide concentrations.

increasing hydroxide concentration. The measured Tc concentration dependence on hydroxide concentration is slightly greater than that observed in the pH range from 9 to 12.5. However, there are not enough data to suggest that a third Tc-gluconate species is present.

Note in Figure 3 that at hydroxide concentrations greater than 1 M of NaOH, the ratio of Tc(IV) to Tc(VII) decreases significantly, although Tc(IV) concentrations are still greater than the Tc(VII) concentrations. This result could suggest that we are experiencing difficulty in maintaining rigorous reducing conditions at high hydroxide concentrations. Difficulty in maintaining reducing conditions might be expected since the Eh-pH boundary for Tc(VII) has a negative slope so that increasingly reducing conditions are required to stabilize reduced Tc species with increasing pH. However, we believe that the high Tc(VII) values do not represent the failure to maintain reducing conditions but are the result of reduced Tc species partitioning in to the organic phase as has been noted by other researchers (*11*). We are currently exploring alternative chemical conditions in which to conduct the solvent extractions for oxidation state determinations at high hydroxide concentrations.

The solubility of $TcO_2 \cdot xH_2O$(am) was measured at fixed pH = 10.5 as a function of gluconate concentration to elucidate the stoichiometry of the Tc-gluconate complex. At this pH, the measured Tc concentrations display only a weak dependence on gluconate concentration, as shown in Figure 4. In fact, no measurable increase in Tc concentration is observed for the highest three gluconate concentrations (0.2, 0.3, and 0.5 M gluconate). At gluconate concentrations between 0.01 and 0.2 M, the Tc concentration increases with increasing gluconate concentration with a slope of approximately 0.3. Such a modest dependence on ligand concentration suggests that the Tc-gluconate species may be polymerizing at this pH. Studies of $TcO_2 \cdot xH_2O$(am) solubility as a function of gluconate concentration are planned at fixed pH of 3 and 8.5 to determine if different dependencies are observed.

Raman investigations were carried out to aid in determining the nature of the Tc-gluconate solution complex. In all samples, the Tc-gluconate complex is present at far lower concentration than "free" or uncomplexed gluconate; therefore, the Raman spectra of free gluconate at identical chemical conditions was subtracted from the Raman spectra of the Tc-gluconate complex leaving only the Raman signature of the gluconate complexed to Tc. In Figure 5, the observed Raman spectrum of Tc-gluconate complex at pH = 10.5 is compared to Th-gluconate complex at pH = 9.5 in the presence of 0.4 M of $NaNO_3$ and 0.5 M of free gluconate at 10.5 in the region where vibrations assigned to the carboxylate group and carbon background occur (*12*). In Figure 6, an analogous comparison is made in the ethylenic stretching region. The Tc and Th gluconate complexes share similar vibrational features in both regions suggesting that the nature of the gluconate bonding is comparable. Our unpublished Raman studies on Th-EDTA and Th-citrate complexes also display similar Raman features in

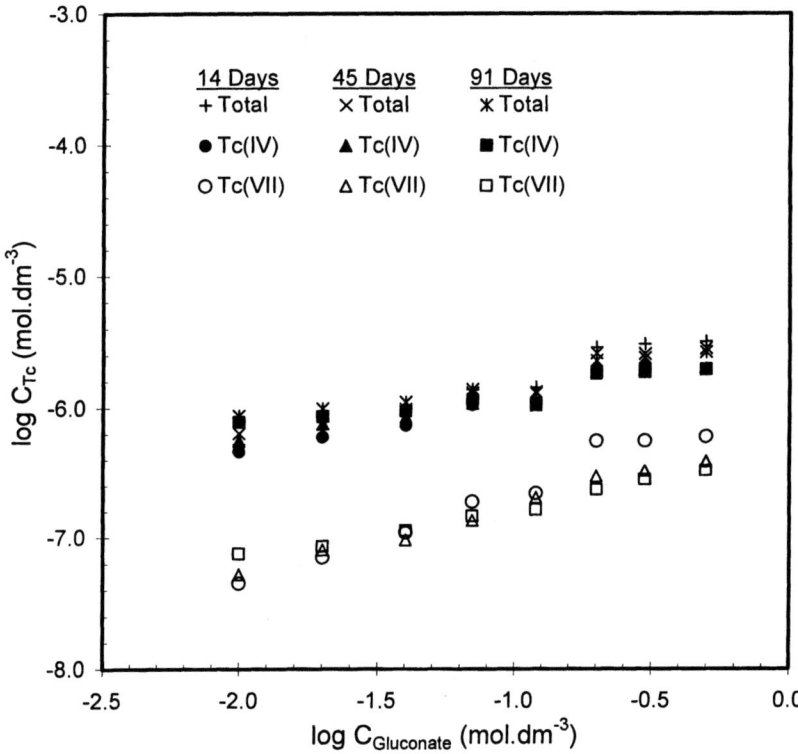

Figure 4. $TcO_2 \cdot xH_2O(am)$ solubility as a function of gluconate concentration at fixed pH = 10.5.

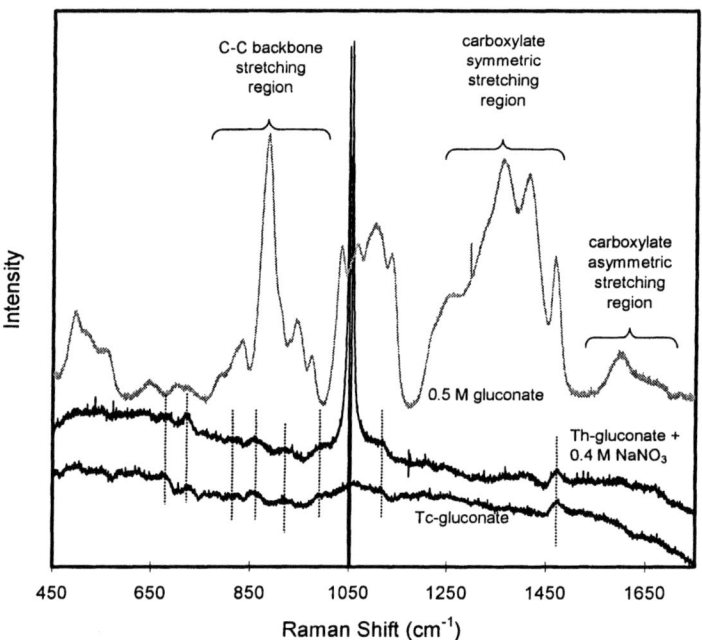

Figure 5. Raman spectra from 450 to 1700 cm^{-1} of Tc- and Th-gluconate complexes at ca. pH = 10. No distinctive features indicating metal-gluconate bonding are observed. The sharp feature in the Th-gluconate spectrum at 1050 cm^{-1} is from incomplete subtraction of $NaNO_3$.

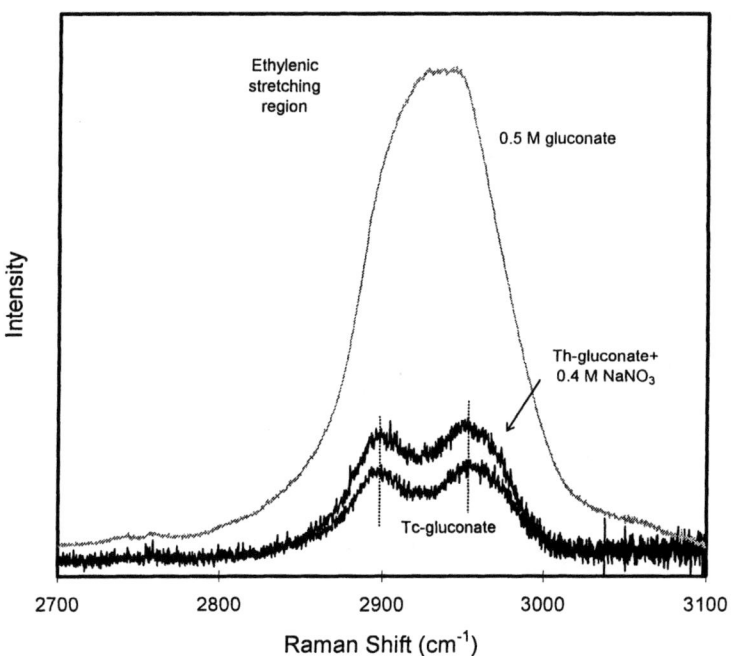

Figure 6. Raman spectra in the ethylenic stretching region from 2700 to 3100 cm^{-1} of Tc- and Th-gluconate complexes at ca. pH = 10. The Raman intensity of the 0.5 gluconate has been reduced by a factor of 5 so that the features in the Raman spectra of the complexed species can be more easily compared.

the lower frequency region where the vibrations are assigned to the carboxylate group and carbon backbone despite differences in the bonding schemes. The apparent similarity suggests that that the observed Raman vibrations in the lower frequency region may not be able to distinguish subtle differences in the bonding schemes of the carboxylate group. However, the Th-EDTA complex did exhibit differences in the ethylenic stretching region which may be attributed to a larger number of ethylene-like groups in EDTA.

At hydroxide concentrations greater than 4 M of NaOH, the Raman spectra of the Tc-gluconate complexes differ significantly from those at lower hydroxide concentrations in both the intensity and the sharpness of the vibrational features. These features are compared to the solid Na Gluconate starting material in Figures 7 and 8. Visually the solutions at 4 M of NaOH are clear, yet a white precipitate is observed at the bottom of the sample tube. At 5 M of NaOH, the Tc-gluconate solution is saturated with white precipitates. In the absence of Tc, solutions of 0.01 M of gluconate at both 4 M and 5 M of NaOH are clear and no precipitate is formed. However, Raman spectra of these Tc-free gluconate solutions show changes in both the low frequency and the ethylenic regions compared to the Raman spectra of Tc-free gluconate solutions at ca. pH = 10 indicating that the gluconate ligand itself is undergoing changes as a function of increasing hydroxide concentration. These changes may reflect deprotonation reactions along the carbon backbone or of the carboxylate group with increasing hydroxide concentration. With the addition of Tc, such reactions could result in further polymerization of the Tc-gluconate complex at hydroxide concentrations greater than 4 M of NaOH.

Acknowledgments

The authors thank Dr. Dhanpat Rai for useful discussions and Mr. Dean Moore for assistance with the solubility experiments. This work was supported by the Environmental Management Science Program (EMSP) and by the Division of Materials Science and Engineering, Office of Basic Energy Sciences, DOE. Pacific Northwest National Laboratory (PNNL) is operated by Battelle for DOE under Contract DE-AC06-76RLO 1830.

References

1. Schroeder, N. C.; Ashley, K. R.; Blanchard, D. L. U.S. Department of Energy Environmental Science Management Project Final Report **2000**, Project Number 59990.
2. Pecsok, R. D; Sandera, J. *J. Am. Chem. Soc.* **1955**, *77*, 1489–1494.

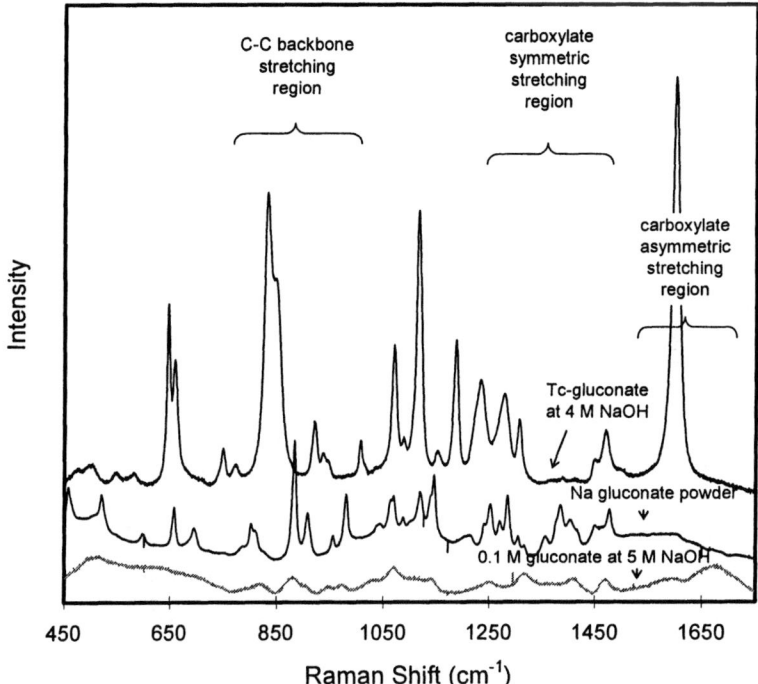

Figure 7. Raman spectra from 450 to 1700 cm^{-1} of Tc-gluconate complex at 4.0 M of NaOH. The Raman spectrum of the Tc-complex at 4 M of NaOH is compared to solid Na gluconate starting material and "free" gluconate at 5 M of NaOH. At 4 M of NaOH, the vibrational features Raman spectrum of the Tc-gluconate complex are much sharper than those observed at lower hydroxide concentrations, shown in Figure 5.

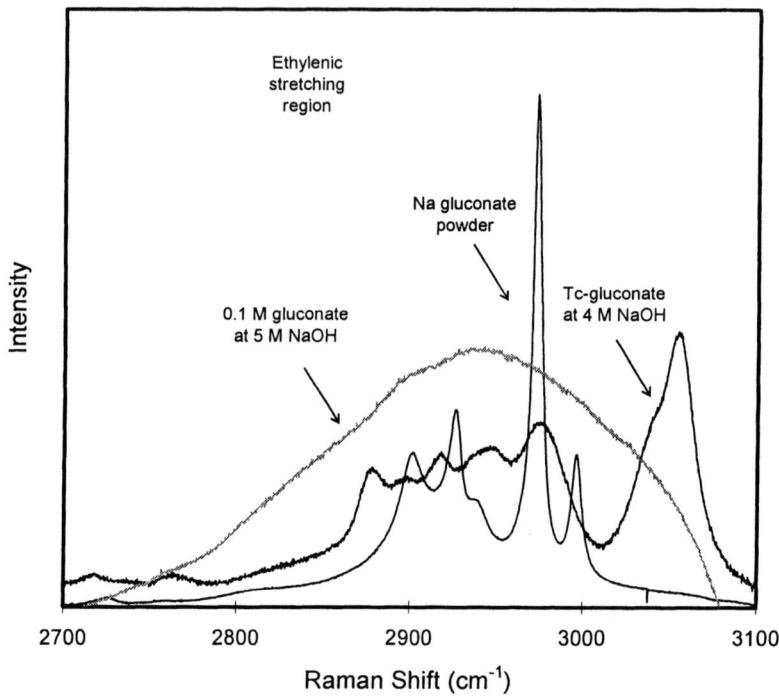

Figure 8. Raman spectra in the ethylenic stretching region of Tc-gluconate complex at 4.0 M of NaOH. The Raman spectrum of the Tc-complex at 4 M of NaOH is compared to solid Na gluconate starting material and "free" gluconate at 5 M of NaOH. As observed in the low frequence region, the vibrational features for Tc-complex at 4 M of NaOH differ from those observed at lower hydroxide concentrations (compare to Figure 6).

3. Bodini, M. E.; Willis, L. A.; Riechel, T. L.; Sawyer, D. T. *Inorg. Chem.* **1976,** *15*, 1538–1543.
4. Bodini, M. E; Sawyer D. T. *J. Am. Chem. Soc.* **1976,** *98*, 8366–8371.
5. Lis, T. *Acta. Cryst. B* **1979,** *B35*, 1699–1701.
6. Ramos, M. L.; Caldeira, M. M.; Gil, V. M. S. *Carb. Res.* **1997,** *304*, 97–109.
7. Hwang, L. L.-Y.; Ronca, N.; Solomon, N. A.; Steigman, J. *Int. J. Appl. Radiat. Isot.* **1985,** *36*, 475–480.
8. De Kievet, W. *J. Nucl. Med.* **1981,** *22*, 703–709.
9. Meyer, R. E.; Arnold, W. D.; Case, F. I.; O'Kelley, G. D. *Radiochim. Acta* **1991,** *55*, 11–18.
10. Eriksen, T. E.; Ndalamba, P.; Bruno, J.; Caceci, M. *Radiochim. Acta* **1992,** *58/59*, 67–70.
11. Ianovici, E.; Kosinski, M.; Lerch, P.; Maddock, A. G. *J. Radio. Chem.* **1981,** *64*, 315–326.
12. McConnell, A. A.; Nuttall, R. H.; Stalker, D. M. *Talanta* **1978,** *25*, 425–434.

Chapter 17

Behavior of Technetium in Alkaline Solution: Identification of Non-Pertechnetate Species in High-Level Nuclear Waste Tanks at the Hanford Reservation

Wayne W. Lukens[1], David K. Shuh[1], Norman C. Schroeder[2], and Kenneth R. Ashley[3]

[1]Actinide Chemistry Group, Chemical Sciences Division, Lawrence Berkeley National Laboratory, Berkeley, CA 94720
[2]Chemical Science Technology, Los Alamos National Laboratory, Los Alamos, NM 87545
[3]Department of Chemistry, Texas A&M University at Commerce, Commerce, TX 75429

Technetium is a long-lived (^{99}Tc: 213,000 year half-life) fission product found in nuclear waste and is one of the important isotopes of environmental concern. The known chemistry of technetium suggests that it should be found as pertechnetate, TcO_4^-, in the extremely basic environment of the nuclear waste tanks at the Hanford site. However, other chemical forms of technetium are present in significant amounts in certain tanks, and these non-pertechnetate species complicate the treatment of the waste. The only spectroscopic characterization of these non-pertechnetate species is a series of X-ray absorption near edge structure (XANES) spectra of actual tank waste. To better understand the behavior of technetium under these conditions, we have investigated the reduction of pertechnetate in highly alkaline solution in the presence of compounds found in high-level waste. These results and the X-ray absorption fine structure (XAFS) spectra of these species are compared to the chemical behavior and XANES spectra of the actual non-pertechnetate species. The identity of the non-pertechnetate species is surprising.

Introduction

The Hanford Reservation in eastern Washington State is the site of one of the largest and most costly remediation efforts in the United States. Years of plutonium production have generated 53 million gallons of high-level nuclear waste, which is now stored in 177 underground tanks (*1*). The waste consists of three distinct fractions: supernate, saltcake, and sludge (*2, 3*). The compositions of the these fractions vary greatly depending upon which reprocessing method was used to separate the plutonium and later, uranium, from the spent nuclear fuel. The waste contained in the tanks that are the subject of this manuscript is a mixture of REDOX (reduction-oxidation) and PUREX (plutonium uranium extraction) wastes that have been further processed to remove some of the ^{90}Sr and ^{137}Cs (*3*). In these tanks, the supernate is an aqueous solution of sodium nitrate, nitrite, and hydroxide, and various organic compounds including citrate, gluconate, formate, oxalate, ethylenediaminetetraacetate (EDTA), and nitrilotriacetate (NTA); in addition, appreciable quantities of ^{137}Cs, ^{90}Sr, ^{127}I, ^{237}Np, and ^{99}Tc are present in the supernate. Saltcake consists of water-soluble salts, mainly sodium nitrate and nitrite, that have precipitated during reduction of supernate volume by evaporation. Sludge consists of the waste components that are insoluble under strongly alkaline conditions and includes most of the fission products and actinides plus large quantities of aluminum and iron oxides and aluminosilicates.

The current plan for immobilizing this waste requires separating it into high- and low- activity streams, which will be vitrified separately to form high- and low- activity waste glasses (*1*). The low-activity waste stream mainly consists of supernate and dissolved saltcake; the high-activity waste stream is mainly sludge. Due to the previous performance requirements for the low-activity glass, almost all of the ^{137}Cs and ^{90}Sr and approximately 80% of the technetium (^{99}Tc) needed to be removed from the low-activity waste stream and sent to the high-activity waste stream, as illustrated in Scheme 1 (*4, 5*). This technetium separation was to be accomplished by ion exchange of pertechnetate, TcO_4^-, the most thermodynamically stable form of technetium at high pH. Although ion exchange was effective for many tanks, work by Schroeder showed that it failed for Complexant Concentrate (CC) waste tanks, including tanks SY-101 and SY-103, which contain significant amounts of organic complexants including NTA, EDTA, citrate, and gluconate (*4, 6*). In these tanks, the vast majority of technetium is present as a soluble, lower-valent, non-pertechnetate species (NPS) that is not removed during pertechnetate ion exchange (*4, 6*).

The identity of the NPS is unknown, and its behavior has hampered removal efforts. It is not readily removed by ion exchange, and although the NPS is air-sensitive (it slowly decomposes to pertechnetate), it is difficult to oxidize in practice (6, 7). The only spectroscopic characterization of the NPS is a series of Tc K-edge X-ray absorption near edge structure (XANES) spectra of CC samples reported by Blanchard (Figure 1) (7). Although its identity cannot be determined directly from these spectra, the NPS was believed to be Tc(IV) based on the energy of its absorption edge, 7.1 eV lower than that of TcO_4^-. This edge shift is similar to that of TcO_2, 6.9 eV lower than that of TcO_4^- (8). The presence of soluble, lower-valent technetium species is unexpected in light of the known chemistry of technetium; under these conditions, insoluble $TcO_2 \cdot 2H_2O$ would be expected rather than soluble complexes. The work described here identifies the potential candidates for the non-pertechnetate species and identifies technetium complexes that have XANES spectra identical to that of the NPS shown in Figure 1.

Experimental Section

Procedures

Caution: ^{99}Tc *is a β-emitter* ($E_{max} = 294$ *keV,* $\tau_{1/2} = 2 \times 10^5$ *years*). All operations were carried out in a radiochemical laboratory equipped for handling this isotope. Technetium, as $NH_4^{99}TcO_4$, was obtained from Oak Ridge National Laboratory. The solid $NH_4^{99}TcO_4$ was contaminated with a large amount of dark, insoluble material. Prolonged treatment of this sample with H_2O_2 and NH_4OH did not appreciably reduce the amount of dark material. Ammonium pertechnetate was separated by carefully decanting the colorless solution from the dark solid. A small amount of NaOH was added to the colorless solution, and the volatile components were removed under vacuum. The remaining solid was dissolved in water, and the colorless solution was removed from the remaining precipitate with a cannula. The concentration of sodium pertechnetate was determined spectrophotometrically at 289 nm ($\varepsilon = 2380$ M l^{-1} cm^{-1}). UV-visible spectra were obtained using an Ocean-Optics ST2000 spectrometer.

X-ray absorption fine structure (XAFS) spectra were acquired at the Stanford Synchrotron Radiation Laboratory (SSRL) at Beamline 4-1 using a Si(220) double crystal monochromator detuned 50% to reduce the higher order harmonic content of the beam. All ^{99}Tc samples were triply contained inside

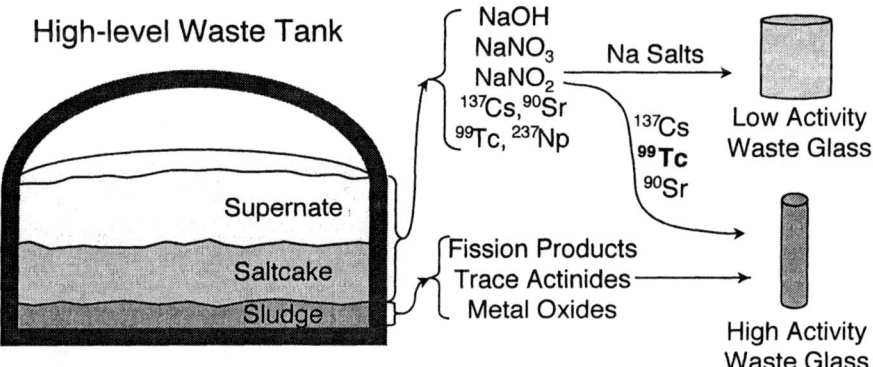

Scheme 1. Simplified Illustration of Immobilization of high-Level Nuclear Waste at the Hanford Site Illustrating the Role of ^{99}Tc Separation.

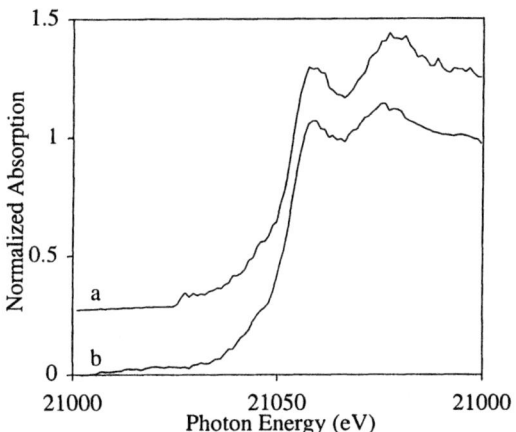

Figure 1. Tc K-edge XANES spectra of the non-pertechnetate species (NPS) in tanks (a) SY-103, and (b) SY-101 (Reported by Blanchard in reference 7.) (Reproduced with permission from Environ. Sci. Technol. **2004**, 33, 229.*)*

sealed polyethylene vessels. XAFS were obtained in the transmission mode at room temperature using Ar filled ionization chambers or in fluorescence yield mode using a multi-pixel Ge-detector system (9). The spectra were energy calibrated using the first inflection point of the pre-edge peak from the Tc K-edge spectrum of an aqueous solution of NH_4TcO_4 defined as 21044 eV. To determine the Tc K-edge charge state shifts, the energies of the Tc K-edges at half height were used. Extended XAFS (EXAFS) analysis and radiolysis experiments were carried out as previously described (10).

All operations were carried out in air, except as noted. Water was deionized, passed through an activated carbon cartridge to remove organic material, and then distilled. Iminodiacetic acid was recrystallized three times from water. All other chemicals were used as received. The $Tc(CO)_3(H_2O)_3^+$ stock solution was prepared from $TcOCl_4(n-Bu_4N)$ (11) by the procedure developed by Alberto (12) then dissolving the reaction product in 0.01-M triflic acid. The ^{99}Tc concentration was determined by liquid scintillation.

Solutions for NMR spectroscopy were prepared by addition of 0.10-mL aliquots of the $Tc(CO)_3(H_2O)_3^+$ stock solution to 0.90 mL of 1.1-M NaOH in D_2O with and without 0.11-M organic complexant. NMR samples were contained inside a Teflon tube inside a 10-mm screw cap NMR tube. Solutions for XAFS spectroscopy were prepared by addition of 0.20-mL aliquots of the $Tc(CO)_3(H_2O)_3^+$ stock solution to 0.80 mL of D_2O solutions of 1.1-M NaOH with and without 0.11-M organic complexant.

The Tc(IV) gluconate complex was prepared by reducing a solution of TcO_4^- (2 mM, 1 mL, 2 μmol) in 0.1-M potassium gluconate and 1-M NaOH with sodium dithionite (2 M, 10 μL, 20 mol). Solutions for EXAFS spectroscopy were sealed under Ar inside 2-mL screw-cap centrifuge tubes, which were placed inside two consecutive heat sealed, heavy walled polyethylene pouches. Pouches were stored under Ar in glass jars sealed with poly(tetrafluoroethylene) tape until their spectra were recorded.

Tc(IV) gluconate was also prepared by treating $TcO_2 \cdot 2H_2O$ (10) with a solution of 0.1-M potassium gluconate in 1-M NaOH in D_2O forming a pale, pink solution. The solution was treated with CO for 5 min then heated to 85 °C in a closed NMR tube with air in the head-space. After 1 h the sample contained a large amount of TcO_4^- as determined by ^{99}Tc NMR. After 18 h at 85 °C, the solution was colorless and contained mainly TcO_4^- as determined by ^{99}Tc NMR.

Results and Discussion

Tc(IV) Alkoxide Complexes

As a first step in investigating the behavior of technetium in highly alkaline solutions relevant to high-level waste, solutions of TcO_4^- in alkaline solution containing organic compounds, including complexants, were irradiated to reduce the TcO_4^-, and the lower-valent technetium products produced were identified (10). The use of irradiation in these experiments does not imply a similar mechanism for reduction of TcO_4^- in high-level waste tanks. Both chemical (13) and radiolytic (14) pathways exist for reduction of TcO_4^- under these conditions, but the radiation-chemical pathway is different from the pathway that is operative here—reduction of TcO_4^- by hydrated electrons from the radiolysis of water and by the reducing radicals produced by hydrogen atom abstraction from alcohols (14).

The initial results of the radiolysis experiments showed that none of the carboxylate complexants, citrate, EDTA, or NTA, form stable complexes with Tc(IV) in alkaline solution. Under these conditions, only $TcO_2 \cdot 2H_2O$ is produced. However, soluble, lower-valent complexes are produced by the radiolytic reduction of TcO_4^- in alkaline solution containing the alcohols, ethylene glycol, glyoxylate, and formaldehyde. Although glyoxylate and formaldehyde are aldehydes, they exist as geminal diols in aqueous solution, and therefore can act as alkoxide ligands. The EXAFS spectrum and proposed structure of the Tc(IV) glyoxylate complex are shown in Figure 2, and the structural parameters are given in Table I. The proposed structure is very similar to that of the well known $(H_2EDTA)_2Tc_2(\mu-O)_2$ complex with the EDTA ligands replaced by glyoxylate ligands, presumably acting as diolate ligands.

These radiolysis experiments clearly show that soluble Tc(IV) alkoxide complexes can be formed in highly alkaline solution under conditions similar to those found in high-level waste. However, none of the potential ligands examined are present in high-level waste in sufficient concentrations to account for the formation of the NPS (3). The alkoxide ligand present in large quantities in CC waste is gluconate (15). Moreover, gluconate can act as a tridentate alkoxide ligand (using the carboxylate group and hydroxyl groups on carbon atoms 2 and 3). The resulting $Tc(gluconate)_2^{2-}$ complex would presumably be similar to an analogous complex of Tc(IV) with two tridentate alkoxide ligands, $Tc[(OCH_2)_3CN(CH_3)_3]_2$, described by Anderegg (16). This complex is the most hydrolytically stable of the Tc(IV) alkoxide complexes. While most Tc(IV) alkoxide complexes are stable only above pH 10, $Tc[(OCH_2)_3CN(CH_3)_3]_2$ is stable towards hydrolysis down to pH 4, and a related, Tc(IV) alkoxide complex with a different tridentate ligand, 1,2,4-butanetriolate, is stable down to pH 8 (17). Consequently, a Tc(IV) gluconate complex would be expected to be quite hydrolytically stable although less stable than $Tc[(OCH_2)_3CN(CH_3)_3]_2$ in which the cationic ammonium group increases the acidity of the hydroxyl groups.

The light pink Tc(IV) gluconate complex was prepared in situ by reducing TcO_4^- with dithionate in a solution of 0.1-M gluconate and 1-M NaOH. The EXAFS spectrum and Fourier transform of Tc(IV) gluconate are shown in

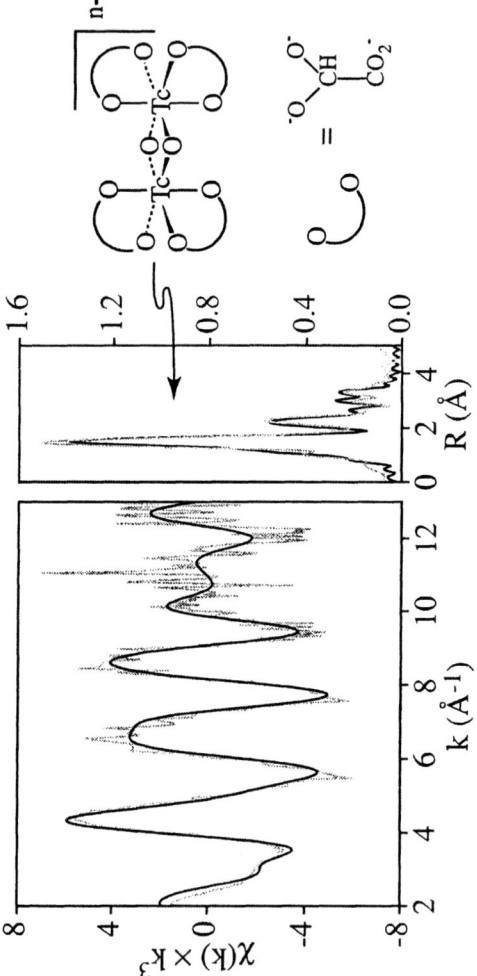

Figure 2. EXAFS spectrum and Fourier transform of the Tc(IV) species formed by radiolysis of TcO_4^- in a solution of 0.1 glyoxylic acid in 1-M NaOH; data are shown in gray and the fit in black. The structure of the complex consistent with the EXAFS spectrum is shown on the right.

Table I. Structural Parameters of Soluble Radiolysis Product Derived from EXAFS[a]

Scattering Path	Coordination Number[b]	Distance (Å)	Debye-Waller Parameter (Å^2)[b]	ΔE_0 (eV)[c]
Tc-O	6.7(3)	2.008(3)	0.0058(5)	-7.9
Tc-Tc	0.7(1)	2.582(4)	0.003*	-7.9
Tc-O-Tc-O[d]	6*	4.06(2)	0.002(3)	-7.9

a. Numbers in parentheses are the standard deviation of the given parameter derived from least-squares fit to the EXAFS data. The standard deviations do not indicate the accuracy of the numbers; they are an indication of the agreement between the model and the data. In general, coordination numbers have an error of ± 25%, and bond distances have an error of ± 0.5%, when compared to data from crystallography.
b. Parameters with an asterisk were not allowed to vary during analysis.
c. E_0 was refined as a global parameter for all scattering paths. The large negative value results from the definition of E_0 in EXAFSPAK.
d. This scattering path is a 4-legged multiple scattering path between two mutually *trans* oxygen atoms in the technetium coordination sphere.

Table II. Structural Parameters of Tc(IV) Gluconate Derived from EXAFS[a]

Scattering Path	Coordination Number[b]	Distance (Å)	Debye-Waller Parameter (Å^2)[b]	ΔE_0 (eV)[c]
Tc-O	6*	2.010(1)	0.0045(1)	-5.2(3)
Tc-C	6*	3.37(2)	0.015(3)	-5.2
Tc-O-Tc-O[d]	6*	4.03(2)	0.008(3)	-5.2

a. Numbers in parentheses are the standard deviation of the given parameter derived from least-squares fit to the EXAFS data. The standard deviations do not indicate the accuracy of the numbers; they are an indication of the agreement between the model and the data. In general, coordination numbers have an error of ± 25%, and bond distances have an error of ± 0.5%, when compared to data from crystallography.
b. Parameters with an asterisk were not allowed to vary during analysis (scale factor was varied instead; $S_0^2 = 1.38(3)$).
c. E_0 was refined as a global parameter for all scattering paths. The large negative value results from the definition of E_0 in EXAFSPAK.
d. This scattering path is a 4-legged multiple scattering path between two mutually *trans* oxygen atoms in the technetium coordination sphere.

Figure 3; fit parameters are given in Table II. The coordination environment of the Tc center is simple: 6 O neighbors at 2.01 Å and 6 C neighbors at 3.37 Å. The bond distances are similar to the aforementioned $Tc[(OCH_2)_3CN(CH_3)_3]_2$, in which the average Tc-O bond length is 1.996(9) Å (*16*). Although the coordination geometry of the coordinated gluconate ligand cannot be determined directly from the EXAFS data, the gluconate ligand is believed to be coordinated to the Tc center by a carboxylate and two hydroxyl groups, as illustrated in Figure 3.

Although the EXAFS experiment establishes the existence of Tc(IV) gluconate, it does not establish whether Tc(IV) gluconate is the NPS observed in CC waste. In fact, as shown in Figure 4, Tc(IV) gluconate is not the NPS in Tanks SY-101 and SY-103. Although the XANES spectra of Tc(IV) gluconate and the NPS are superficially similar, the energies of their absorption edges differ by 1.6 eV. More importantly, no combination of the XANES spectra of Tc(IV) gluconate, Tc(V) gluconate, and TcO_4^- will fit the XANES spectrum of the NPS. Not only is the NPS not Tc(IV) gluconate, the NPS cannot be any kind of Tc(IV) alkoxide complex. The energies of the Tc-K edges of Tc(IV) alkoxide complexes and other Tc(IV) complexes with oxygen neighbors, including $TcO_2 \cdot 2H_2O$, fall within a very narrow range around 5.5 eV below the energy of the TcO_4^- absorption edge (*10*). For comparison, the Tc-K edge of the NPS occurs at 7.1 eV below the TcO_4^- absorption edge.

The fact that Tc(IV) gluconate is not the NPS is surprising. The radiolysis experiments clearly show that Tc(IV) alkoxides can be formed and are hydrolytically stable in highly alkaline solution. In fact, the Tc(IV) gluconate complex is so hydrolytically stable that it can be prepared by dissolving $TcO_2 \cdot 2H_2O$ in 1-M NaOH containing 0.1-M gluconate, which may have implications for treating high-level waste. If CC waste containing gluconate is added to tanks containing $TcO_2 \cdot 2H_2O$, soluble Tc(IV) gluconate will be formed. Like the NPS, Tc(IV) gluconate would not be removed by the ion exchange resins that remove TcO_4^-. However, like other Tc(IV) compounds (*10*), Tc(IV) gluconate is fairly air-sensitive in alkaline solution (it decomposes to TcO_4^- in < 18 hr at 85 °C when exposed to air), and should be easy to oxidize, unlike the NPS. This sensitivity to oxidation could be the reason that Tc(IV) gluconate is not observed in CC waste. Another possibility is that the thermodynamic stability of the NPS is greater than that of Tc(IV) gluconate.

fac-Tc(CO)$_3$ Complexes

A different approach was taken since the systematic investigation described above seemed unable to yield the identity of the NPS. Theoretical XANES spectra were calculated (*18*) for a variety of lower valent technetium complexes regardless of whether the ligands were present in high-level waste and disregarding the oxidation state of the technetium center. The complex that

311

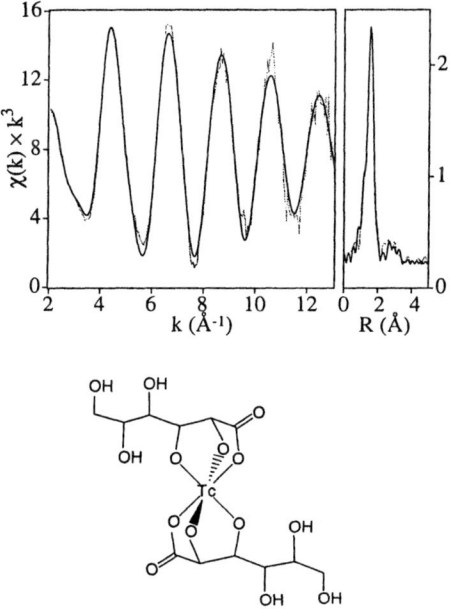

Figure 3. EXAFS spectrum and Fourier transform of the Tc(IV) gluconate complex; data are shown in gray and the fit in black. The structure of the complex consistent with the EXAFS spectrum is shown on the right.

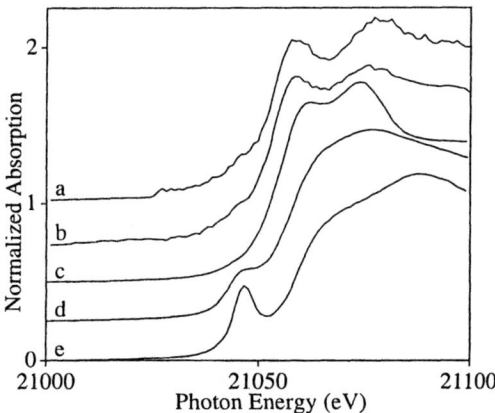

Figure 4. Tc K-edge XANES spectra of (a) NPS in tank SY-103, (b) NPS in tank SY-101, (c) Tc(IV) gluconate, (d) Tc(V) gluconate, and (e) TcO_4^-. The spectra of the non-pertechnetate species in tanks SY-101 and SY-103 are from reference 7. (Reproduced with permission from Environ. Sci. Technol. **2004**, *33, 229.*

had a calculated XANES spectrum most similar to that of the NPS was $fac\text{-Tc(CO)}_3(H_2O)_3^+$ (*12, 19–21*). Since the crystal structure of this complex has not been reported, the Tc-C and Tc-O distances for the carbonyl and water ligands were taken from the crystal structure of $fac\text{-Tc(CO)}_3\{[OP(OCH_3CH_3)_2]_3Co(C_5H_5)\}$ (*22*). The synthesis and chemistry of $fac\text{-Tc(CO)}_3$ complexes are the subject of extensive research, largely due to Alberto, since $fac\text{-Tc(CO)}_3$ complexes are potentially useful as ^{99m}Tc radiopharmaceuticals (*12, 19–21*). Of particular relevance is the fact that $fac\text{-Tc(CO)}_3$ complexes can be prepared from TcO_4^- in alkaline solution at low CO concentrations. This characteristic, along with the fact that CC waste tanks contain CO (the head space gas consists of 0.25 to 0.5 mol % CO) (*23*), suggests that $fac\text{-Tc(CO)}_3$ could have formed in the Hanford high-level waste tanks.

The behavior of $fac\text{-Tc(CO)}_3$ complexes in alkaline solution has previously been investigated, and the species formed at different hydroxide concentrations have been identified by Gorskov using ^{99}Tc-NMR spectroscopy (*24*). Figure 5 shows the ^{99}Tc-NMR spectra of the $fac\text{-Tc(CO)}_3$ species produced by adding $fac\text{-Tc(CO)}_3(H_2O)_3^+$ to different alkaline solutions. In 1-M NaOH, the only species present is $fac\text{-Tc(CO)}_3(H_2O)_2(OH)$, which has a chemical shift of –1060 ppm. As with Tc(IV), the carboxylate complexants do not form complexes with $fac\text{-Tc(CO)}_3$ in alkaline solution. Only $fac\text{-Tc(CO)}_3(H_2O)_2(OH)$ was observed in 1-M NaOH solutions containing 0.1-M EDTA, NTA, or citrate. However, gluconate does form a complex with $fac\text{-Tc(CO)}_3$, which is indicated by the presence of a new peak in the ^{99}Tc-NMR spectrum at –1240 ppm of $fac\text{-Tc(CO)}_3$ in a solution of 0.1-M gluconate and 1-M NaOH. When $fac\text{-Tc(CO)}_3(H_2O)_3^+$ is added to an SY-101 simulant, both $fac\text{-Tc(CO)}_3(H_2O)(OH)$ and $fac\text{-Tc(CO)}_3(\text{gluconate})^{2-}$ are observed. However, after one week, the only observable technetium species are $fac\text{-Tc(CO)}_3(\text{gluconate})^{2-}$ and TcO_4^-. These NMR experiments demonstrate that $fac\text{-Tc(CO)}_3$ species are stable in alkaline solutions approximating the composition of high-level waste. For comparison, alkaline solutions of Tc(IV) alkoxides are more air-sensitive, and will oxidize to TcO_4^- in less than a week if exposed to air (*10*).

Several of the $fac\text{-Tc(CO)}_3$ complexes were also characterized by EXAFS spectroscopy as shown in Figure 6, and the structural details are summarized in Table III. It is not surprising that their spectra and the parameters derived from fitting their spectra are very similar. The main differences among these complexes are Tc-C and Tc-O distances of the carbonyl and water, hydroxide, or gluconate ligands. Since the scattering atoms are identical and the bond distances change little among the complexes, the spectra differ only slightly. The change in bond distances is systematic and consistent with the nature of the ligands. The Tc-O distances for the first shell oxygen decrease in order from $fac\text{-Tc(CO)}_3(H_2O)_3^+$ to $fac\text{-Tc(CO)}_3(H_2O)_2(OH)$ to $fac\text{-Tc(CO)}_3(\text{gluconate})^{n-}$, in agreement with the observation that gluconate

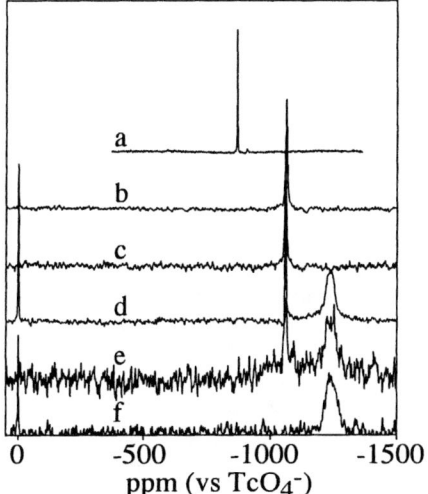

Figure 5. ^{99}*Tc-NMR spectra of fac-Tc(CO)$_3$(H$_2$O)$_3^+$ dissolved in (a) dilute triflic acid, (b) 1- M NaOH, (c) 1- M NaOH with 0.1- M EDTA, (d) 1-M NaOH with 0.1- M gluconate, (e) SY-101 simulant, and (f) SY-101 simulant after 1 week in air.*

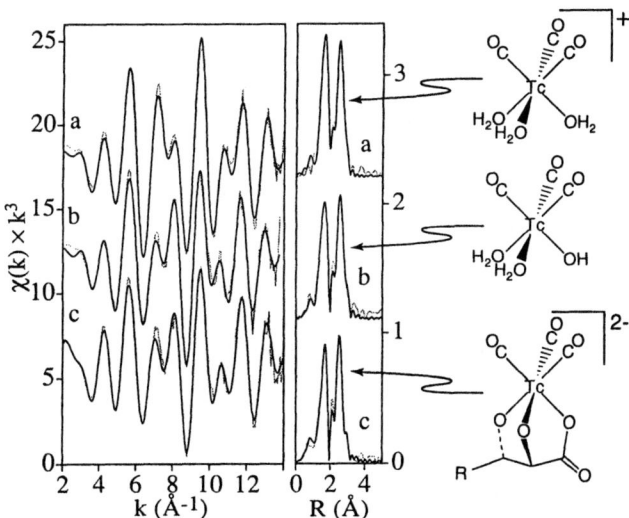

Figure 6. EXAFS spectra and Fourier transforms of (a) fac-Tc(CO)$_3$(H$_2$O)$_3^+$, (b) fac-Tc(CO)$_3$(HO)(H$_2$O)$_2$, and (c) fac-Tc(CO)$_3$(gluconate)$^{2-}$. Data are illustrated in gray, and the least squares fits are black. The structures of the complexes consistent with the EXAFS spectra are to the right of the spectra.

forms the most stable complex with fac-Tc(CO)$_3$, followed by hydroxide, and then followed by water. In addition, the CO distance is shorter in fac-Tc(CO)$_3$(H$_2$O)$_3{}^+$ than in the other two complexes as water is a weaker π–donor than the other two ligands. Overall, the EXAFS data are consistent with the known stabilities of these three complexes and clearly show that these are three distinct complexes, in agreement with the ^{99}Tc-NMR data.

The results described above show that fac-Tc(CO)$_3$ complexes are stable under conditions found in high-level waste, but do not establish whether they are actually the NPS. As shown in Figure 7, the XANES spectra of the fac-Tc(CO)$_3$ complexes are very similar, if not identical, to those of the NPS. The Tc K-edge energies of the XANES spectra of fac-Tc(CO)$_3$ complexes occur at 7.5 eV below that of TcO$_4{}^-$, in excellent agreement with the observed edge shift of 7.1 eV for the NPS. Most convincing is the fact that the spectrum of the NPS in tank SY-103 can be fit using only the spectrum of fac-Tc(CO)$_3$(gluconate)$^{2-}$, and the spectrum of the NPS in tank SY-101 can be fit using the spectrum of fac-Tc(CO)$_3$(gluconate)$^{2-}$ containing 7% TcO$_4{}^-$, presumably due to oxidation. The spectra of the NPS can also be fit using the spectrum of fac-Tc(CO)$_3$(H$_2$O)$_2$(OH), but the fit is of slightly poorer quality. Given the similarity of the EXAFS spectra of the fac-Tc(CO)$_3$ complexes, it is not possible to definitively assign the spectra to a specific fac-Tc(CO)$_3$ complex, but XANES spectra in Figure 1 are clearly assignable to fac-Tc(CO)$_3$ species.

The identity of the NPS explains some of its behavior. Simplest to explain is the fact that it is not removed by the cationic resins used to separate TcO$_4{}^-$ from the waste. The most weakly solvated anions (TcO$_4{}^-$ in this case) are the most strongly bound by the resins used to separate TcO$_4{}^-$. Although fac-Tc(CO)$_3$(gluconate)$^{2-}$ is anionic, it should be more strongly solvated than NO$_3{}^-$, which is present in much higher concentrations, consequently anionic Tc(CO)$_3$(gluconate)$^{2-}$ cannot be separated in the presence of excess nitrate using these resins. Of course, if neutral fac-Tc(CO)$_3$(H$_2$O)$_2$(OH) is present, it would not be removed by ion exchange.

The seemingly strange behavior of the NPS with regard to oxidation is largely explained by its identity. The fac-Tc(CO)$_3$ complexes are not thermodynamically stable with respect to oxidation to TcO$_4{}^-$; however, they are kinetically inert due to their low-spin d^6 electronic structure. As a result, they will react slowly with potential oxidizing agents, such as oxygen. The kinetic inertness of these complexes also affects oxidation by strong oxidizers. Since fac-Tc(CO)$_3$ complexes will react relatively slowly with strong oxidizers (although presumably much faster than they react with oxygen), the strong oxidizers will preferentially react with other waste constituents, such as nitrite or organic molecules, that are more reactive and present in much higher concentrations.

One aspect of the chemistry of fac-Tc(CO)$_3$ in CC waste that has not been addressed is the mechanism of its formation. As noted above,

Table III: EXAFS Fitting Results for *fac*-Tc(CO)₃ Complexes

Scattering Path		$Tc(CO)_3$ $(H_2O)_3^+$	$Tc(CO)_3$ $(H_2O)_2(OH)$	$Tc(CO)_3$ $(gluconate)^{2-}$
Tc-$\underline{C}$O	N	3	3	3
	R(Å)	1.904(2)	1.886(3)	1.911(2)
	$\sigma^2(\text{Å}^2)$	0.0041(2)	0.0058(3)	0.0062(2)
Tc-$\underline{O}$	N	3	3	3
	R(Å)	2.163(2)	2.155(3)	2.137(2)
	$\sigma^2(\text{Å}^2)$	0.0052(2)	0.0047(5)	0.0068(3)
Tc-C$\underline{O}^a$	N	3	3	3
	R(Å)	3.045(9)	3.083(8)	3.09(3)
	$\sigma^2(\text{Å}^2)$	0.0050(2)	0.0046(2)	0.0015(2)
Tc-O$\underline{C}$	N			3
	R(Å)			3.44(2)
	$\sigma^2(\text{Å}^2)$			0.011(2)
4 leg MS path with *trans* ligands coordinated to Tc	N	6	6	6
	R(Å)	3.96(2)	4.01(2)	3.96(1)
	$\sigma^2(\text{Å}^2)$	0.017(4)	0.010(3)	0.019(3)
ΔE_0		-14.8(4)	-11.6(6)	-11.1(3)
Scale Factor (S_0^2)		1.39(4)	1.19(6)	1.68(7)

fac-Tc(CO)₃ complexes can be prepared from TcO₄⁻ in alkaline solution at elevated temperature at low CO concentration; however, the formation of *fac*-Tc(CO)₃ complexes from TcO₄⁻ in waste simulants remains to be investigated.

Acknowledgments

This work was supported by the Environmental Management Science Program of the Office of Science and Technology of the U.S. Department of Energy (DOE) and was performed at the Lawrence Berkeley National Laboratory, which is operated by the DOE under Contract No. DE-AC03-76SF00098. Part of this work was performed at the Stanford

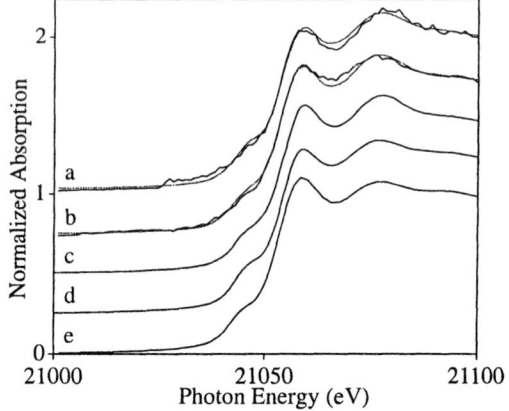

Figure 7. Tc K-edge XANES spectra of (a) NPS in tank SY-103 (black) and Tc(CO)$_3$(gluconate)$^{2-}$ (gray), (b) NPS in tank SY-101 (black) and 93% Tc(CO)$_3$(gluconate)$^{2-}$ with 7% TcO$_4^-$ (gray), (c) Tc(CO)$_3$(gluconate)$^{2-}$, (d) Tc(CO)$_3$(HO)(H$_2$O)$_2$, and (e) Tc(CO)$_3$(H$_2$O)$_3^+$. The spectra of the NPS in tanks SY-101 and SY-103 are from reference 7. (Reproduced with permission from Environ. Sci. Technol. **2004**, 33, 229.)

Synchrotron Radiation Laboratory, which is operated by the Director, DOE, Office of Science, Office of Basic Energy Sciences, Division of Chemical Sciences.

References

1. National Research Council. *Research Needs for High-Level Waste Stored in Tanks and Bins at U.S. Department of Energy Sites: Environmental Management Science Program*; National Academy Press: Washington, DC, 2001.
2. Gephart, R. E.; Lundgren, R. E. *Hanford Tank Clean Up: A Guide to Understanding the Technical Issues*; Battelle Press: Columbus, OH, 1998.
3. Agnew, S. F.; Boyer, J.; Corbin, R. A.; Duran, T. B.; Fitzpatrick, J. R.; Jurgensen, K. A.; Ortiz, T. P.; Young, B. L. *Hanford Tank Chemical and Radionuclide Inventories: HDW Model Rev. 4*; LA-UR-96-3860; Los Alamos National Laboratory: Los Alamos, NM, 1996.
4. Schroeder, N. C.; Radzinski, S. D.; Ball, J. R.; Ashley, K. R.; Cobb, S. L.; Cutrel, B.; Whitener, G. *Technetium Partitioning for the Hanford Tank Waste Remediation System: Anion Exchange Studies for Partitioning Technetium from Synthetic DSSF and DSS Simulants and Actual Hanford*

Waste (101-SY and 103-SY) Using Reillex^{TM} -HPQ Resin; LA-UR-95-4440; Los Alamos National Laboratory: Los Alamos, NM, 1995.
5. The separation requirements for high level waste at Hanford are currently being revised and may eliminate technetium separation.
6. Schroeder, N. C.; Radzinski, S. D.; Ashley, K. R.; Truong, A. P.; Sczcepaniak, P. A. In *Science and Technology for Disposal of Radioactive Tank Wastes*; Schulz, W. W., Lombardo, N. J., Eds.; Plenum Press: New York, 1998, p 301.
7. Blanchard, D. L.; Brown, G. N.; Conradson, S. D.; Fadeff, S. K.; Golcar, G. R.; Hess, N. J.; Klinger, G. S.; Kurath, D. E. *Technetium in Alkaline, High-Salt, Radioactive Tank Waste Supernate: Preliminary Characterization and Removal*; PNNL-11386; Pacific Northwest National Laboratory: Richland, WA, 1997.
8. Almahamid, I.; Bryan, J. C.; Bucher, J. J.; Burrell, A. K.; Edelstein, N. M.; Hudson, E. A.; Kaltsoyannis, N.; Lukens, W. W.; Shuh, D. K.; Nitsche, H.; Reich, T. *Inorg. Chem.* **1995**, *34*, 193.
9. Bucher, J. J.; Allen, P. G.; Edelstein, N. M.; Shuh, D. K.; Madden, N. W.; Cork, C.; Luke, P.; Pehl, D.; Malone, D. *Review of Scientific Instruments* **1996**, *67*, 4.
10. Lukens, W. W.; Bucher, J. J.; Edelstein, N. M.; Shuh, D. K. *Env. Sci. Tech.* **2002**, *36*, 1124.
11. Davison, A.; Trop, H. S.; DePamphilis, B. V.; Jones, A. G. *Inorg. Synth.* **1982**, *21*, 160.
12. Alberto, R.; Schibli, R.; Egli, A.; Schubiger, A. P.; Herrmann, W. A.; Artus, G.; Abram, U.; Kaden, T. A. *J. Organomet. Chem.* **1995**, *493*, 119.
13. Bernard, J. G.; Bauer, E.; Richards, M. P.; Arterburn, J. B.; Chamberlin, R. M. *Radiochim. Acta* **2001**, *V89*, 59.
14. Lukens, W. W.; Bucher, J. J.; Edelstein, N. M.; Shuh, D. K. *J. Phys. Chem. A* **2001**, *105*, 9611.
15. Golcar, G. R.; Colton, N. G.; Darab, J. G.; Smith, H. D. *Hanford Waste Tank Simulants Specification and Their Applicability for the Retrieval, Pretreatment, and Vitrification Process*; PNWD-2455; Pacific Northwest National Laboratory: Richland, WA, 2000.
16. Alberto, R.; Albinati, A.; Anderegg, G.; Huber, G. *Inorg. Chem.* **1991**, *30*, 3568.
17. Alberto, R.; Anderegg, G.; May, K. *Polyhedron* **1986**, *5*, 2107.
18. Zabinsky, S. I.; Rehr, J. J.; Ankudinov, A.; Albers, R. C.; Eller, M. J. *Phys. Rev. B* **1995**, *52*, 2995.
19. Alberto, R.; Schibli, R.; Egli, A.; Schubiger, A. P.; Abram, U.; Kaden, T. A. *J. Amer. Chem. Soc.* **1998**, *120*, 7987.
20. Alberto, R.; Schibli, R.; Waibel, R.; Abram, U.; Schubiger, A. P. *Coord. Chem. Rev.* **1999**, *192*, 901.

21. Alberto, R.; Ortner, K.; Wheatley, N.; Schibli, R.; Schubiger, A. P. *J. Amer. Chem. Soc.* **2001**, *123*, 3135.
22. Kramer, D. J.; Davison, A.; Jones, A. G. *Inorg. Chim. Acta* **2001**, *312*, 215.
23. Johnson, G. D. *Flammable Gas Program Report*; WHC-SP-1193; Westinghouse Hanford Company: Richland, WA, 1996.
24. Gorshkov, N. I.; Lumpov, A. A.; Miroslavov, A. E.; Suglubov, D. N. *Radiochemistry* **2000**, *43*, 231.

Chapter 18

Effects of Sodium Hydroxide and Sodium Aluminate on the Precipitation of Aluminum Containing Species in Tank Wastes

Shas V. Mattigod[1], David T. Hobbs[2], Kent E. Parker[1], David E. McCready[1], and Li-Qiong Wang[1]

[1]Pacific Northwest National Laboratory, P.O. Box 999, Richland WA 99352
[2]Savannah River National Laboratory, Aiken, SC 29808

Aluminosilicate deposit buildup experienced during the tank waste volume-reduction process at the Savannah River Site (SRS) required an evaporator to be shut down. Studies were conducted at 80 °C to identify the insoluble aluminosilicate phase(s) and to determine the kinetics of their formation and transformation. These tests were carried out under conditions more similar to those that occur in HLW tanks and evaporators. Comparison of our results with those reported from the site show very similar trends. Initially, an amorphous phase precipitates followed by a zeolite phase that transforms to sodalite and which finally converts to cancrinite. Our results also show the expected trend of an increased rate of transformation into denser aluminosilicate phases (sodalite and cancrinite) with time and increasing hydroxide concentrations.

High-level wastes (HLW) from fuel-reprocessing operations are evaporated at Savannah River Site (SRS) to concentrate the waste to about 30–40% of its original volume before it is discharged into a holding tank. The 2H-Evaporator system at SRS consists of a feed tank (Tank 43H) that receives liquid wastes primarily from fuel-reprocessing operations and the Defense Waste Processing Facility (DWPF), the evaporator, and the concentrate receipt tank (currently Tank 38H). After evaporation, the concentrated wastes are transferred via a gravity drain line (GDL) to the concentrate receipt tank. Frequently, the concentrated wastes from the concentrate receipt tank are transferred back into the evaporator feed tank for further volume evaporation.

For about four decades, SRS evaporators operated successfully with only occasional minor problems such as $NaNO_3$ salt buildup and clogging of the drain lines from the evaporators. Because these deposits were water-soluble, the drain lines were unclogged easily by flushing with water. In 1997, the 2H-Evaporator feed tank began receiving silicon-rich wastes from the DWPF recycle stream. The DWPF recycle waste stream is more dilute (contains less soluble salts) than fuel-reprocessing wastes, and thus requires a higher degree of volume reduction (typically 90%) to reach the same salt concentrations as are found in the fuel-reprocessing wastes. The higher concentration requirement for DWPF waste and the existing operational problems resulted in significant increases in the residence time of these wastes in the 2H Evaporator.

Beginning in 1997, the silicon-rich DWPF waste stream was mixed with the aluminum-rich stream from the fuel-reprocessing operation in Tank 43H, and this mixture was fed to the 2H-Evaporator. Soon after, a sodium-aluminosilicate (sodalite) deposit of limited solubility began to form in the evaporator (*1*) necessitating the shutdown of this evaporator for cleaning. Deposits of the zeolite and a sodium uranate phase were found in the GDL (*2*), which was back flushed, and evaporation operations were resumed. From June 1998, operation of the 2H Evaporator became progressively more difficult due to the more frequent buildup of limited solubility aluminosilicate compounds, resulting in the shutdown of the evaporator in October 1999. An inspection revealed significant accumulations of deposits on most of the exposed surfaces of the evaporator. Analysis revealed (*3*) that the deposits in the evaporator and the drain lines consisted mainly of a sodium aluminosilicate compound, sodalite [$Na_8Al_6Si_6O_{24}(NO_3)_2 \cdot 4H_2O$], and sodium diuranate ($Na_2U_2O_7$). Based on the extent of the low-solubility deposits in the evaporator and in the drain lines, and the criticality concerns from the aggregated sodium-diuranate compound, it was decided to stop evaporator operations until a solution for this problem was found.

One of the solutions to this clogging problem was the evolution of a method that uses nitric acid to dissolve the low-solubility compounds in the 2H Evaporator (*4*). However, a complete amelioration of this problem requires a process control tool that can be used to predict critical mixing ratios for aluminum-rich and silicon-rich waste streams that will prevent the formation of the limited-solubility aluminosilicate compound (zeolite-sodalite) in the evaporator and the drain lines.

A brief review of literature that is pertinent to the aluminosilicate phases that have been observed to form in the 2H-Evaporator at SRS indicates that zeolite A, $Na_{12}Al_{12}Si_{12}O_{48} \cdot 27H_2O$, is a phase related structurally to both sodalite and nosean (5). The alumina:silica ratio of the basic cage structure is the same in both the zeolite A and sodalite. Basically, zeolite A consists of a double-unit cell of sodalite without the NaCl, Na_2SO_4, or NaOH groups attached. Recent work (6) on waste forms found that zeolite A formed as a precursor to the formation of sodalite, $Na_8Al_6Si_6O_{24}(Cl_2)$. The zeolite A transformed to NaCl-containing sodalite under increased temperature and pressure along with a small amount of nepheline ($NaAlSiO_4$).

From these data we can assume that it is very likely that in the 2H-Evaporator, sodalite also forms from the zeolite A precursor. The formation of zeolite A is well studied, and zeolite A is known to be kinetically a fast former. Well-crystallized zeolite A has been reported to form when a mixture of sodium-aluminate gel (87 wt % $NaAlO_2$ and 13 wt % NaOH commercially available as Alfloc) and 1 M colloidal silica sol (particles of 250A) has been reacted at temperatures between 85 and 110 °C and at pH values ≥ 10 for 2 or 3 h (5, 7). These investigators also reported slower crystallization with an increase in the silica content of the gel, and faster crystallization in the presence of excess NaOH (5, 7).

Studies by Gasterger et al. (8) on spent pulping liquor evaporators indicated that a mixture of sodalite and hydroxysodalite [$Na_8Al_6Si_6O_{24}Cl_2$ and $Na_8Al_6Si_6O_{24}(OH)_2$] precipitated at 95 °C in the Al/Si range between 0.076 and 3 in which $Al(OH)_4^-$ and $HSiO_4^{3-}$ were the predominant aqueous species. These aqueous alkaline solutions had ionic strengths between 1.0 and 4.0 mol/kg with the corresponding OH^- concentrations of ≤ 0.09 and ≤ 3.8 mol/L, respectively. He also observed that a sodium-aluminosilicate (NAS) gel was a precursor phase leading to the formation of zeolite A. Similarly, Ejaz et al. (9) reported the initial formation of a metastable aluminosilicate precursor phase of higher solubility that with time alters to less soluble crystalline zeolite A. Based on an extensive literature review, Barnes et al. (10) reported that reactions occurring in Bayer liquor at temperatures below 80 °C initially form a metastable amorphous phase that subsequently crystallizes into zeolite A.

Zeolite A ($Na_{12}Al_{12}Si_{12}O_{48} \cdot 27H_2O$) has also been found to form in the SRS M-Area waste tanks (11). Zeolite was found to form preferentially in tanks under high pH (12–12.8) conditions when solid $Al(OH)_3$ was present. Experiments were performed to determine how zeolite A got into the tanks because no zeolite had been used in any M-Area process. These experiments demonstrated that a zeolite phase identified as sodalite [$Na_8Al_6Si_6O_{24}(Cl)_2$] could form rapidly (within 29 h) at room temperature from the interaction of high surface area aluminosilicates (perflo and diatomaceous earth) in the tank with a solution of 6 M NaOH solution *(12)*.

The main focus of this study was to obtain data on the characteristics of solid and liquid phases that would help verify the thermodynamic stability of aluminosilicate compounds under waste tank conditions as predicted by supersaturation calculations *(12)*. Such verification would enhance the utility and reliability of activity diagrams and supersaturation indices as predictive tools.

Experimental

A test matrix was designed to cover a range of hydroxide and salt concentrations and the reaction temperature and time encountered in evaporator operation (Table I).

Table I. Solution Compositions Used in the Test Matrix (M)

Solution	OH	Al	Na
1	0.1	0.2	3.31
2	0.1	0.5	3.61
3	1.0	0.2	4.21
4	1.0	0.5	4.51
5	4.5	0.2	7.71
6	4.5	0.5	8.01

Si and NO_3 concentrations in all experiments were 0.01M and 3M respectively.

Four reaction temperatures were selected (40 °C, 80 °C, 120 °C, and 175 °C) to cover the range of temperatures encountered during the storage and evaporation of waste solutions. In this study, data obtained from experiments conducted at 80 °C will be presented. The reactions were carried out over varying periods of time to study the initial precipitation and the crystallization (transformation) process.

The precipitation experiments were conducted as follows. Required quantity of NaOH was dissolved in deionized distilled H_2O in a polypropylene beaker. Appropriate quantity of sodium nitrate was added to the NaOH solution and stirred. Weighed quantity of $NaAlO_2$ and a sufficient volume of deionized distilled water was added to the NaOH solution to bring the total solution to the required concentration and stirred for 30 min. These steps provided 0.5 L to 6 L of solution containing 0.2 M Al, 4.5 M NaOH, and 3 M $NaNO_3$. Next, an appropriate amount of sodium silicate (containing 14 wt % silica, SiO_2) was added to the solution. The addition of sodium silicate was designed to produce solutions that contain 10^{-2} M silica concentrations. The test mixtures were stirred and kept in an oven set to 80 °C. At appropriate times, samples were removed from the oven for analysis. Samples were centrifuged at 3000 rpm for 5 min to separate the solid and liquid phases. An aliquot of the liquid phase was filtered and analyzed for dissolved constituents. The solid phase was washed free of salts using deionized distilled water, air-dried before analyses.

The solid samples were analyzed by XRD to identify the crystalline phases containing aluminum and silicon. The XRD apparatus was a Philips X'Pert MPD system (Model PW3040/00) with a Cu Kα X-ray source operated at 45 kV, 45 mA (1.8 kW). The scan range was 5–75° and the typical scan rate was ~ 2°/min. The XRD data were analyzed using the program *JADE* (V5.0, V6.0, and V6.1, Materials Data Inc., Livermore, CA) and reference data from the Powder Diffraction File Database (PDF-2, International Centre for Diffraction Data, Newtown Square, PA). Semi-quantification of crystalline phase content was based on the ratio of the peak height above background for each identified phase to the combined peak heights of all phases present. Using JADE (Materials Data Inc., Livermore, California) software, this was accomplished by scaling PDF reference data against the background-subtracted experimental patterns. Normally, the 100% peak for each phase was used for scaling. Where this was not possible due to superposition, the strongest resolved peak was utilized. The sum of the phase scale factors was taken to be the total diffracted intensity, and the concentration of each phase was determined from the ratio of its scale factor to the sum. The semiquantitative XRD analyses were all unstandardized; therefore, the mass estimates are less precise than values obtained from quantitative analyses conducted with appropriate standards. Therefore, the numerical values in graphs represent trends in phase transformation. The solutions were analyzed for dissolved aluminum and silicon using inductively-coupled plasma (ICP) spectrometry.

Results and Discussion

Solid phase analyses indicated that upon mixing, an amorphous phase formed in Solutions 1 through 4 (Figures 1 and 2). The precipitate from

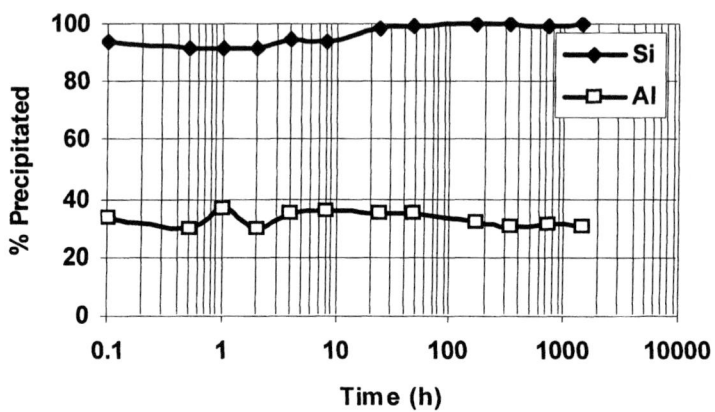

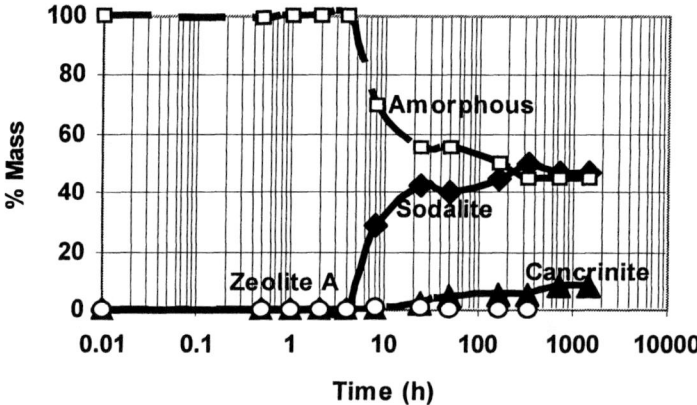

Figure 1. Fractions of Al and Si precipitating out of solution (1 and 2) and solid phases identified as a function of reaction time.

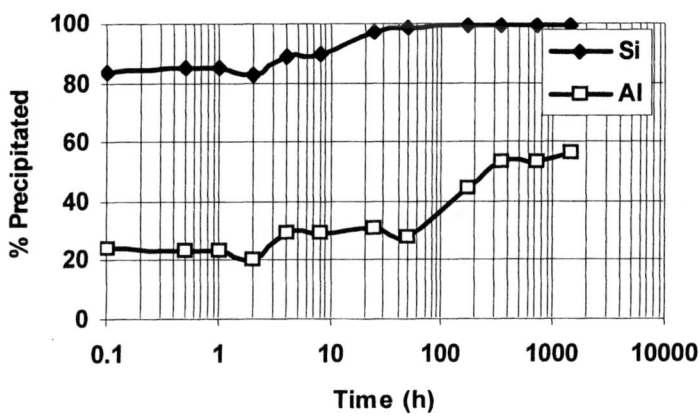

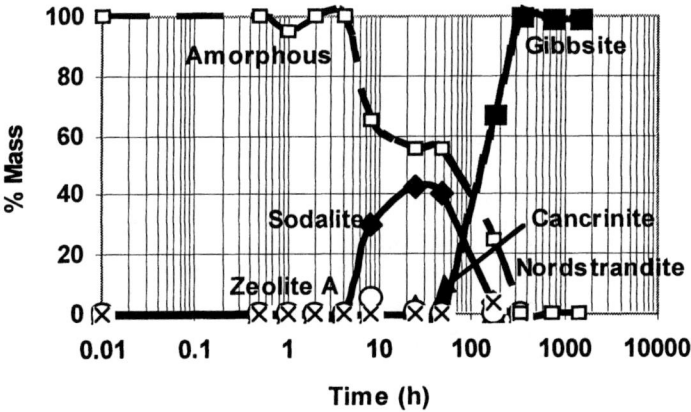

Figure 1. Continnued.

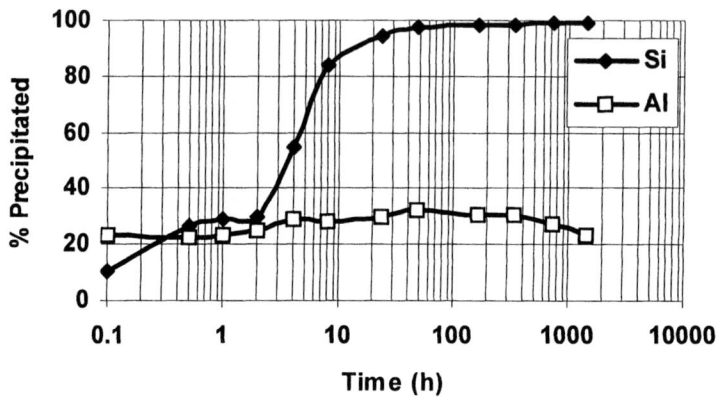

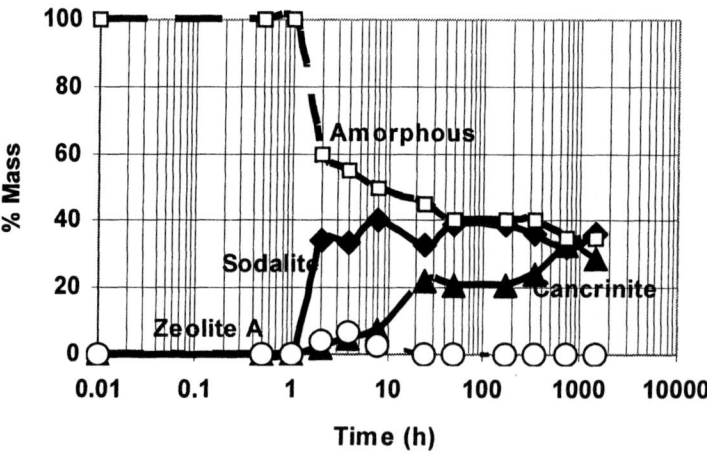

Figure 2. Fractions of Al and Si precipitating out of Solution (3 and 4) and solid phases identified as a function of reaction time.

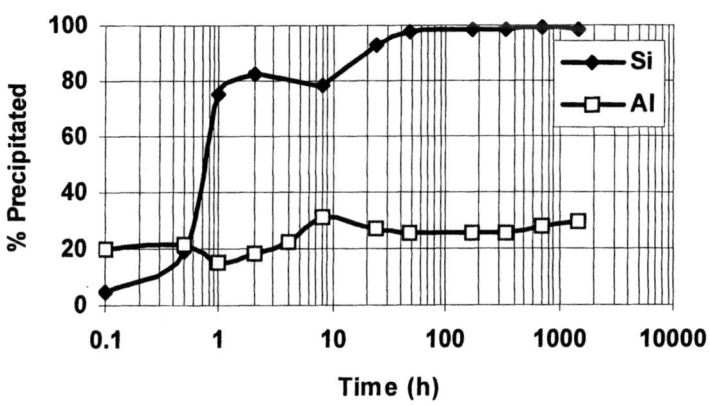

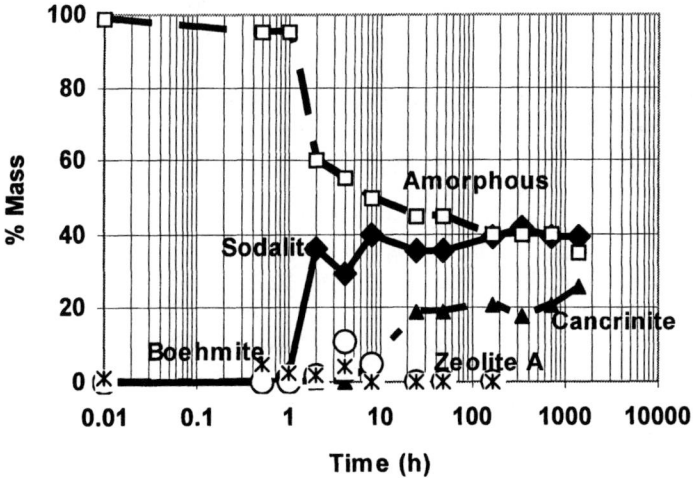

Figure 2. Continued.

Solution 1 (Figure 1) remained amorphous for about 8 h before formation of crystalline phases were detected. Concentrations of dissolved species indicated that about 30–35% of Al in Solution 1 precipitated when the reaction was initiated (Figure 1). Initially, about 90–95% of dissolved Si precipitated, and complete precipitation occurred after about 24 h of reaction time. As the reaction proceeded further, no additional Al appeared to precipitate. The crystalline phases consisted mainly of sodalite with trace amount of zeolite A. With increasing reaction time, the quantities of sodalite increased with a concomitant decrease in the quantity of amorphous material. The solution data indicated that increased Si precipitation was associated with an increased rate of formation of sodalite (Figure 1). After 24 h, in addition to the amorphous phase and sodalite, trace quantities of cancrinite were detected. Continued sampling of precipitates indicated continual decrease in quantities of amorphous material coupled with increasing quantities of crystalline phases, namely sodalite and cancrinite. Apparently, equilibrium existed in this system because after 14 days of reaction there were no substantial changes in relative quantities of these phases.

Initially, about 20% of the Al and about 80% of Si present in solution 2 precipitated, and after 4 h of reaction, continued Si precipitation culminated in almost complete removal of Si from solution. The initial precipitate from this solution was amorphous for the initial 4-h reaction period (Figure 1). Solution data indicated that additional Al precipitation occurred after about 4 h and 24 h of reaction. Again, additional Si precipitation seemed to be triggered by the formation of sodalite. Samples obtained after 8 h showed that about a third of the solid mass consisted of sodalite with a trace of zeolite A. Also, precipitation of additional dissolved Al after about 48 h of reaction coincided with the formation of gibbsite. Solid phases at the end of 1 and 2 days contained decreasing amounts of amorphous materials with increasing quantities of sodalite and minor amounts of cancrinite. Samples obtained after 7 days showed that gibbsite constituted a major fraction (~ 70%) of the solid mass with trace quantities of sodalite and nordstrandite. Reaction that continued for periods of 14, 30, and 60 days revealed no detectable quantities of amorphous material, with gibbsite being the major crystalline phase with trace quantities of boehmite. At the end of 60 days of reaction, about 55% of dissolved Al had precipitated. Although the solution data indicated that crystalline aluminosilicates persisted at the end of the experiment, these phases were not detected by XRD because they constituted a minor fraction of the solid mass that was dominated by gibbsite forming from additional Al precipitation that occurred after 48 h.

In Solution 3, during the initial reaction period extending up to 1 h, only the amorphous phase was present (Figure 2). However, at the end of 2 h, sodalite constituted about a third of the solid mass with minor amounts of zeolite A and cancrinite. Additional samples obtained periodically from 1 to 60 days showed a continual decline in the quantity of amorphous material with increasing quantities of the crystalline fraction containing sodalite and cancrinite. Solid materials obtained after 30- and 60-day reaction times showed that the mass constituted roughly equal proportions of sodalite, cancrinite, and amorphous material. The solid phases and their relative transformations with reaction progress in Solution 4 were substantially similar to what was observed in Solution 3, except that during the early phases of the reaction (up to 4 h) traces of boehmite were detected in the predominantly amorphous matrix (Figure 2). The Al precipitation reactions in Solutions 3 and 4 seemed to be similar in that initially about 20% of added Al precipitated (Figure 2). After the reaction had progressed for about 4–8 h, an additional ~ 10% of the Al was observed to precipitate. Precipitation of Si in both solutions was initially low (~ 5–10%), and additional precipitation seemed to occur after about .5-h and 2-h reaction times. In both solutions, almost complete precipitation of Si had occurred after 24 h of reaction. The solid phase characterization data (Figure 2) suggested that the onset of additional Si removal in solution may be related to rapid formation of crystalline aluminosilicate phases such as sodalite and cancrinite.

The solids in Solution 5 during the initial stages of the reaction (< 2 h) were mainly colloidal in nature and therefore could not be recovered in sufficient quantities for characterization. Therefore, a larger volume of solution (6 L) was reacted for 2 h to obtain sufficient quantities of precipitates for characterization. The precipitate consisted of an amorphous phase (about one-half the mass) and crystalline components such as sodalite, cancrinite, and boehmite (Figure 3). In this sample, a trace amount of an unidentifiable crystalline component and some quartz were also present, likely a result of contamination of the sample. With increasing reaction time, the quantities of amorphous material decreased, while the mass of crystalline fraction increased. After 4 h, about one-half of the solid mass consisted of sodalite, with the remaining mass consisting of cancrinite and amorphous material. As the reaction proceeded, the quantities of both sodalite and amorphous material decreased, with a concomitant increase in the mass of cancrinite. Samples obtained at the end of the 60-day reaction period indicated that the solid mass consisted of cancrinite, sodalite, and an amorphous phase (Figure 3).

Solids obtained at the end of 2 h from Solution 6 (total volume 6 L) also showed equal proportions of the amorphous phase with crystalline materials (mainly sodalite with minor amounts of cancrinite) (Figure 3). Continued reaction generated increasing fractions of crystalline material that consisted of mainly cancrinite with decreasing amounts of sodalite. At the end of 60 days, about 70% of the material was crystalline in nature with cancrinite being the dominant crystalline phase.

The solid phases obtained from all solutions at the end of the 60-day reaction period are listed in Table II. The data showed that except in Solution 2, rapid formation of sodalite and cancrinite occurred in all solutions. Although

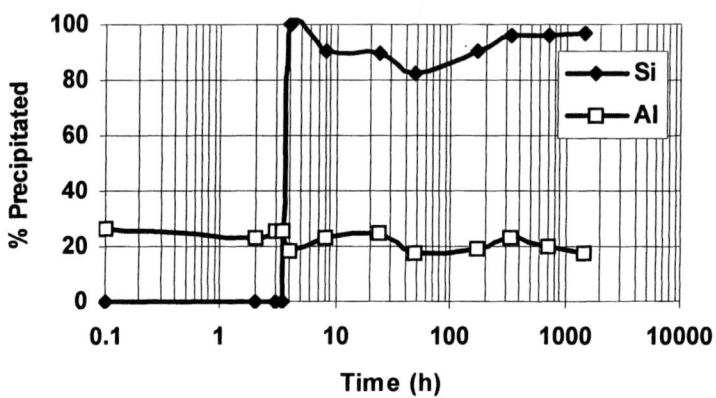

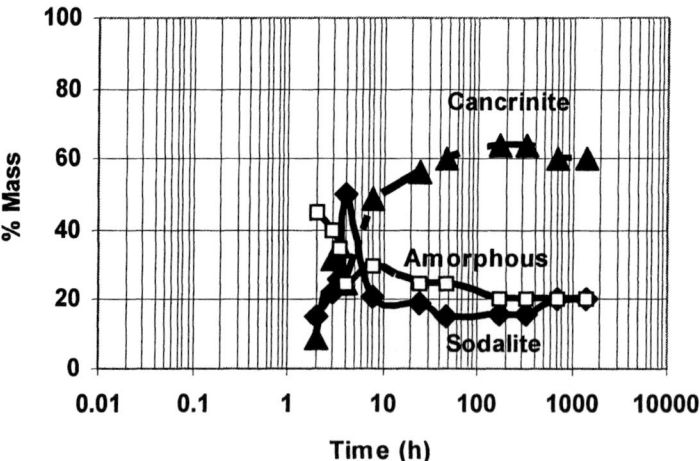

Figure 3. Fractions of Al and Si precipitating out of solution (5 and 6) and solid phases identified as a function of reaction time.

331

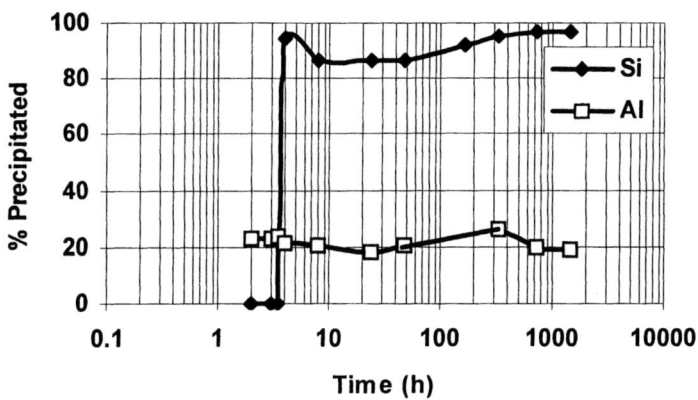

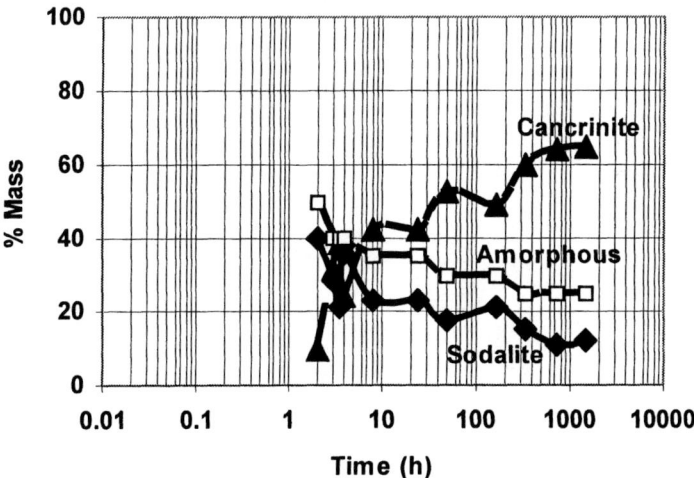

Figure 3. Continued.

Table II. Phases Identified at the End of 60-Day Recreation at 80 °C

Solution	Mass (wt %)				
	100–80	80–60	60–40	40–20	20–<10
1			Sodalite, Amor. Phase		Cancrinite
2	Gibbsite				
3				Amor. Phase Sodalite Cancrinite	Boehmite
4				Amor. Phase Sodalite Cancrinite	
5		Cancrinite			Sodalite Amor. Phase
6		Cancrinite		Amor. Phase	Sodalite

Sodalite: $Na_8[AlSiO_4]_6(NO_3)_2$ (PDF#50-0248)
Cancrinite: $Na_8[AlSiO_4]_6(NO_3)_2 \cdot 4H_2O$ (PDF#38-0513)
Gibbsite: $Al(OH)_3$ (PDF#33-0018)
Amor. Phase: Amorphous Phase

minor amounts of zeolite A were initially detected in some cases, the higher reaction temperatures seemed to promote very rapid transformation of this phase into more stable phases. Also, increasing hydroxide concentrations appeared to initiate kinetically fast crystallization of sodalite and cancrinite.

Solution data indicated that initial reactions precipitated about 20–25% of Al in Solutions 5 and 6. Following this removal, no additional Al appeared to precipitate from either of these solutions during the remaining reaction period (Figure 3). However, in both solutions, measurable precipitation of Si did not occur during the first 3.5 h of reaction. Following this apparent quiescent period, extremely rapid removal of Si (~ 80–100%) was observed. Such rapid Si precipitation following the initially nonreactive stage appeared to be triggered by rapid crystal growth of aluminosilicate phases (sodalite and cancrinite) that may have initially formed from homogeneous nucleation. These data suggested that removal of Si from solution was influenced strongly by hydroxide concentrations (Figure 3). Higher hydroxide concentrations appeared to delay the onset of formation of crystalline aluminosilicate phases. At present, the reason for the extended quiescent period (no Si precipitation) under high-hydroxide conditions is not known.

Previous testing, conducted primarily in support of the aluminum production industry, indicates sodium aluminosilicate phases form upon mixing alkaline solutions of aluminate and silicate. At temperatures below 80 °C, a number of studies reported formation of zeolite A (*13*). With time, zeolite A transforms into sodalite (*14–16*), and other studies indicate that sodalite is also a metastable phase and therefore will transform into cancrinite (*18–20*). The rate of transformation is influenced by many factors, including temperature and the presence of other components such as hydroxide and carbonate ions. Based on these studies, Barnes et al. (*10*) proposed the following order of precipitation at temperatures of less than 160 °C:

[aluminosilicate precursor species] → amorphous → zeolite A → sodalite → cancrinite

More recent testing in support of the HLW evaporator plugging issue has shown similar trends in the formation of aluminosilicate phases (*19–21*). These tests were carried out under conditions more similar to those that occur in HLW tanks and evaporators.

Comparison of our results with those reported above show very similar trends. Initially, an amorphous phase precipitates followed by a zeolite phase followed by formation of sodalite and finally cancrinite. Our results also show the expected trend of an increased rate of transformation into denser aluminosilicate phases (sodalite and cancrinite) with increasing hydroxide concentration. Under high Al and low hydroxide concentrations, only gibbsite, $Al(OH)_3$, is the identified crystalline material at equilibrium or near equilibrium conditions. Initially, a small amount of aluminosilicate solids precipitate and after depletion of the available silicon, gibbsite slowly crystallizes from solution and effectively dilutes the aluminosilicate phases to a concentration below that which can be detected by XRD.

References

1. Wilmarth, W. R.; Coleman, C. J.; Hart, J. C.; Boyce, W. T. *Characterization of Samples from the 242-16H Evaporator Wall*; WSRC-TR-2000-00089; Westinghouse Savannah River Company: Aiken, SC, 2000.
2. Wilmarth, W. R.; Fink, S. D.; Hobbs, D. T.; Hay, M. S. *Characterization and Dissolution Studies of Samples from the 242-16H Evaporator Gravity Drain Line*; WSRC-TR-97-0326, Rev 0; Westinghouse Savannah River Company: Aiken, SC, 1997.
3. Wilmarth, W. R.; Colemen, C. J.; Jurgenson, A. R.; Smith, W. M.; Hart, J. C.; Boyce, W. T.; Missimer, D.; Conley, C. M. *Characterization and Dissolution Studies of Samples from the 242-16H Evaporator;* WSRC-TR-2000-00038, Rev 0; Westinghouse Savannah River Company: Aiken, SC, 2000.
4. Boley, C. S.; Thompson, M. C.; Wilmarth, W. R. *Technical Basis for the 242-16H Evaporator Cleaning Flowsheet*; WSRC-TR-2000-00038, Rev 0; Westinghouse Savannah River Company: Aiken, SC, 2000.
5. Barrer, R. M.; Baynham, J. W.; Bultitude, F. W.; Meier, W. M. *J. Chem. Soc. London.* **1959**, 195–208.
6. Johnson, S. G.; Ebert, W. L. *Extension of C1285-97 to the ANL Ceramic Waste Form*; ASTM C26.13; American Society of Testing Materials: Philadelphia, PA, 2000.
7. Milton, R. M. U.S. Patents 2,882,243 and 2,882,244, 1959.
8. Gasteiger, W.; Frederick, J.; Streisel, R. C. *Ind. Eng. Chem. Res.* **1992**, *31*, 1183–1190.
9. Ejaz, T.; Jones, A. G.; Graham, P. *J. Chem. Eng. Data.* **1999**, *44*, 574–576.
10. Barnes M. C.; Addai-Mensah, J.; Gerson, A. R. *Micro. Meso. Mat.* **1999**, *31*, 287–302.
11. Jantzen, C. M. *Vitrification of M-Area Mixed (Hazardous and Radioactive) Wastes: I. Sludge and Supernate Characterization;* WSRC-TR-94-0234; Westinghouse Savannah River Company: Aiken, SC, 1989.
12. Jantzen, C. M.; Laurinat, J. E. *Thermodynamic Modeling of Deposition in Savannah River Site (SRS) Evaporators*; WSRC-TR-2000-00293, Rev 0; Westinghouse Savannah River Company: Aiken, SC, 2001.
13. Wehrli, J. T.; Aguila, D. D. *Preparation of DSP at Low Temperatures*; 22.12; Queensland Alumina Ltd.: Australia, 1993.

14. Subotic, B.; Skritic, D.; Smit, I.; Sekovanic, L. *J. Cryst. Growth* **1980**, *50*, 498.
15. Subotic, B.; Sekovanic, L. *J. Cryst. Growth* **1986**, *75*, 561.
16. Grujic, B.; Subotic, B.; Despotovic, L. J. A. In *Studies in Surface Science and Catalysis No. 24, Zeolites – Synthesis, Structure, Technology and Application;* Jacobs, P. A., Van Santen, R. A., Eds.; Elsevier: Amsterdam, 1989; p 261.
17. Addai-Mensah, J.; Gerson, A. R.; Zheng, K.; O'Dea, A.; Smart, RS-C. *Light Metals* **1997**, 23.
18. Zheng, K.; Addai-Mensah, J.; Gerson, A. R.; Smart, RS-C. *J. Chem. Eng. Data* **1998**, *43*, 31.
19. Barnes, M. C.; Addai-Mensah, J.; Gerson, A. R. *J. Cryst. Growth* **1999**, *200*, 246.
20. Rosencrance, S.; Herman, D.; Healy, D. *Formation and Deposition of Aluminosilicates in Support of the 2H-Evaporator Fouling Problem;* WSRC-TR-2001-00464, Rev. 0; Westinghouse Savannah River Company: Aiken, SC, 2001.
21. Hu, M. Z.; DePauli, D. W.; Bostick, D. T. *Dynamic Particle Growth Testing: Phase I Studies*; ORNL/TM-2001/100; Oak Ridge National Laboratory: Oak Ridge, TN, 2001.

Indexes

Author Index

Allain, Leonardo R., 223
Ashley, Kenneth R., 302
Babain, Vasily A., 171
Bao, Lili, 53
Barnes, Craig E., 223
Barrans, Richard E., Jr., 250
Bond, L. J., 100
Bonnesen, Peter V., 12, 146
Bramlett, J. Morris, 223
Brennecke, Joan F., 250
Brodsky, A., 100
Brown, Gilbert M., 12
Buchanan, A. C., III, 146
Burgess, L., 100
Busch, Daryle H., 186
Bushan, K. Mani, 186
Chen, Tianniu, 223
Cheng, Meng-Dawn, 240
Chiarizia, Renato, 250
Dabestani, Reza, 12
Dai, Sheng, 34, 53, 146, 223
Dietz, Mark L., 250
Dzielawa, Julie A., 250
Fagan, Bryan C., 223
Felmy, Andrew R., 286
Givens, Richard S., 186
Goretzki, Gudrun, 12
Greenwood, M. S., 100
Gu, Baohua, 53
Hassan, Mansour M., 186
Herbst, R. Scott, 171
Herlinger, Albert W., 250
Hess, J. N., 250
Hess, Nancy J., 286
Hirsch, Roland F., 2
Hobbs, David T., 319

Hoefnagels, Johan P. M., 133
Im, Hee-Jung, 223
Lee, Doh-Won, 240
Lee, Jong-ill, 186
Li, Liyu, 268
Li, Zuojiang, 34
Liu, Jun, 64
Lubbers, Christopher, 250
Lukens, Wayne W., 302
Luo, Huimin, 146
Luther, Thomas A., 171
Macdonald, Digby D., 64
Mahurin, Shannon M., 34, 53
Marx, Brian M., 64
Mattigod, Shas V., 319
McAlister, Daniel R., 250
McCasland, Anne, 186
McCready, David E., 319
Meuse, Curtis W., 133
Navrotsky, Alexandra, 268
Nenoff, Tina M., 268
Nyman, May D., 268
Parker, Kent E., 319
Peterman, Dean R., 171
Pipino, Andrew C. R., 133
Roecker, Lee E., 223
Rubas, Audris V., 250
Schroeder, Norman C., 302
Scurto, Aaron M., 250
Sepaniak, Michael J., 223
Shuh, David K., 302
Silin, Vitalii, 133
Smirnov, Igor V., 171
Stepinski, Dominique, 250
Stoyanov, Evgenii S., 171
Su, Yali, 268

Wai, Chien M., 161
Wang, Li-Qiong, 319
Wang, Paul W., 2
Woodward, John T., 133
Xia, Yuanxian, 286
Xu, Hongwu, 268
Xue, Zi-Ling, 223
Yan, Hui, 53
Yang, Yihui, 223
Yost, Terry L., 223
Yu, Xianghua, 223
Zachry, Tiffany, xi
Zalupski, Peter R., 250
Zhang, Chi, 186

Subject Index

A

[3-({3-[1-(Acetyl-pyridin-2-yl)-ethylamino]-propyl}-methylamino)-propyl]-carbamic acid benzyl ester, synthesis, 190 191

[3-({3-[1-(6-Acetyl-pyridin-2-yl)-ethylideneamino]-propyl}-methylamino)-propyl]-carbamic acid ester, synthesis, 190

Actinide and lanthanide extraction with supercritical carbon dioxide, overview, 163 164

Alkaline solution, technetium behavior, 302 318. *See also* pH effect

1-(2-Alkoxy-2-oxoethyl)-3-methyl imidazolium, general synthesis, 150

Alkyl chain length effect on strontium ion extraction by imidazolium-based ionic liquids, 154, 156t-157

Alkylenediphosphonic acids with silicon-containing functional groups, 252 265

TMSP-esterified, 257 263

Alumina particle formation in surface decontamination by laser plasma, 241 249

Aluminosilicate deposit buildup in tank waste volume-reduction, sodium hydroxide and sodium aluminate effect, 319 335

Aluminum containing species in tank wastes, precipitation, 319 335

Ambient-temperature ionic liquids. *See* Room-temperature ionic liquids

[241]Americium extraction in CCD/PEG system, nitric acid dependence, 177 179

Americium(III) ion extraction in TMSP-esterified alkylenediphosphonic acids, 257 259

Amino-anchored silica sol-gels in copper(II) removal, 225 227

{3-[(3-Aminopropyl)-methylamino]-propyl}-carbamic acid benzyl ester, synthesis, 189 190

1-[6-(1-{3-[(3-Aminopropyl)-methylamino]-propylamino]-ethyl)-pyridin-2-yl]-ethanone
formation constants, metal complexes, 198, 218
kinetic studies, Ni(II) complexes, 201 203, 204f, 206 208, 209f-211f, 218 219
protonation, 197 198, 217 218
rate constants for complex formation with Ni(II), 203, 205 206
specific rate constants for nickel(II) reacting with ligand species, 217t
synthesis, 191, 195 196

Anion effect on strontium ion extraction by imidazolium-based ionic liquids, 152, 155t

Anthracene-fluorophore, 17, 20f-21, 22f, 23f, 29 30

Aqueous durability, perovskites, strontium release rate, ASTM-PCT (product consistency test), 272, 277 280, 281t

ASTM-PCT (product consistency test) for strontium release rates, 272, 277 280, 281t

Avalanche photodiode detection, fluorescence measurement, 38

B

Barrier oxide layer formed on iron, physicochemical processes, 82, 84f–96f

1,1-Bis(2-alkoxy-2-oxoethyl)-imidazolium bis[(trifluoromethyl)sulfonyl]-imide, general synthesis, 150

1,1-Bis(2-alkoxy-2-oxoethyl)-imidazolium bromide, general synthesis, 149 150

25,27-Bis(benzyloxy)-26,28-bis(5-chloro-3-oxapentyloxy)calix[4]arene, 1,3 alternate, synthesis, 29

25,27-Bis(benzyloxy)calix[4]arene azacrown-5, 1,3-alternate, synthesis, 29 30

1,3-Bis(2-tert-butoxy-2-oxoethyl)-imidazolium bis[(trifluoromethyl)sulfonyl]-amide, synthesis, 151, 152, 153f

1,3-Bis(2-tert-butoxy-2-oxoethyl)-imidazolium bromide, synthesis, 150, 152, 153f

6,7-Bis-[2-(2-chloroethoxy)-ethoxy]-4-methylcoumarin, synthesis, 30

1,3-Bis(2-ethoxy-2-oxoethyl)-imidazolium bis[(trifluoromethyl)sulfonyl]-imide, synthesis, 151, 152, 153f

1,3-Bis(2-ethoxy-2-oxothyl)-imidazolium bromide synthesis, 150, 152, 153f

6,7-Bis-[2-(iodoethoxy)-ethoxy]-4-methylcoumarine, synthesis, 30

Borate buffer solutions, electrochemical impedance, data for iron, 64 99

Borosilicate glass containing uranium lifetime fluorescence distribution, 43
spectroscopic properties and redox chemistry, 34 52

1-(2-tert-Butoxy-2-oxoethyl)-3-methylimidazolium bis[(trifluoromethyl) sulfonyl]-imide, synthesis, 151, 152, 153f

1-(2-tert-Butoxy-2-oxoethyl)-3-methylimidazolium bromide, synthesis, 159, 152, 153f

C

^{13}C NMR spectra
perchloric acid contacted with CMPO, 183, 184t
polyethylene glycol complexes, 182t

Calix[4]arene-bis(tert-octylbenzo-crown-6) effect on strontium ion extraction by imidazolium-based ionic liquids, 159

Calix[4]arene fluorescent ligands, preparation and metal ion complexation, 12 33

Cancrinite formation upon mixing sodium aluminate and sodium silicate, 323 334

Carbon dioxide, supercritical. See Supercritical carbon dioxide

Cavity ring-down spectroscopy (CDRS) in chemical detection, 133 144

CCD/PEG-400 systems, cesium(I) and strontium(II) extraction, 175 181, 182t

CCD systems, cesium(I) selective extraction, 174 175
See also Hexachlorocobalt dicarbollide anion, structure

CDRS. See Cavity ring-down spectroscopy

Ceramic waste forms from strontium loaded SOMS-1. See Perovskite, strontium loaded

Cesium and strontium extraction CCD/PEG-400 systems, 175 181

supercritical CO_2, dicyclohexano-18-crown-6 ether with counteranions, 168 169
Cesium complexation, fluorescence quenching, 24, 26f–27
Cesium extraction by imidazolium-based ionic liquids, alkyl chain length effect, 154, 156t–157
Cesium extraction by supercritical carbon dioxide, 168 169
^{137}Cesium extraction CCD/PEG system, nitric acid dependence, 177 179
^{137}Cesium in Hanford waste tanks, 13
Cesium selective extraction in chlorinated cobalt dicarbollide systems, 174 175
Cesium separation agent, crystalline silicotitanate, 269 270
Chelation enhanced fluorescence effects in photo-induced electron transfer sensor, 17, 20f–21, 22f, 23f
Chelation to metal ion, switch-binding by linear ligand with transformation to macrocycle, 186 222
Chemosensors, fluorophores based on calix[4]arenes, 12 33
Chlorinated cobalt dicarbollide systems. *See* CCD systems
Cis-dicyclohexano-18-crown-6 in imidazolium-based ionic liquids, strontium extraction, 146 160
See specific compounds for syntheses
CMPO, perchloric acid complexation in polar diluent systems, 181 184
Complexant Concentrate waste tanks, 303
Concrete particle formation in surface decontamination by laser plasma, 241 249
CONTIN, continuous model analysis program, 39
Copper(II) complexes

chelation enhanced fluorescence effects, 21, 22f
non-cyclized β–aminoketone, 220
Copper(II) removal by silica sol-gels anchored with amino and mercapto ligands, 225 227
Correlation analysis, laser fluence *vs.* generated particle concentration, 245, 247 248
Courmarin-fluorophore, 21, 24 27, 30 31
Critical frequency calculations, grating equation, 106 108, 109f
Crown ether ligands in supercritical carbon dioxide, cesium and strontium extraction, 168 169
Crown ethers, encapsulated, ligand-anchored silica sol-gels in strontium(II) separation, 228 232f
Crown ethers, room-temperature ionic liquids for fission-product separation, 146 160
Crystalline silicotitanate, separation agent for cesium, 269 270

D

Dansyl fluorophores, 14 17, 18f, 19f, 27 29
Data acquisition system for ultrasonic diffraction grating spectroscopy, 102, 104f–106
DataFit software, transfer coefficient and standard rate constant determinations, 92, 94f
DCE. *See* 1,2-Dichloroethane
Deactivation and decommission problem, science research agenda, 5 6
Defense Waste Processing Facility, 320
Di-[3-(trimethylsilyl)-1-propyl] alkylenediphosphonic acids. *See* TMSP-esterified

1,2-Dichloroethane (DCE), diluent, 179 184
Dicyclohexano-18-crown-6 ether with counteranions, cesium and strontium extraction in supercritical carbon dioxide, 168 169
Diphenyl-*N,N*-di-n-butylcarbamoylmethylphosphine oxide. See CMPO
gem-Diphosphonates, bridge-substituted tetraalkyl, 254 256*f*
Distribution coefficient fission products, enhancement with ionic liquids, 147
Dithioacetal derivatives grafted on silica gels in mercury(II) separation, 232 234
Dynamic flow in solubilities determination, TMSP-esterified alkylenediphosphonic acids in supercritical carbon dioxide, 259 263

E

EDTA in borate buffer solution, electrochemical impedance, passive iron, 65
Electrochemical cell, three-electrode polytetrafluoroethylene (PTFE), description, 65 68
Electrochemical impedance data analysis, iron in borate buffers, 64 99
Environmental Management Science Program (EMSP), mission, 2 3
Environmental Remediation Sciences Program, 3, 9
Esaki tunneling, 73
1-(2-Ethoxy-2-oxoethyl)-3-methylimidazolium bis[(trifluoromethyl) sulfonyl]-imide, synthesis, 151, 152, 153*f*
1-(2-Ethoxy-2-oxoethyl)-3-methylimidazolium bromide, synthesis, 149, 152, 153*f*
Ethylenediamine-anchored silica sol-gels in copper(II) removal, 226 227
Ethylenediaminetetraacetic acid, disodium salt, See EDTA
^{157}Europium extraction CCD/PEG system, nitric acid dependence, 177 179
Evanescent wave cavity ring-down spectroscopy (EW-CRDS), 138 140
Evanescent wave spectra, TCE, cis-DCE, and trans-DCE, 140 141*f*
Evaporator failure in fuel-reprocessing operations at Savannah River Site, 319 322
EW-CRDS. *See* Evanescent wave cavity ring-down spectroscopy
EXAFS spectra and Fourier transforms, *fac*-technetium-carbonyl complexes, 312 314, 315*t*
EXAFS spectra and structure technetium(IV) gluconate complex, 307, 309 311*f*
technetium(IV) glyoxylate complex, 307 309
Extended X-ray absorption fine structure. *See* EXAFS spectra

F

Finkelstein conditions, 24
Fission-product separation by room-temperature ionic liquids, 146 160
Fluorescence, probe for local environment investigations, glass matrices, 35 36, uranium glass
Fluorescence properties, uranium doped borosilicate glass, 39 45
Fluorescent calix[4]arenes as chemosensors, 12 33

calix[4]arene scaffold and crown or azacrown ether binding sites, syntheses, 27 31
Fluorescent chemosensor design, logic, 13
Formation voltage effect in electrochemical impedance spectroscopy, passive iron in borate buffer, 72 73, 75f–76f
Frequency sweep effect in electrochemical impedance spectroscopy, passive iron in borate buffer, 68 69, 70f
Frit-165 uranium-doped glass
  steady-state fluorescence, 43 45
  UV-Vis-NIR absorption spectrum, 45 46f
Frit-202 uranium-doped glass, lifetime fluorescence distribution, 43
FS-13 (Phenyltrifluoromethyl sulfone) diluent, 172, 174 175, 177 182

G

Glass composition, effect on uranium valances in immobilized glass, 47 50
Gluconate effect on technetium(IV) oxide aqueous solubility, 286 301
Gluconate ligand bonding sites, 287 288, 289f
Gold nanoparticle surface-plasmon resonance response, 136 138f
Grating equation, critical frequency calculations, 106 108, 109f
Green solvent, supercritical carbon dioxide, overview, 251 252
Green technology for nuclear waste management, 161 170

H

^{1}H NMR spectra
  acid peak in ^{137}cesium extraction by chlorinated cobalt dicarbollide, 175
  imidazoliun ring, 154t
  perchloric acid contacted with CMPO, 182t–183
  tri-n-butylphosphate–nitric acid complex in supercritical fluid carbon dioxide, 166 167f
Hanford Site
  borosilicate glass, HLW immobilization, 269
  density sensor on pipeline, 131
  gluconate use in waste stream treatment, 287
  high-level liquid waste, characteristics, 3 4
  non-pertechnetate species in high-level nuclear waste tanks, 302 318
  waste tanks, cesium and strontium isotope ratios, 13
Hexachlorocobalt dicarbollide anion, structure, 173f
  See also CCD systems
High Field Model, 73
High-level waste reprocessing, evaporator failure at Savannah River Site, 319 322
High-level waste science research agenda on, 3 4
High-level waste tanks, non-pertechnetate species, 302 318
Humic acid, surface-enhanced Raman scattering (SERS) spectrum, 61
Hydroxide effect upon solid phase formation upon mixing sodium silicate and sodium aluminate, 323 334

I

Idaho National Laboratory
  high-level liquid waste, characteristics, 3 4

UNEX development, 172–174
Imidazolium-based ionic liquids in fission-product separations, 146–160
Impedance model development based on Point Defect Model, 82, 84f–96f
Interfacial Control Model, 88
IR spectra, CCD/PEG-400 extraction system with strontium and barium cations, 179–180
Iron in borate buffer solutions, electrochemical impedance data analysis, 64–99

J

Japanese demonstration project, uranium and plutonium extraction from mixed oxide fuel in supercritical carbon dioxide, 169

K

Khlopin Radium Institute, UNEX development, 172
Kinetic measurements, nickel(II) chelation, 193–194
Kramers-Kronig transforms, electrochemical impedance data verification, 77, 81–82, 83f
Kroger-Vink notation, 82

L

Lanthanide and actinide extraction with supercritical carbon dioxide, 163–164
Laplace inversion, measured decay curves, 39, 40f
Laser fluence vs. generated particle concentration, correlation analysis, 245, 247–248
Laser plasma in surface decontamination, nanoparticle formation, 240–249
Laser wavelength effects on particle generation, 244–245, 246f
Lifetime fluorescence distribution functions, uranium-doped glasses, 39, 40f–45
Linear Systems Theory, electrochemical impedance data verification, 77, 81–82, 83f
Lux-Flood acid-base concept, application to glass melts, 36

M

Matlab program for performing Kramers-Kronig transforms, 81–82, 83f
Matlab software in shear wave reflection technique, 118
Mercapto-ligand anchored silica sol-gels
  copper(II) removal, 225–227
  mercury(II) removal, 227, 232–234
Mercury(II) complex, chelation enhanced fluorescence effects, 21, 22f, 23f
Mercury(II) complexation in fluorescence quenching, 14, 17–19f
Mercury(II) reactions with thioether carboxylic acids, 232–234
Mercury(II) removal by silica sol-gels anchored with mercapto ligands, 227, 232–234
Metal ion chelation, switch-binding by linear ligand with transformation to macrocycle, 186–222
Metal ion extractants, supercritical carbon dioxide-soluble, 250–267
Metal separation by organofunctional sol-gels, 223–237
25,27-{4-Methylcoumarin-6,7-diylbis[2-(2-oxoethoxy)ethoxy]}-

26,28-bis-benzyloxy-calix[4]arene, synthesis, 31
Methylenebis(siloxypropylmethylphosphonate) esters, 254 256f
Methylenediphosphonic acids, SC-CO_2 solubility variations with structure, 261t–264
Mine wastes containing uranium, extraction by supercritical carbon dioxide, 163
Monolithic folded resonator, experimental configuration, 138 140

N

N-(9-anthrylmethyl)-25,27-bis-(benzyloxy)-calix[4]arene azacrown-5, 1,3-alternate, synthesis, 30
N-dansyl-5,11,17,23-tetrakis(1,1-dimethylethyl)-26,28-dihydroxy-calix[4]arene azacrown-5, synthesis, 28 29
N-tosyl 25,27-bis(benzyloxy)calix[4]arene azacrown-5, synthesis, 29
Nanoparticle formation during surface decontamination by laser plasma, 240 249
experimental setup, 241, 243f–244
National Research Council, National Academy of Sciences, reports, 4 6
Natural and Attenuated Bioremediation Research Program, 3
Nickel(II) ion chelation, switch-binding by linear ligand with transformation to macrocycle, 186 222
Niobate based ion exchangers. See SOMS
Nitric acid concentration in CCD/PEG extraction system, 175 179
NMR spectra. See specific nuclei
Non-pertechnetate species in high-level nuclear waste tanks, 302 318
Nuclear facilities, decontamination and decommissioning, 241
Nuclear waste management with radionuclide supercritical fluid extraction, 161 170

O

Office of Environmental Management (US-DOE), establishment, 2
Oleic acid as sacrificial cation exchanger in strontium ion extractions, 156 158f
Oligo(dimethylsiloxane)-substituted tetraalkyl gem-diphosphonates, structures, 253t
On-line instrument for density and viscosity measurements, 129, 131
Optical basicity
application to glass melts, 36 37
uranium redox equilibria in silicate glasses, 47 49
Optical resonator design for evanescent wave cavity ring-down spectroscopy, 138 140
Organofunctional sol-gels in metal separation, 223 237
Oxygen fugacity, effect on uranium valances in immobilized glass, 47

P

^{31}P NMR spectra, perchloric acid contacted with CMPO, 181 183
Particle concentration, generated, vs. laser fluence, correlation analysis, 245, 247 248
Particle removal from surface, required energy, 248 249
PCE. See Perchloroethylene

Pechmann synthesis, 24
Pentadecafluoro-n-octanoic acid, strontium extraction in supercritical carbon dioxide, 168
Perchloric acid complexation with CMPO in polar diluent systems, 181, 182t–183
Perchloroethylene (PCE) detection by cavity ring-down spectroscopy, 133 144
Perfluoro-1-octanesulfonic acid, counteranion, cesium extraction in supercritical carbon dioxide, 169
Perovskite, thermally converted strontium loaded phases, chemical and thermal durability, 274, 276f–280
Perturbation voltage effect in electrochemical impedance spectroscopy, passive iron in borate buffer, 69, 72, 74f
pH effect
  electrochemical impedance spectroscopy, passive iron in borate buffer, 77, 78f–80f
  gluconate effect on technetium(IV) oxide aqueous solubility, 291 294, 295f
  selectivity of SOMS-1 for strontium(II) ion, 273 274, 275f
  See also Alkaline solution
Phenyltrifluoromethyl sulfone. See FS-13
Photo-induced electron transfer sensor with chelation enhanced fluorescence effects, 17, 20f–21, 22f, 23f
Photo-induced internal charge transfer sensor, 21, 24 27
Photon decay time. See Ring-down time
Phthalyltetrathioacetic acid grafted on silica gels in mercury(II) separation, 232 233
Pipeline sensor in ultrasonic diffraction grating spectroscopy, 101, 103f

Planar waveguide technology sensitivity, comparison to EW-CRDS for TCE detection, 142
Point Defect Model, 73, 97 98
  impedance model development, 82, 84f–96f
Polyethylene glycol. See CCD/PEG-400 systems
Portland cement. See Concrete particle formation
Potassium(I) complexation, fluorescence quenching, 24, 26f–27
Potentiometric measurements, nickel(II) chelation, 194 195
Proton as sacrificial cation exchanger in strontium ion extractions, 156, 157, 160
PUREX process, 165, 303
[3-({3-[1-(Pyridin-2-yl)-ethylamino]-propyl}-methylamino)-propyl]-carbamic acid benzyl ester, synthesis, 192
[3-({3-[1-(Pyridin-2-yl)-ethylamino]-propyl}-N-methyl)-1-aminopropane, synthesis, 192 193, 196
  complex formation with Ni(II), kinetics, 208, 212 216f, 219
  formation constants, metal complexes, 198
  protonation constants, 197 198
  specific rate constants for nickel(II) reacting with ligand species, 217t
[3-({3-[1-(Pyridin-2-yl)-ethylideneamino]-propyl}-methylamino)-propyl]-carbamic acid benzyl ester, synthesis, 191 192

R

Radioactive wastes, Universal Extraction process chemistry, 171 185

Raman spectral features, technetium-gluconate solution complexes, 294, 296f–298, 299f–300f
See also Surface-enhanced Raman scattering
Reaction kinetics, switch-binding by linear ligand with transformation to macrocycle upon metal chelation, 188 222
Redox chemistry, uranium species in molten glass, 45 50
Reflection coefficient and viscosity, relationship, 125 126
Reflection coefficient at quartz-liquid interface, 119, 125
Reflection technique for viscosity, sensor calibration and reliability, 127 129t, 130
Research Needs for High-Level Waste Stored in Tanks and Bins at U.S. Department of Energy sites: Environmental Management Science Program, 4
Research Opportunities for Deactivating and Decommissioning Department of Energy Facilities, 5 6
Ring-down time, definition, 134
Room-temperature ionic liquids for fission-product separations, 146 160

S

Sacrificial cation exchangers in strontium ion extractions, 147, 155t, 156 157
Safety note, ^{99}technetium, beta-emitter, 304
Savannah River Site
evaporator failure in high-level waste reprocessing, 319 322
frit composition, 37 38
high-level liquid waste, characteristics, 3 4
Savannah River Defense Waste Processing Facility, 269
Scanning electron microscopy, silver-doped sol-gel film, 55, 57, 58f–59f
Science research agenda on high-level waste, 3 4
Self-calibrating feature, reflection technique, definition, 126
Sensor calibration and reliability, reflection technique, 127 129t, 130
SERS. *See* Surface-enhanced Raman scattering
Shear wave reflection techniques in viscosity measurement, 118 131
Silica sol-gels in metal separation, 223 237
Silicotitanate-based ion exchangers. *See* SOMS
Siloxane-functionalized *gem*-diphosphonates, 252 256f
Silver-doped sol-gel film, characterization, 55 57, 58f–59f
Silver-doped sol-gel substrate, preparation, 55
Slope and reflection coefficient, relationship, 124f, 125
Slurry properties characterization
shear wave reflection techniques, 118 131
ultrasonic diffraction grating spectroscopy, 102 118
Sodalite formation in waste evaporator, 321
Sodalite formation upon mixing sodium aluminate and sodium silicate, 323 334
Sodium ion as sacrificial cation in strontium ion extractions, 147, 155t, 157 159
Sodium ion effect on selectivity of SOMS-1 for strontium(II) ion, 273 274, 275f

Sodium tetraphenylborate as sacrificial cation exchanger in strontium ion extractions, 147, 155*t*, 157 159
Sol-gel glass, uranyl-doped, lifetime fluorescence distribution, 41, 42*f*–43
Sol-gel substrate, silver-doped, in Raman scattering, uranyl–humic complexes, 53 63
Solid phase formation upon mixing sodium silicate and sodium aluminate, 323 334
Solubility, aqueous technetium(IV) oxide in presence of gluconate, 291 295*f*
Solubility experiments, aqueous technetium(IV) oxide in presence of gluconate, method, 288, 290 291
SOMS and related perovskites as ion exchangers, 268 284
SOMS-1, selectivity for strontium(II) ion, 273 274, 275*f*
South Carolina. *See* Savannah River Site
Speed of sound. *See* Velocity of sound
Stainless steel 316 particle formation in surface decontamination by laser plasma, 241 249
Strontium and cesium extraction in CCD/PEG-400 systems, 175 181
Strontium complex, chelation enhanced fluorescence effects, 21, 22*f*, 23*f*
^{90}Strontium extraction CCD/PEG system, nitric acid dependence, 177 179
^{90}Strontium in Hanford waste tanks, 13
Strontium ion distribution coefficient measurements and selectivity, SOMS-1, 271 275*f*
Strontium ion extraction by imidazolium-based ionic liquids, 152, 154 159
Strontium ion extraction by supercritical carbon dioxide, 168 169
Strontium loaded phases, perovskite, chemical and thermal durability, 274, 276*f*–280
Strontium release rates by ASTM-PCT (product consistency test), 272, 277 280, 281*t*
Strontium selective separation, silica sol-gels with encapsulated crown ether ligand, 228 232*f*
Sugar water, liquid in ultrasonic diffraction grating spectroscopy and reflection techniques investigation, 100 131
Supercritical carbon dioxide
 lanthanide and actinide extraction, overview, 163 164
 radionuclide separations in nuclear waste management, 161 170
 soluble metal ion extractants, 250 267
TMSP-esterified alkylenediphosphonic acids, solubilities, 259 263
Surface decontamination by laser plasma, nanoparticle formation, 240 249
Surface-enhanced Raman scattering (SERS)
 uranyl–humic complexes, spectra, 53 63
 uranyl ions, spectra, 57, 60*f*
 See also Raman spectral features
Surface particle removal, threshold energy, 248 249
Surface-plasmon-resonance-enhanced cavity ring-down detection, 135 138*f*
Switch-binding definition, 188
Switch-binding dynamics, linear ligand transformation to macrocycle, 186 222
Symposium and book organization, 6 9

T

Tank wastes, aluminum containing species, precipitation, 319 335
^{99}Tc NMR spectra, *fac*-technetium-carbonyl complexes, 312 313*f*
TCE. *See* Trichloroethylene
fac-Technetium carbonyl complexes, 310, 312 316
Technetium gluconate solution complexes, Raman spectra features, 294, 296*f*–298, 299*f*–300*f*
Technetium(IV) alkoxide complexes, radiolysis experiments, 307 310
Technetium(IV) gluconate complex, EXAFS spectrum, coordination environment, 307, 309 311*f*
Technetium(IV) glyoxylate complex, EXAFS spectrum and structure, 307 309
Technetium(IV) oxide, aqueous solubility in presence of gluconate, 286 301
Technetium(IV) tetra-L-alamine complex, cyclization, 219 220
5,11,17,23-Tetrakis(1,1-dimethylethyl)-25,27-bis(5-chloro-3-oxapentyloxy)-26,28-dihydroxycalix[4]-arene, synthesis, 28
5,11,17,23-Tetrakis(1,1-dimethylethyl)-25,27-bis(2-dansylamidoethoxy)-26,28-dihydroxycalix[4]arene, synthesis, 27 28
1,4,10,13-Tetraoxa-7,16-diazacyclooctadecane-7,16-bis(malonate), selective strontium(II) separation, 228 232
TGA-DTA analysis, SOMS-1 conversion to Perovskite form, 272, 274, 275*f*–276*f*
Thenoyltrifluoroacetone, chelating agent in supercritical fluid carbon dioxide extraction, actinides and lanthanides, 162
Thermogravimetric analysis-differential thermal analysis. *See* TGA-DTA
Thorium and uranyl ions, extraction by supercritical carbon dioxide containing fluorinated β-diketones, 163
Threshold energy for particle removal from surface, 248 249
TMSP-esterified alkylene-diphosphonic acids
  extraction behavior, 257 259
  solubility in supercritical carbon dioxide, 259 263
Tri-n-butylphosphate in supercritical carbon dioxide, actinides and lanthanides extraction, 163 164
Tri-n-butylphosphate–nitric acid complex
  characterization in supercritical carbon dioxide, 165 166, 167*f*
  uranium(IV) oxide dissolution in supercritical carbon dioxide, 163 165
Trichloroethylene (TCE) detection by cavity ring-down spectroscopy, 133 142

U

Ultrasonic diffraction grating spectroscopy (UGDS), slurry properties characterization, 102 118
  experimental measurements, 108, 110*f*–116, 117*f*
Universal Extraction (UNEX) process, fundamental chemistry, 171 185
Uranium as probing atom measuring concentration in borosilicate glass, 41, 42*f*
Uranium coordination in borosilicate glasses, 49 50

Uranium in borosilicate glass, 34 52
  fluorescence properties, 39 45
  frits from Savannah River Site, 37 38
Uranium redox equilibria in silicate glasses, 49
Uranium species in molten glass, redox chemistry, 45 50
Uranium trioxide, fluorescence decay in uranium doped glass, 39 45
Uranium(IV) oxide, direct dissolution in supercritical carbon dioxide with tri-n-butylphosphate–nitric acid complex, 164 165
Uranyl and thorium ions, extraction by supercritical carbon dioxide containing fluorinated β-diketones, 163
Uranyl-humic complexes, surface-enhanced Raman scattering (SERS), 53 63
Uranyl ions, surface-enhanced Raman scattering (SERS) spectrum, 57, 60f
UV spectra, imidazolium-based ionic liquids with cis-dicyclohexano-18-crown-6 ether, 158f
UV-Vis-NIR spectra, frit-165 uranium-doped glass, 45 46f
UV-Vis spectra, silver-doped sol-gel film, 55 56f

V

Velocity of sound determination by UDGS, 108 118
Viscosity and reflection coefficient, relationship, 125 126
Viscosity effect on UDGS velocity of sound determination, 116, 118
Viscosity measurements by shear wave reflection techniques, 118 131
Viscosity sensor calibration and reliability, reflection technique, 127 129t, 130

W

Washington. *See* Hanford Site
Waste at Hanford Site from REDOX and PUREX processes, immobilization plans, 303 304, 305f
Waste immobilization efficiency, radionuclides in borosilicate glasses, 35
Waste management, green technology, 161 170
Waste reprocessing, evaporator failure at Savannah River Site, 319 322
Waste tanks SY-101 and SY-102, 303, 310
Waste unified processing, UNEX process, 171 185
Wastes from uranium mine, supercritical carbon dioxide extraction, 163

X

X-ray diffraction, thermally converted strontium loaded perovskite, 274, 276f–277, 278f
XANES (X-ray absorption near edge structure) spectra in identification, *fac*-technetium carbonyl complexes 304, 305f, 310, 312, 314

Z

Zeolite A formation
  in waste evaporator, 321
  upon mixing sodium aluminate and sodium silicate, 323 334

Bestsellers from ACS Books

The ACS Style Guide: A Manual for Authors and Editors (2nd Edition)
Edited by Janet S. Dodd
470 pp; clothbound ISBN 0-8412-3461-2; paperback ISBN 0-8412-3462-0

Reagent Chemicals: Specifications and Procedures: Tenth Edition
By ACS Committee on Analytical Reagents
816 pp; clothbound ISBN 0-8412-3945-2

Advances in Arsenic Research: Integration of Experimental and Observational Studies and Implications for Mitigation
Edited by Peggy A. O'Day, Dimitrios Vlassopoulos, Xiaoguang Meng, and Liane G. Benning
446 pp; clothbound ISBN 0-8412-3913-4

Chemical Activities (student and teacher editions)
By Christie L. Borgford and Lee R. Summerlin
330 pp; spiralbound ISBN 0-8412-1417-4; teacher edition,
ISBN 0-8412-1416-6

Chemical Demonstrations: A Sourcebook for Teachers, Volumes 1 and 2,
Second Edition
Volume 1 by Lee R. Summerlin and James L. Ealy, Jr.
198 pp; spiralbound ISBN 0-8412-1481-6
Volume 2 by Lee R. Summerlin, Christie L. Borgford, and Julie B. Ealy
234 pp; spiralbound ISBN 0-8412-1535-9

The Internet: A Guide for Chemists
Edited by Steven M. Bachrach
360 pp; clothbound ISBN 0-8412-3223-7; paperback ISBN 0-8412-3224-5

Laboratory Waste Management: A Guidebook
ACS Task Force on Laboratory Waste Management
250 pp; clothbound ISBN 0-8412-2735-7; paperback ISBN 0-8412-2849-3

Metal-Containing and Metallosupramolecular Polymers and Materials
Edited by Ulrich S. Schubert, George R. Newkome, and Ian Manners
598 pp; clothbound ISBN 0-8412-3929-0

For further information contact:
Order Department
Oxford University Press
2001 Evans Road
Cary, NC 27513
Phone: 1-800-445-9714 or 919-677-0977

More Bestsellers from ACS Books

Microwave-Enhanced Chemistry: Fundamentals, Sample Preparation, and Applications
Edited by H. M. (Skip) Kingston and Stephen J. Haswell
800 pp; clothbound ISBN 0–8412–3375–6

Fire and Polymers IV: Materials and Concepts for Hazard Prevention
Edited by Charles A. Wilkie and Gordon L. Nelson
436 pp; clothbound ISBN 0–8412–3948–7

Ionic Liquids as Green Solvents: Progress and Prospects
Edited by Robin D. Rogers and Kenneth R. Seddon
614 pp; clothbound ISBN 0–8412–3856–1

Fermentation Biotechnology
Edited by Badal C. Saha
300 pp; clothbound ISBN 0–8412–3845–6

Chemometrics and Chemoinformatics
Edited by Barry K. Lavinex
216 pp; casebound ISBN 0–8412–3858–8

Polymeric Drug Delivery I: Particulate Drug Carriers
Edited by Sönke Svenson
352 pp; clothbound ISBN 0–8412–3918–5

Polymeric Drug Delivery II: Polymeric Matrices and Drug Particle Engineering
Edited by Sönke Svenson
390 pp; clothbound ISBN 0–8412–3976–2

Food Lipids: Chemistry, Flavor, and Texture
Edited by Fereidoon Shahidi and Hugo Weenen
248 pp; clothbound ISBN 978–0–8412–3896–1

Herbs: Challenges in Chemistry and Biology
Edited by Mingfu Wang, Shengmin Sang, Lucy Sun Hwang, and Chi-Tang Ho
384 pp; clothbound ISBN 978–0–8412–3930–2

For further information contact:
Order Department
Oxford University Press
2001 Evans Road
Cary, NC 27513
Phone: 1-800-445-9714 or 919-677-0977